AF393803

Trends in Atomic
and Molecular Physics

Trends in Atomic and Molecular Physics

Edited by

Krishan K. Sud

and

Upendra N. Upadhyaya

Department of Physics
College of Science Campus
M. L. Sukhadia University
Udaipur, India

Springer Science+Business Media, LLC

Library of Congress Cataloging-in-Publication Data

Trends in atomic and molecular physics / edited by Krishan K. Sud and Upendra N. Upadhyaya.
    p. cm.
    Includes bibliographical references and index.
    ISBN 978-1-4613-6912-7   ISBN 978-1-4615-4259-9 (eBook)
    DOI 10.1007/978-1-4615-4259-9
       1. Collisions (Nuclear physics)--Congresses. 2. Scattering (Physics)--Congresses. 3.
    Atoms--Congresses. 4. Molecules--Congresses. I. Sud, Krishan K. II. Upadhyaya,
    Upendra N. III. National conference on Atomic and Molecular Physics (12th :
    1998-1999 : Udaipur, India)
QC794.6.C6 . T7454 2000
539.7'57--dc21
                                                                    99-088822

Proceedings of the XII National Conference on Atomic and Molecular Physics, held 29 December 1998–
2 January 1999, in Udaipur, India.

ISBN 978-1-4613-6912-7

© 1999 Springer Science+Business Media New York
Originally published by Kluwer Academic / Plenum Publishers in 1999
Softcover reprint of the hardcover 1st edition 1999
10 9 8 7 6 5 4 3 2 1

A C.I.P. record for this book is available from the Library of Congress

Attainable neither through speech
nor through mind
nor through eyes is It,
Apart from the one who utters,
'It is what exists' how is it obtainable!

Kaṭha Upaniṣad

Preface

Contemporary research in atomic and molecular physics concerns itself with studies of interactions of electron, positron, photons, and ions with atoms, molecules, and clusters; interactions of intense ultrashort laser interaction with atoms, molecules, and solids; laser assisted atomic collisions, optical, and magnetic traps of neutral atoms to produce ultracold and dense samples; high resolution atomic spectroscopy and experiments by using synchrotron radiation sources and ion storage rings. In recent years, important advances have been made in the experimental as well as theoretical understanding of atomic and molecular physics. The advances in atomic and molecular physics have helped us to understand many other fields, like astrophyics, atmospheric physics, environmental science, laser physics, surface physics, computational physics, photonics, and electronics.

XII National Conference on Atomic and Molecular Physics was held at the Physics Department, M. L. S. University, Udaipur from 29th Dec. 1998 to 2nd Jan. 1999 under the auspices of the Indian Society of Atomic and Molecular Physics. This volume is an outcome of the contributions from the invited speakers at the conference. The volume contains 24 articles contributed by the distinguished scientists in the field. The contributors have covered a wide range of topics in the field in which current research is being done. This also reflects the trend of research in this field in Indian universities and research institutes.

We are grateful to the national programme committee, national, and local organizing committees, and members of the Physics Department and Computer Centre, M. L. S. University Udaipur, for giving their time and energy. We would also like to express our gratitude to the University Grant Commission (New Delhi), Department of Science and Technology (New Delhi) and Department of Atomic Energy (Mumbai), and the Physics Department (MLSU, Udaipur) for the financial support which made this conference possible.

We would like to thank Prof. A. K. Singh, Vice-Chancellor, M. L. S. University for his kind support and encouragement in this academic venture and our colleagues Dr. V. K. Agarwal, Dr. K. Venugopalan, and Dr. B. L. Ahuja for all possible help. Mr. A. S. Bhullar and Mr. C. Selvaraju, members of the theory group of the Physics Department, M. L. S. University deserve our sincere appreciation and thanks for the time and energy spent in organizing this conference and in the preparation of this volume.

Finally, we are grateful to the authors of the invited talks for their stimulating presentation, and interaction with the participants of the conference. We believe that this volume represents the direction of current research in a vast and active field of atomic and molecular physics.

Krishan K. Sud
Upendra N. Upadhyaya

Contents

Intense, Ultrashort, Laser-Solid Interactions

Sudeep Banerjee, G. Ravindra Kumar and Lokesh C. Tribedi

Tata Institute of Fundamental Research
Mumbai (Bombay) – 400 005, India

The physics of intense laser–solid interaction is briefly reviewed. The emission of x-ray pulses from laser produced 'solid' like plasmas under intense, picosecond and femtosecond excitation is discussed. Experimental results are presented for metal plasmas. A new, one step method is presented for the estimation of the *spectrally resolved* absolute x-ray yields using simple solid state detectors. In addition, gamma ray emission upto 500 keV is also reported at intensities of 10^{15} W cm^{-2}.

1.   INTRODUCTION

The study of light-matter interaction continues to throw up exciting discoveries even though it is a field that has attained considerable maturity over the past few decades. In the post-laser era, many new phenomena have been discovered, that pertain to the nature of coherent light-matter interaction. Lasers, particularly pulsed lasers, have also opened up new avenues of creating matter in highly excited states with applications ranging from ablation to inertial confinement fusion. The peak intensities achieved have also increased in a dramatic fashion, with the current state of the art being in the 10^{20} W cm^{-2} range [1, 2]. This has been made possible by the advent of pulse compression methods whereby femtosecond pulses can be routinely generated [3] and an amplification procedure known as Chirped Pulse Amplification (CPA) [4]. Till the advent of CPA, intense laser-matter interactions were studied at a few big sized and expensive, high-energy laser facilities. Now compact *tabletop* Terawatt lasers capable of achieving the same or even larger intensities are available to hundreds of small laboratories around the world.

Trends in Atomic and Molecular Physics,
Edited by Sud and Upadhyaya. Kluwer Academic/Plenum Publishers, New York, 2000.

2

The application of intense, ultrashort laser pulses to excite matter has revealed certain fundamentally different features of interaction not present in moderately strong fields [5, 6]. Specific examples of these include above threshold ionization (ATI), very high order harmonic generation (HHG), tunnelling ionization, multiple ionization and coulomb explosions, softening of molecular bond, particle acceleration to keV and MeV energies [7] and so on. Many of these phenomena have been observed in gaseous as well as solid systems. The excitation of solids in particular has thrown up possibilities of studying the properties of plasmas under unusual conditions. Though laser produced plasmas have been investigated for over three decades now, the recent spurt in interest is due to the high density and excitation levels that can be produced under femtosecond excitation. Such plasmas are not only rich in their physics, but also are efficient sources of a) VUV and X-ray radiation (both coherent and incoherent), b) high energy electrons and c) high energy, highly charged ions. The physics of such plasmas is interesting because of the extremely small timescales of the formation of the plasma and subsequent interaction of the plasma with the laser pulse. For the same peak intensity, in the long pulse, nanosecond regime 1) the mechanism of ionization and plasma formation is mainly due to cascade collisions of the initial laser ionized electrons with other atoms in the solid, 2) the plasma gradient is small and a significant portion of the laser pulse interacts with the plasma, rather than the solid, 3) the plasma has enough time to redistribute the excitation into different internal modes and 4) the emitted x-ray pulses also have long duration. In contrast, in the ultrashort regime, 1) the plasma is formed mainly by the direct ionization of atoms in the solid by the laser, 2) the plasma has a steep gradient and most of the laser pulse interacts with the solid, 3) there is not much time for the redistribution of the energy and the onset of instabilities while the laser pulse is incident on the solid and 4) the x-ray emission is also of ultrashort duration. These fundamental differences demarcate the physics in the two regimes. From the point of view of this paper, it is interesting to know that the emission of x-rays is much more efficient from ultrashort pulse produced plasmas than from those created by long pulses. In addition, the availability of ultrashort x-ray pulses in copious fluxes from femtosecond laser excited plasmas opens up exciting applications in technology [8, 9] and in basic research – notable among the latter being real time studies of the dynamics of atoms in crystals and biological systems, time resolved ionization of core level electrons in atoms and so on (for example, see ref. [10]). The biological applications are really promising because of the existence of a window in water absorption around 200 eV, which should facilitate deeper penetration of x-rays than other radiation. Lastly, the gamma rays produced from these plasmas could be used for nuclear excitation.

In a sense the study of real time dynamics of atomic motion parallels the revolution in the study of real time electron dynamics in atoms and molecules made possible by femtosecond lasers in the infrared-visible-ultraviolet region of the electromagnetic spectrum. It may not be an exaggeration to say that table top terawatt lasers in the optical region can act as tabletop synchrotrons and particle accelerators!

We will not attempt a review of the physics of ultrashort pulse laser produced plasmas because a number of excellent reviews exist in the literature [5, 11-14]. We move on to describe our experiments and discuss our results. We have recently proposed a new method for estimating the absolute yields of x-rays from ultrashort pulse excited plasmas [15, 16]. We present this method and the resulting analysis of our data. Later we discuss the results of our measurements on the dependence of x-ray emission on various laser parameters. Finally we present observations of gamma ray emission from these plasmas.

2. STUDIES OF X-RAY EMISSION

In our laboratory, we study femtosecond and picosecond intense laser interactions with solids using, at the present time, x-ray emission as the major diagnostic. Previous studies on x-ray emission have focussed on high resolution spectra and their interpretation. X-ray yields have also been estimated in these studies, but as we will show below, the procedures involved are rather cumbersome and perhaps inaccurate. There is a great need to correctly establish the total x-ray yields from the point of view of comparison of the x-ray emission characteristics under different excitation conditions (and the physics behind the dependence) and in view of the applications mentioned above. Our studies provide a unique method of measuring the yields in a simple and inexpensive manner. As a major additional feature, we highlight the fact that the same measurements also yield important information on the photon time correlation aspects of the x-ray source, an aspect which has deep implications for characterising the coherence of the source. Interestingly, it turns out that the photon correlation aspect is a critical input in the estimation of x-ray yields from ultrashort-pulse produced plasmas.

We present here, measurements of the absolute x-ray yields from different metal plasmas created by focussed picosecond and femtosecond laser pulses. We show that, unlike in normal dispersive methods used to study x-ray emission which rely on a VUV (vacuum ultraviolet) grating monochromator [17, 18] or a crystal monochromator [19, 20], it is possible in a simple manner to get reasonable information on the spectral content

4

alongwith the absolute x-ray yields. To do so, we model the photon statistics of the x-ray source assuming a thermal distribution. This model is essential to make a correct estimate of the emitted x-ray flux. This also has important implications for understanding the coherence properties of the x-ray source. We point out the advantages of our method over the existing methods.

2.1   EXPERIMENTAL SETUP

The experimental details are depicted in Figs. 1 and 2. Radiation from picosecond Nd-YAG and femtosecond Ti-sapphire lasers has been used to produce plasmas in our experiments. The picosecond laser, which has a passive-active modelocked cavity, produces 35 ps pulses at 1064 nm, 532 nm and 355 nm at a repetition rate of 10 Hz (Continuum 'PY61C-10'). The maximum pulse energy at 1064 nm is 60 mJ. The femtosecond laser is a custom built Chirped Pulse Amplification (CPA) system (Continuum) that uses seed pulses from a Coherent 'Mira Seed' Ti-Sapphire laser which is pumped by a Coherent 'Innova-90/6 plus' argon ion laser. The seed pulses at 806 nm, having a bandwidth of about 22 nm, are stretched by a grating stretcher and fed to a regenerative amplifier and later to a multipass amplifier, both of which are pumped by 532 nm radiation from a 10 Hz, 'Surelite III-10' (Continuum) nanosecond Nd-YAG laser. The amplified pulse can finally be compressed to a duration of 100 fs. The maximum output from the laser is about 50 mJ/pulse. The final pulse duration is measured by an autocorrelator (Femtochrome). The amplified femtosecond pulse has a diameter of 15 mm. The 1064 nm radiation is elliptically polarised (s-polarization fraction = 0.7) while radiation at other wavelengths is linearly polarised (s-polarisation). Figure 1 shows the pulse width and the associated bandwidth of the femtosecond laser.

The laser pulses were focussed onto the metal target using lenses of focal length f = 10 and 20 cm. The focussed spot diameter was estimated to be 15 μm for f = 10 cm and 25 μm for f = 20 cm. The target was mounted in a chamber evacuated to 10^{-3} torr. The solid target, of high purity, was in the form of a disc and was continuously rotated and translated using a stepper motor drive. X-ray emission from the plasma was collected by a cooled (77 K) Si(Li) x ray detector with a Be or mylar window of thickness 25 μm or a NaI detector with an aluminium window. The signal from the detector is amplified and fed to a multichannel analyser through an ADC. The Si(Li) detector is good for the range 1-20 keV with a resolution of 160 eV at 5.9 keV and the calculated resolution is about 80 eV at 1.5 keV, the energy of aluminium K x-ray. Its maximum efficiency is 100 % in the range 4-20 keV.

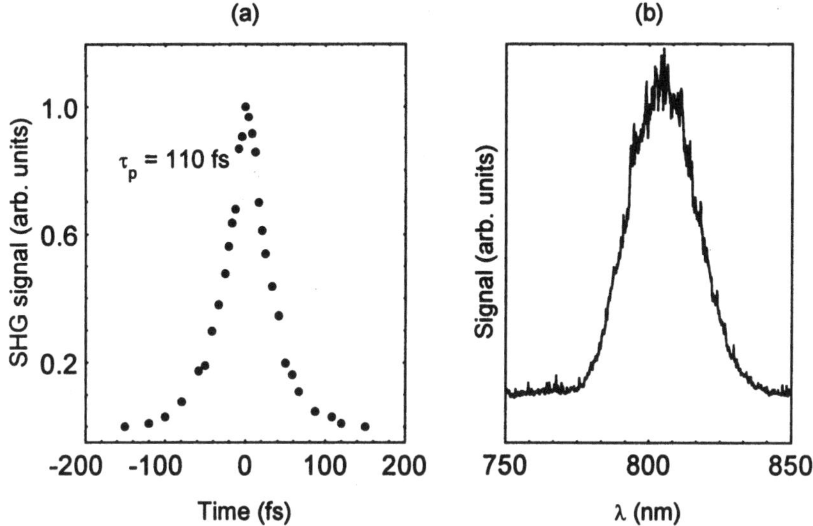

Figure 1. Characteristics of the femtosecond pulse used in the experiments (a) pulse duration as reflected in the autocorrelation (b) bandwidth of the amplified output.

The NaI detector is good for the range 20 keV-10 MeV and has a maximum efficiency (~100%) in the range 50-200 keV. The x-rays pass through appropriate windows mounted on the vacuum chamber before entering the detector. Spectra were typically collected over 20000-40000 shots. The detector was placed in the direction of the plasma plume. The target was mounted at an angle of 45° with respect to the input laser beam. The x-ray yield was corrected for the efficiency of the detector arising due to the absorption in both the windows.

A typical x-ray spectrum from an aluminium plasma measured under excitation by 532 nm radiation at an intensity of 3.0×10^{13} W cm^{-2} is shown in Fig. 3. The raw spectrum consists of a series of successively decreasing peaks. The first peak occurs at 1.64 keV (see below) and the successive peaks occur at multiples of this energy. There is a smaller peak to the left of the main peak, the origin of which is not clearly established and may arise from noise and pickup in the detection system. Detailed examination of such spectra taken over many runs and with both picosecond and femtosecond excitation [15, 16] revealed that the multiple peak structure is a consequence of the simultaneous arrival of more than one photon at the detector. This is a peculiar aspect of this kind of detectors in experiments where many photons can arrive simultaneously at the detector, and hence care must be exercised in estimating the total yields of x-rays measured using such a detection system. To explain further, after excitation by a particular laser shot, two or more x-ray photons may simultaneously arrive

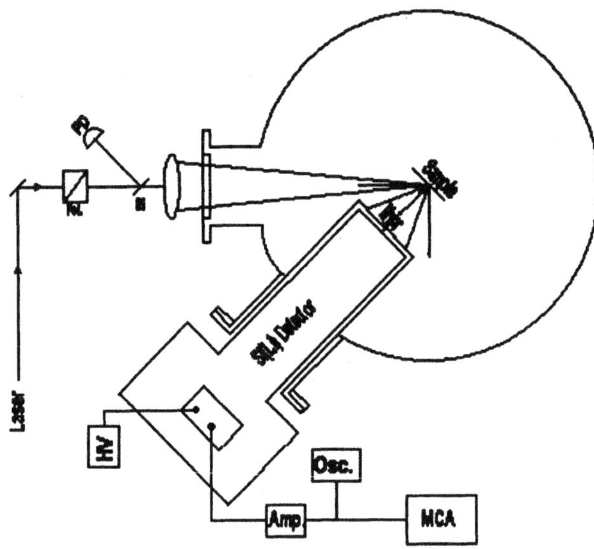

Figure 2. Schematic of the experimental setup. In case of γ ray measurements the Na(I) detector was placed at varying distances as described in the text.

at the detector [21]. Such bursts are common from high luminosity sources like laser produced plasmas, but unlikely to be encountered in a typical ion-atom collision experiment where the x-ray production is not so copious. Though we have not measured the duration of the emitted x-ray pulse, it has been reported [11, 22] to be of the order of the input pulse duration (1-100 ps), a time which is a small fraction of the detector response time (ns). The detector responds to two such closely timed photons as if they were one photon of twice the energy, resulting in a peak at twice the basic x-ray photon energy. For instance, in Fig. 3, the peak at 11.4 keV corresponds to the near simultaneous arrival of 7 x-ray photons of energy 1.64 keV. The total x-ray yield is obtained by integrating the area under each peak and adding up these areas after multiplying by the appropriate number (2 for the second peak, 3 for the third peak and so on).

2.2   PHOTON STATISTICS

The measured data should therefore be analysed keeping the probabilities for the arrival of multiple photons in mind and a proper choice must be made of the kind of statistics relevant to the emission. This property pertains to the temporal coherence characteristics of the source and the arrival times of photons at the detector are described by the 'photon statistics' of the source. It is well known that the arrival times of photons from a coherent source (laser) are described by Poisson statistics, while those for a chaotic, thermal source are described by the Bose-Einstein distribution [23]. The laser produced plasma source, being an incoherent source, clearly belongs to the latter category and the photon statistics are given by [23]

$$P(n) = \frac{1}{1+N}\left(\frac{N}{N+1}\right)^n$$

P(n) is the probability of the simultaneous arrival of *n* photons at the detector and N is the average number of photons arriving at the detector. This distribution leads to successively decreasing probabilities as *n* increases for all N. For the Poissonian distribution

$$P(n) = \frac{1}{N!} N^n \exp(-N)$$

P(n) peaks at N (Fig. 4). The number N is a measure of the source luminosity and larger N indicates a larger x-ray flux.

Using this equation, we analyse the data shown in Fig. 3. It turns out that the best fit is obtained for N = 2. We have obtained consistent fits to different sets of data using the above distribution, shown in Fig. 3(a). Once the fits are obtained, we can calculate the total x-ray yield.

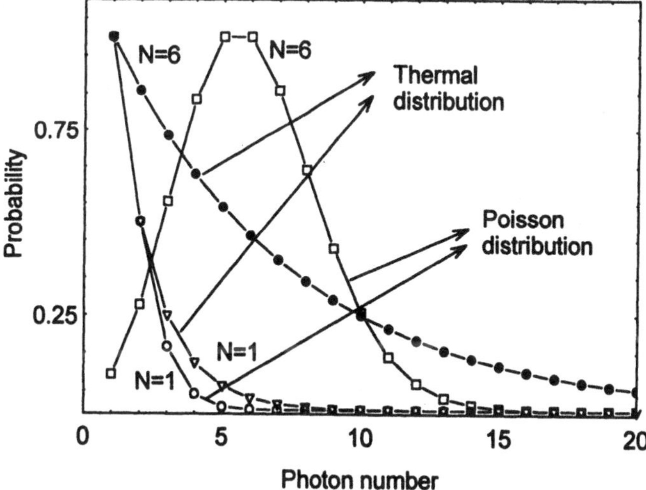

Figure 3. Thermal and Poisson distribution for N = 1 and N = 6 and varying. Note that the thermal distribution is always maximum at n = 1 while for the Poisson case the peak occurs at n = N.

8

We have obtained fits to data at different x-ray luminosity and find that good fits are obtained for almost all sets of data. We have also noticed that under low luminosity, the higher peaks do vanish, as expected.

Let us now compare our measurement with those obtained using monochromators [17-20] or p-i-n photodiodes [19, 20]. In these methods, there are uncertainties associated with the spectrometer (crystal or grating reflectivities), efficiency of the channeltron, photographic film density and so on, as shown in the detailed analysis of Stearns *et al* [17] and Meyerhofer and coworkers [19, 20]. In the other method, which involves using p-i-n photodiodes, only **spectrally integrated yield** is obtained [24] after correcting for filter transmission losses. Besides, different filters are needed for different spectral ranges. In contrast, our method eliminates all these intermediate stages and hence provides the most accurate yields. It is also surprising that none of these studies have taken into account the statistics of the source which is an essential input **irrespective** of the method of detection. Stearns *et al* used Poissonian statistics in their analysis, which, as we demonstrated above is not the correct choice.

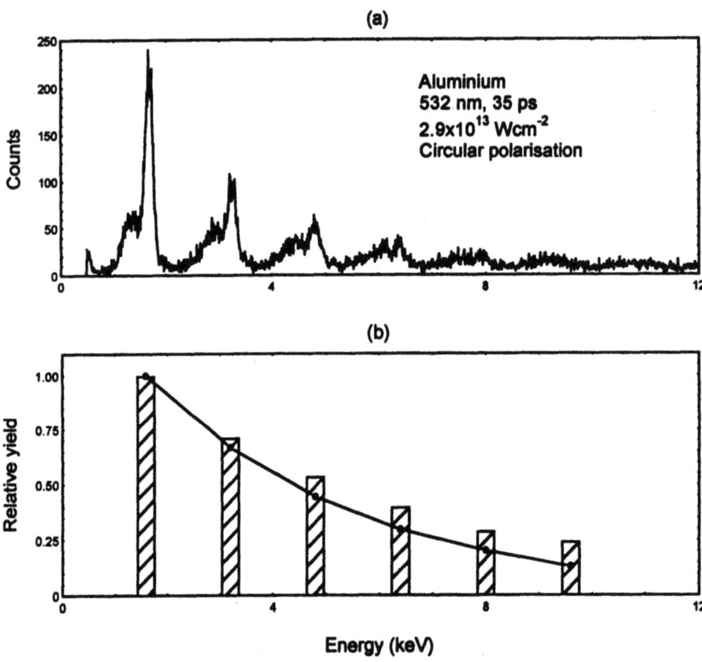

Figure 4. X-ray spectrum of aluminium showing K x-ray emission. (a) Measured line emission. The small peak close to zero is low level electronic noise. (b) Thermal distribution of emitted x-rays. The histograms represent area under the peaks with appropriate multipliers (see text). The solid line is the theoretical fit assuming a thermal distribution with N = 2.

The expected peak position for Al K x-ray is 1.486 keV arising from an Al atom with a single hole in K-shell. If this photon is emitted in presence of multiple vacancies in higher shells (or in K-shell) the x-ray peak is shifted to higher energy side, a fact well known from the study of heavy ion-atom collisions. Therefore, the shift of this peak to a higher energy is due to the emission from highly ionized Al ions in the plasma. The observed shift $\Delta E \sim 170$ eV may correspond charge states as high as Al^{10+} to Al^{12+} and was estimated from the ionization energies calculated by Godwal *et al* [25]. Thus we demonstrate that the Si(Li) detector has good enough resolution to provide information on the charge states in intense, ultrashort laser produced plasmas.

An additional important point to be noted about making a correct choice of statistics is that it helps identify the proper spectral features. As an example, if a 'true' peak is present at a higher energy, where there is a higher order contribution from a lower energy peak, the correct strength of the 'true' peak can be ascertained by subtracting the contribution of the lower energy peak from the total. It is in the estimate of the higher order contribution that the statistics assume significance. We should also note that, the distribution allows us to estimate the higher order contributions that are buried in noise. We can even get an idea of the number of peaks that we really need to consider, in order to get a good signal to noise ratio and estimate the errors associated with the contributions we ignore. As an example, let us consider Fig. 4 again and estimate the error associated with a measurement when neglecting different higher energy peaks. Thus the error turns out to be less than 10 % if we neglect the peaks higher than n = 5, in the case of the aluminium data (Fig. 4(a)). We thus get a reliable estimate of the errors involved in measuring the absolute yields, particularly when the higher energy peaks are barely above the noise level of the detector and cannot be distinguished easily. We draw attention to one interesting aspect of the photon statistics here. While characterising x-ray sources for applications, considerable emphasis is placed on the coherence properties. Photon statistics provide a clear way to demarcate coherent x-ray emission (Poissonian statistics) from incoherent x-ray radiation, which follows thermal statistics. It is thus possible to use solid state detectors to monitor the coherence properties of the x-ray source.

Fig. 4 shows the ratio of the aluminium total K x-ray energy emitted over 2π solid angle as a function of input laser intensity at 1064 nm (assuming an isotropic angular distribution of x-rays). Note that we have calculated the ratios of the energy in the K x-ray beam with that of the laser pulse energy. The emission appears to have been saturated leading to a fall in the efficiency as the laser intensity is increased. The maximum efficiency we measured was 1.0×10^{-9} at 3.0×10^{13} W cm^{-2}. These values compare very favourably with the recent reported efficiencies for x-ray emission from rare gas clusters [21], data which were obtained at intensities three orders of magnitude larger than our values. Similarly, we can attempt a comparison of our yields with similar experiments on aluminium, although these other experiments used much higher intensities. It should also be noted that these experiments used a number of calibration procedures for arriving at the reported yields, unlike in our case, where the yield estimation is straightforward. We note that the efficiency of aluminium K x-ray emission varies from 10^{-6} at 10^{16} W cm^{-2} to 0.001 at 10^{18} W cm^{-2} [26-28]. These intensities are 3 to 5 orders of magnitude larger than those used for the data in Fig 4.

Going back to Fig. 4, we notice that the laser pulses in this measurement have circular polarization. We have done detailed measurements of the dependence of the x-ray yield on the ellipticity of laser light. We have observed that linear polarization produces more yields for the K x-rays than circular polarization. Further, shorter wavelengths produced higher yields for the same x-ray. These studies will be reported elsewhere [29].

2.3   Bremsstrahlung radiation

Apart from the characteristic x-rays, we have also observed bremsstrahlung radiation in the same spectral region (Fig. 5) in considerable abundance. Note that photon 'pile up', as in the case of the data obtained for the characteristic K x-ray, should be avoided in this case because of the continuous nature of the spectrum. The count rates were thus reduced by either inserting small apertures before the detector or moving the detector. The spectra have the typical, exponentially decaying behaviour of bremsstrahlung. From such a dependence, we estimate an electron temperature 4.4 keV for aluminium and 6.1 keV for Magnesium at an intensity of 10^{15} Wcm^{-2}. The latter, understandably has more abundance because of its higher Z number.

3. γ–RAY EMISSION MEASUREMENTS:

Fig. 6 shows the emission of photons in the tens to hundreds keV region from copper plasmas under femtosecond excitation. These data were obtained with the NaI detector placed at a distance of 1 m from the plasma. A temporal gating scheme was devised to collect data only within a well defined time window (10-50 μs) after the laser trigger. This was necessary to eliminate the cosmic ray background. We could suppress the background almost completely. Fig. 5 shows that emission could be observed almost upto 500 keV. The electron temperature corresponding to this emission is estimated to be 7.7 keV. There is a recent report of similar emission from copper, with a 'weak pre-pulse' to create a plasma and further excite it with the main laser pulse [30]. Observations of similar emission are pertinent because we do not use any 'pre-pulse' and our intensities are about an order of magnitude lower.

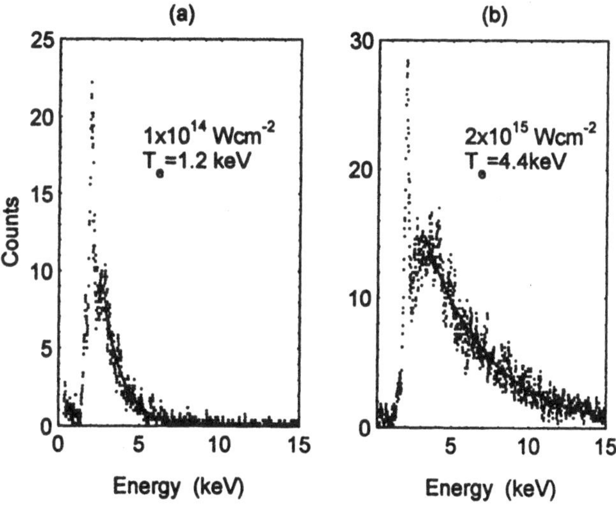

Figure 5. Bremsstrahlung spectra of Al. The solid line is an exponential fit with corresponding electron temperature of $\sim 10^6$ K.

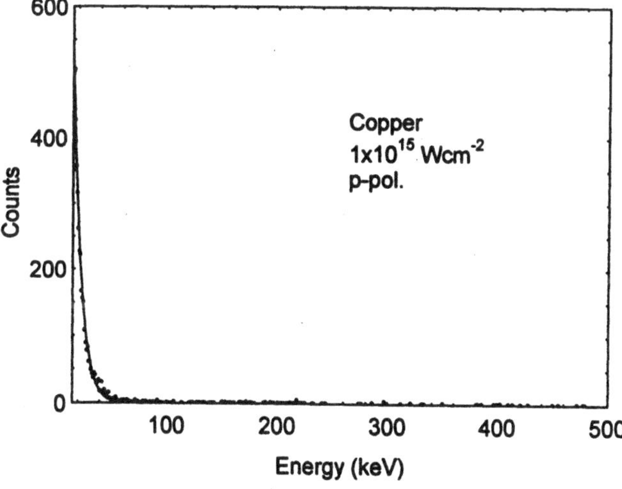

Figure 6. γ-ray emission from copper. The spectrum was recorded with the detector at a distance of 1 m from the target. The electron temperature is 80×10^6 K.

In conclusion, we have presented studies on x-ray emission from intense, ultrashort laser excited solid plasmas. The absolute yields of characteristic x-rays were estimated using simple solid state detectors. The role of photon statistics is crucial in the estimation of absolute yields. Such measurements also provide a reasonable amount of spectral information apart from characterising the coherence properties of the x-ray source. In addition, continuum bremsstrahlung radiation from these plasmas was also measured.

We believe that intense, ultrashort laser produced solid plasmas will continue to provide an experimentally rich situation for the study of matter under extreme excitation.

ACKNOWLEDGMENTS

We thank Prof. Surendra Singh for discussions on photon statistics. Amal Saha and Riju Isaac are thanked for help with some of the experiments. The high energy, femtosecond laser facility has been generously funded by the Department of Science and Technology, Government of India.

REFERENCES

1.   M. D. Perry and G. Mourou, Science **264**, 917 (1994).

2.   See IEEE J. Selected Topics in Quant. Electron. **4** No. 2, (1998).

3.   J. C. Diels and W. Rudolf, *Ultrashort Laser Pulse Phenomena*, Academic Press (1996).

4.   D. Strickland and G. Mourou, Opt. Comm. **56**, 219 (1985).

5.   *Laser Interactions with Atoms, solids and Plasmas,* ed. R. M. More, NATO ASI Series B: **327**, Plenum Press, New York (1994).

6.   *Superintense Laser-Atom Physics*, ed. B. Piraux, Plenum, New York, (1995).

7.   *Laser Interaction with Matter* ed. S. J. Rose, Institute of Physics Conference Series, No. 140, IOP Publishing, Bristol (1997).

8.   *Short Wavelength Lasers and their Applications* ed. C. Yamanaka Springer Proceedings in Physics **30**, Springer-Verlag (1988).

9.   *X-ray Science and Technology* ed. A. G. Michette and C. J. Buckley Institute of Physics Publishing, Bristol (1993).

10. C. Rischel et al., Nature **390**, 490 (1997).

11. M. Murnane, H. Kapteyn, M. D. Rosen and R. Falcone, Science **251**, 531 (1991).

12. E. M. Campbell, Phys. Fluids B **4**, 3781 (1992).

13. E. G. Gamaly, Laser and Particle Beams **12**, 185 (1994).

14. P. Gibbon and E. Forster, Plasma Phys. Control. Fusion **38**, 769 (1996).

15. S. Banerjee, G. Ravindra Kumar, Amal K. Saha and L. C. Tribedi, Opt. Comm. **158**, 72 (1998).

16. S. Banerjee, L. C. Tribedi, Amal K. Saha and G. Ravindra Kumar, Phys. Scripta. (to appear in 1999).

17. D. G. Stearns, O. L. Landen, E. M. Campbell and J. H. Scofield, Phys. Rev. A **37**, 1684 (1988).

18. J. A. Cobble, G. A. Kyrala, A. A. Hauer, A. J. Taylor, C. C. Gomez, N. D. Delamater and G. T. Schappert, Phys. Rev. A **39**, 454 (1989).

19. H. Chen, B. Soom, B. Yakobi, S. Uchida and D. D. Meyerhofer, Phys. Rev. Lett. **70**, 3431 (1993).

20. D. D. Meyerhofer, H. Chen, J. A. Delettrez, B. Soom, S. Uchida and B. Yakobi, Phys. Fluids B **5** 2584 (1993).

21. S. Dobosz, M. Lezius, M. Schmidt, P. Meynadier, M. Perdrix, D. Normand, J. P. Rozet and D. Vernhet, Phys. Rev. A **56**, R2526 (1997).

22. D. W. Phillion and C. J. Hailey, Phys. Rev. A **34**, 4886 (1986).

23. B. E. A. Saleh and M. C.Teich, *Fundamentals of Photonics*, John-Wiley & Sons,Inc. New York, p406 (1991).

24. M. Chaker, J. C. Kieffer, J. P. Matte, H. Pepin, P. Audebert, P. Maine, D. Strickland, P. Bado and G. Mourou, Phys. Fluids. B **3**, 167 (1991).

25. B. K. Godwal, A. Ng, L. DaSilva, Y. T. Lee and D. A. Liberman, Phys. Rev. E **40**, 4521 (1989).

26. J. A. Cobble, G. T. Schappert, L. A. Jones, A. J. Taylor, G. A. Kyrala and R. D. Fulton, J. Appl. Phys. **69**, 3369 (1991).

27. G. Malka, N. Blanchart, D. Desenne, M. Louis-Jacquet, A. Mens, J. L. Miquel, O. Peyrusse, J. Opt. Soc. Amer. B **14**, 2091 (1997).

28. U. Teubner, T. Missalla, I. Uschmann, E. Forster, W. Theubalt, C. Wulker, Appl. Phys. B **62**, 213 (1996).

29. S. Banerjee *et al.* (to be published).

30. P. Zhang et al., Phys. Rev. E **57**, R3746 (1998).

Intense-Field Many-Body S-Matrix Theory of Atomic and Molecular Processes in Femtosecond Laser Pulses

F. H. M. Faisal, A. Becker and J. Muth-Böhm
Fakultät für Physik
Universitat Bielefeld, D-33501 Bielefeld, Germany

A review of the "intense-field many-body S-matrix theory" (or IMST) for investigating highly non-perturbative interaction of strong laser pulses with many-electron atomic and molecular systems is briefly presented. Application of the theory to a number of current problems in the field of interaction of laser light with atoms and molecules are given. They include formation of (a) 'hot electron plateaus' in ATI spectra of He, (b) intensity dependence of double ionization yields for He, and (c) *multiple* ionization yields of Xe, (d) 'enhanced ionization' of a diatomic molecule, H_2, and (e) of a polyatomic molecule benzene C_6H_6, and its cation $C_6H_6^+$.

1.1   INTRODUCTION

A number of new phenomena have been observed recently when intense femtosecond laser pulses interact with atomic and molecular targets such as (e.g. [1, 2, 3]): the observation of the so-called 'hot electron plateau' in the above-threshold ionization (ATI) spectra of atoms, copious production of double *and* multiply charged ions of two and many-electron atoms, and a so-called phenomenon of 'enhanced ionization' of molecules at a specific internuclear distance. They generally require a theoretical method of analysis capable of handling both the highly non-perturbative field interactions and the many-body problem involved. In view of the resulting combined theoretical difficulty, quantitative investigations of these processes are at present still in its beginning phase. For, example, direct simulations of the interaction of many-electron systems with an intense laser field involves a high-dimensional time-dependent Schrödinger equation

Trends in Atomic and Molecular Physics,
Edited by Sud and Upadhyaya. Kluwer Academic/Plenum Publishers, New York, 2000.

which is extremely arduous (even with the 'state-of-the-art' computational hard- and softwares) already for a two-electron (six-dimensional) atomic system like He (e.g. [4, 5, 6]); for many-electron systems this is out of question for the forseeable future. As an alternative, a so-called 'intense-field many-body S-matrix' (IMST) method has been developed recently [7, 8, 9]. In this paper we first briefly sketch the theory and then report the results of its application to the above mentioned problems.

1.2   INTENSE-FIELD MANY-BODY S-MATRIX THEORY (IMST)

The S-matrix series is well-known in atomic collision physics in the so-called 'prior', and 'post' forms [10], both of which are based on a single partitioning of the total Hamiltonian of the system into a reference Hamiltonian and an interaction. In the 'prior' form, the interaction is defined by the difference of the total Hamiltonian and a reference Hamiltonian, chosen such that the initial state is an eigenfunction of the same; in the 'post' form, the reference Hamiltonian is defined with respect to the final state. The reference propagator with respect to which the S-matrix series in these forms is, therefore, restricted to that of either the initial or the final reference Hamiltonian only.

In contrast, the IMST is a systematic reformulation of the many-body S-matrix series that provides a more flexible scheme of expansion based on three different (or more, if desired) partitionings of the total Hamiltonian $H(t)$:

$$H(t) = H_i^0(t) + V_i(t) \qquad \text{(initial state partitioning)} \tag{1.1}$$

$$= H^0(t) + V_0(t) \qquad \text{(intermediate state partitioning)} \tag{1.2}$$

$$= H_f^0(t) + V_f(t) \qquad \text{(final state partitioning).} \tag{1.3}$$

A formal solution of the Schrödinger equation, associated with the total Hamiltonian $H(t)$, is given by the time-dependent generalization of the Lippmann-Schwinger equation [10]:

$$\left| \Psi_i^{(+)}(t) \right\rangle = \left| \Phi_i^0(t) \right\rangle + \int_{-\infty}^{t} dt' \, G(t, t') V_i(t') \left| \Phi_i^0(t') \right\rangle, \tag{1.4}$$

where $G(t, t')$ is the total Green's function and $\Phi_i^0(t)$ is the initial state of the system associated with the reference Hamiltonian H_i^0.

We may first expand $G(t, t')$ in terms of the Green's function associated with the final state Hamiltonian H_f^0:

$$G(t, t') = G_f^0(t, t') + \int_{t'}^{t} dt_1\, G_f^0(t, t_1)\, V_f(t_1)\, G(t_1, t') \qquad (1.5)$$

and then in terms of the Green's function associated with any desired intermediate Hamiltonian H^0,

$$G(t, t') = G^0(t, t') + \int_{t'}^{t} dt_2\, G^0(t, t_2)\, V_0(t_2)\, G(t_2, t'). \qquad (1.6)$$

To get the rearranged S-matrix series we first substitute Eq. 1.6 in Eq. 1.5 and then substitute the resulting expression in Eq. 1.4, which provides the desired expansion of the time-dependent wavefunction of the total system in the form:

$$
\begin{aligned}
\left|\Psi_i^+(t)\right\rangle = {}& \Phi_i^0(t) + \int_{-\infty}^{t} dt'\, G_f^0(t, t')\, V_i(t') \left|\Phi_i^0(t')\right\rangle \\
& + \int_{-\infty}^{t} dt' \int_{t'}^{t} dt_1\, G_f^0(t, t_1)\, V_f(t_1)\, G^0(t_1, t')\, V_i(t') \left|\Phi_i^0(t')\right\rangle \\
& + \int_{-\infty}^{t} dt' \int_{t'}^{t} dt_1 \int_{t'}^{t_1} dt_2\, G_f^0(t, t_1)\, V_f(t_1)\, G^0(t_1, t_2) \\
& \qquad\qquad \times V_0(t_2)\, G^0(t_2, t')\, V_i(t') \left|\Phi_i^0(t')\right\rangle \\
& + \dots
\end{aligned}
$$

$$(1.7)$$

Finally, we project $\Psi_i^+(t)$ onto the final state, $\Phi_f^0(t)$, associated with H_f^0, and obtain the transition amplitude (or the S-matrix) in the form:

$$(S - 1)_{fi} = \lim_{t \to \infty} \sum_{j=1}^{\infty} S_{fi}^{(j)}(t). \qquad (1.8)$$

The first three terms of the above S-matrix expansion (referred to as 'IMST' in the present context) are shown in Fig. 1.1(a). They should be

18

compared with those of the well-known prior or post forms of the many-body S-matrix expansions; the latter are shown in Fig. 1.1(b) and (c).

(a) IMST series

$$S_{fi}^{(1)}(t) = -\frac{i}{\hbar} \int_{-\infty}^{t} dt' \langle \Phi_f^0(t') | V_i(t') | \Phi_i^0(t') \rangle$$

$$S_{fi}^{(2)}(t) = -\frac{i}{\hbar} \int_{-\infty}^{t} dt' \int_{-\infty}^{\infty} dt_1 \langle \Phi_f^0(t_1) | V_f(t_1) G^0(t_1, t') V_i(t') | \Phi_i^0(t') \rangle$$

$$S_{fi}^{(3)}(t) = -\frac{i}{\hbar} \int_{-\infty}^{t} dt' \int_{-\infty}^{\infty} dt_1 \int_{-\infty}^{\infty} dt_2 \langle \Phi_f^0(t_1) | V_f(t_1) G^0(t_1, t_2) V_0(t_2)$$
$$\times\, G^0(t_2, t') V_i(t') | \Phi_i^0(t') \rangle$$

$$\cdots$$

(b) 'prior' form

$$S_{fi}^{(1)}(t) = -\frac{i}{\hbar} \int_{-\infty}^{t} dt' \langle \Phi_f^0(t') | V_i(t') | \Phi_i^0(t') \rangle$$

$$S_{fi}^{(2)}(t) = -\frac{i}{\hbar} \int_{-\infty}^{t} dt' \int_{-\infty}^{\infty} dt_1 \langle \Phi_f^0(t_1) | V_f(t_1) G_f^0(t_1, t') V_i(t') | \Phi_i^0(t') \rangle$$

$$S_{fi}^{(3)}(t) = -\frac{i}{\hbar} \int_{-\infty}^{t} dt' \int_{-\infty}^{\infty} dt_1 \int_{-\infty}^{\infty} dt_2 \langle \Phi_f^0(t_1) | V_f(t_1) G_f^0(t_1, t_2) V_f(t_2)$$
$$\times\, G_f^0(t_2, t') V_i(t') | \Phi_i^0(t') \rangle$$

$$\cdots$$

(c) 'post' form

$$S_{fi}^{(1)}(t) = -\frac{i}{\hbar} \int_{-\infty}^{t} dt' \langle \Phi_f^0(t') | V_f(t') | \Phi_i^0(t') \rangle$$

$$S_{fi}^{(2)}(t) = -\frac{i}{\hbar} \int_{-\infty}^{t} dt' \int_{-\infty}^{\infty} dt_1 \langle \Phi_f^0(t_1) | V_f(t_1) G_i^0(t_1, t') V_i(t') | \Phi_i^0(t') \rangle$$

$$S_{fi}^{(3)}(t) = -\frac{i}{\hbar} \int_{-\infty}^{t} dt' \int_{-\infty}^{\infty} dt_1 \int_{-\infty}^{\infty} dt_2 \langle \Phi_f^0(t_1) | V_f(t_1) G_i^0(t_1, t_2) V_i(t_2)$$
$$\times\, G_i^0(t_2, t') V_i(t') | \Phi_i^0(t') \rangle$$

$$\cdots$$

Figure 1.1 Comparison of the first three terms of the IMST series (panel a) and those of the usual 'prior' (panel b) and the 'post' form (panel c) of the S-matrix theory.

The flexibility of the present IMST expansion is due to the fact that, as indicated already, any desired intermediate interaction, $V_0(t)$, and the associated propagator, $G^0(t, t')$, can be introduced without affecting the choice of either of the end states. Such a formulation is particularly fruitful in suggesting physically useful models of otherwise virtually intractable problems, as will be shown below. This new form of the S-matrix series can be also considered as a 'three-potential' scheme, $V_i(t)$, $V_f(t)$ and $V_0(t)$, in the many-body S-matrix theory, and should have applications for problems other than that of the present interest (e.g. in chemical reaction dynamics). Finally, we note that the IMST series (c.f. fig. 1.1(a)) reduces to the prior form (panel b) for the special choice, $G^0 = G_f^0$ (or $V_0 = V_f^0$), or the (time-reversed) 'post' form (panel c) for the choice $G^0 = G_i^0$ (or $V_0 = V_i^0$) [11].

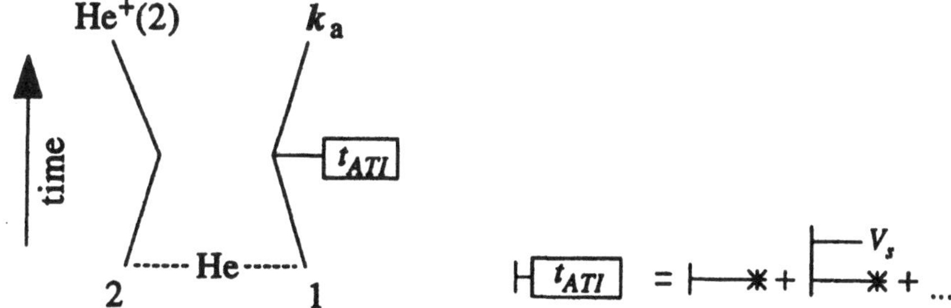

Figure 1.2. Leading Feynman diagram for laser induced single ionization of He. The lines stand for the states of the two atomic electrons, 1 and 2. t_{ATI} is the T-matrix for a virtual ATI-like process, where '⁻*' is the electron-field interaction and V$_s$ is the screened core potential. k$_a$ is the final momentum.

1.3 APPLICATIONS

A number of applications of IMST in connection with the interaction of intense femtosecond laser pulses with two *and* many-electron quantum systems will be given below.

1.3.1 Formation of hot electron plateaus of the ATI spectrum of helium

The leading order diagram for the single ionization of He is shown in fig. 1.2. The expansion of the T-matrix for the ATI process, t_{ATI}, is, as indicated in the Figure, given by the contributions from the successive terms with no-

20

rescattering, single-rescattering, double-rescattering, etc. of the ejected electron with the residual ion. We have evaluated the amplitude approximately by retaining the series upto the single rescattering. The first term with no-rescattering is calculated as in the so-called KFR amplitude [12, 13, 14], and the second term corresponding to the single-rescattering is evaluated approximately as follows (c.f. [15]): all the integrals over the coordinates are calculated analytically. The integrations over the intermediate momentum are carried out by assuming the pole-approximation and taking the contribution along the polarization axis. Finally, the rapidly oscillating mixed terms arising after the modulo squaring of the amplitude are neglected while summing the resulting terms.

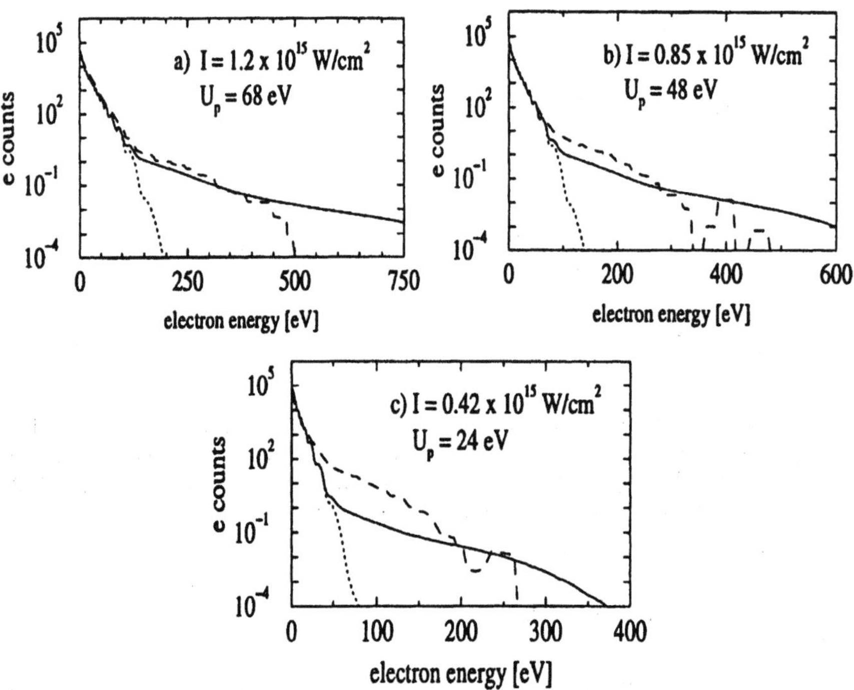

Figure 1.3. Single electron angle integrated ATI rates per He atom as a function of the electron energy at a laser wavelength of 780 nm and the intensities: a) $I = 1.2 \times 10^{15}$ W/cm^2, b) $I = 0.85 \times 10^{15}$ W/cm^2, and c) $I = 0.42 \times 10^{15}$ W/cm^2. The theoretical rates with single-rescattering (solid line) and no-rescattering (dotted line) has been scaled for the purpose of comparison to the experimental data (dashed line, from [16]) at one point, respectively.

In Fig. 1.3(a-c) we compare the results of calculations with no-rescattering (dotted curve) and that including single rescattering (solid curve), for the angle-integrated ATI-spectrum of He at $\lambda = 780$ nm and at three different intensities, (a) $I = 1.2 \times 10^{15}$ W$/$cm^2, (b) $I = 0.85 \times 10^{15}$ W$/$cm^2, and (c) $I = 0.4 \times 10^{15}$ W$/$cm^2. They are also compared with the corresponding experimental data (dashed lines) obtained by Sheehy et al. [16]. For the purpose of comparison we have matched the theoretical rates to the relative experimental data of the single ionization yield from Ref. [16] at one point. Comparisons of the calculated rates show clearly the remarkable increase by many orders of magnitude of the production of 'hot electrons' due to the single rescattering effect (c.f. dotted and solid curves); this is seen to lead to a much improved agreement with the data over a wide range of energies. In particular we note that the rescattering effect in fact gives rise to a 'hot electron plateau' as in the experiment. Note, however, that the present approximate calculation shows a rather slow cut-off compared to a rather sharp cut-off at $10U_p$, as sometimes expected corresponding to the maximum classical energy attainable for a free electron in a laser field, that has initially *zero* kinetic energy (e.g. [17]). Unfortunately, the experimental data near the highest measured energies appear to show rather irregular variations that make it difficult to ascertain the possible existence (or otherwise) of the classically expected sharp cut-off. Clearly, this requires further investigations for a complete clarity.

1.3.2 Intensity dependence of double ionization yields of He

The theory also permits us to investigate the role of electron correlation in the dynamics of laser induced double ionization of He, [7, 8, 9, 15, 18, 19, 20, 21, 22]. Thus, we could first identify a dominant Feynman diagram for the *non-sequential* (NS) double ionization process [15, 20, 21, 22], given in Fig. 1.4(c), which complements the diagram for the single ionization process, shown in Fig. 1.4(a) for He, and in Fig. 1.4(b) for He$^+$. Note that the last two diagrams contribute to the step-wise rates for double ionization of the neutral atom. The actual ion yields that are measured in the experiment as a function of the intensity, in general, contain contributions from both the step-wise and the NS rates.

The NS diagram, Fig. 1.4(c), turns out to be an intense field generalization of the so-called two-step-one (or TS1) diagram, that is well-known in the context of weak field single-photon double ionization by synchrotron radiation [23, 24]. Fig. 1.4(c) also immediately suggests a *physical* picture of the non-sequential process of interest. Thus, one can read it directly to see that at first an 'active' electron absorbs the photon energy from the laser field by a virtual process analogous to ATI and then shares this energy with the other electron via the electron correlation, until both of them have enough energy to escape together from the binding force of the nucleus. Unlike in the case of the step-wise (sequential) process, here the two elementary 'steps' are coupled together in a virtual intermediate state of the two-electron system, which gives rise to *one* transition amplitude for the NS process (hence the name, 'two-step-one' (or TS1)). We may note that the TS1 process differs conceptually from the so-called 'shake-off' picture [30, 31].

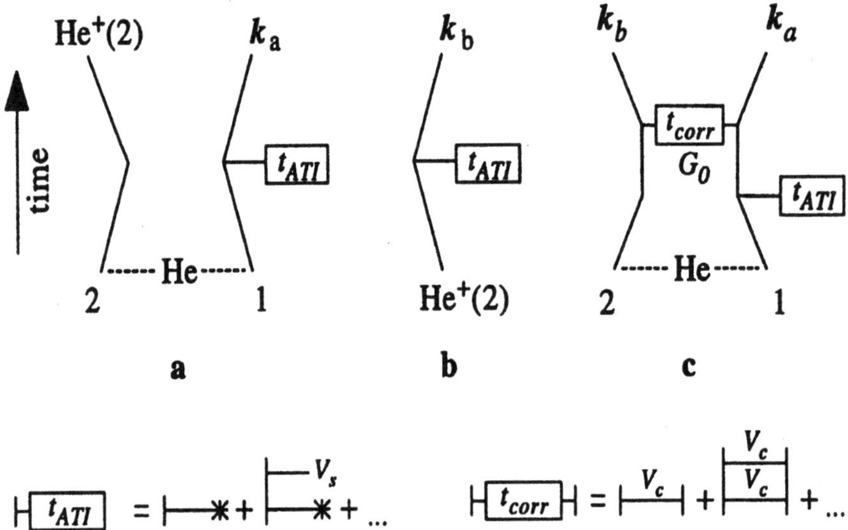

Figure 1.4. Dominant Feynman diagrams for laser induced single ionization of He (panel a) and of He$^+$ (panel b), as well as for non-sequential double ionization of He (panel c). t_{corr} is the T-matrix for e-2e transition and incorporates the e-e correlation V_c. The other symbols have the same significance as in Fig. 1.2.

Evaluation of the NS diagram, including the correlation interaction in all orders, is a practically impossible computational task. We could, however, construct a simplified model from it by an approximate analysis of its structure [15, 20, 21, 22]. This provided an interesting connection between the rate of NS double ionization process and the rates of two other simpler processes, namely, the above-threshold ionization (ATI) and the so-called 'e-2e' collisional ionization process (e.g. [23]). The approximate model formula for the double ionization can be written as [21, 22]:

$$\Gamma^{2+} = \sum_{N=N_0}^{\infty} \Gamma^{(e-2e)}\left(\mathcal{E}_N\right) \frac{\pi k_N}{2} \Gamma^{(ATI)}\left(\mathbf{k}_N \| \varepsilon\right). \qquad (1.9)$$

where $\Gamma^{(ATI)}\left(\mathbf{k}_N \| \varepsilon\right)$ is the N-photon ATI rate; $\mathcal{E}_N = k_N^2/2 + 4 k_N \sqrt{U_p} + 8 U_p$, $k_N^2/2 = N\omega - U_p - E_B$, and $\Gamma^{(e-2e)}\left(\mathcal{E}_N\right)$, is the rate of the 'e-2e' collisional ionization process. $U_p = I/4\omega^2$ is the ponderomotive energy, and E_B is the binding energy of the atom.

It should be noted that the intermediate Volkov electron can have energies corresponding to emission and absorption of an arbitrary number of photons during the propagation in the field. The dominant intermediate energies are of the order of the energy gained by at least a single back-scattering (e.g. [8]). Quantum mechanically, the back-scattering energy occurs significantly up to the order of the Bessel functions comparable to their argument, since beyond that order their contributions fall-off exponentially. The corresponding intermediate energies after the back-scattering are given by $\mathcal{E}_N$ [32].

Physically speaking, the non-sequential rate formula, Eq. 1.9, combines the individual above-threshold rates $\Gamma_{ATI}^{(n)}$ with the individual 'e-2e' rates $\Gamma^{(e-2e)}\left(\mathcal{E}_N\right)$ into a single quantum process that dominates the dynamics of non-sequential double ionization. We note that the physics of this simple formula, derived from a *systematic* S-matrix theory, includes the rather analogous concepts of the 'antenna' [25, 26, 27] and the 'quasi-static tunnelling and classical rescattering [28, 29] pictures mentioned above. However, there are significant differences both physically and technically. Thus, where as the antenna-model is rather similar in the overall structure of Eq. 1.9, the former uses a semiclassical approximation of the propagator as opposed to the quantum on-shell approximation used in Eq. 1.9. More importantly, in the present model the energy to initiate the 'e-2e' process arises *after* the back-scattering with the core. Where as in the antenna-model the energy that initiates the e-2e process is that at which the electron first emerges in the continuum. In the quasi-static rescattering model, the electron emerges with near zero energy and the collision occurs only on the return of the electron after a half-cycle. In contrast, the present model permits dominant (back)-scattering on the *way out* as well. In the present energy-domain representation there is, of course, no explicit signature of the

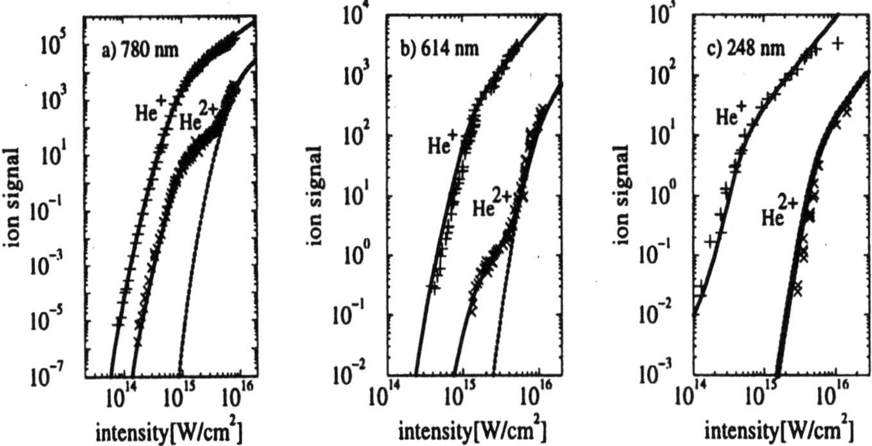

Figure 1.5. Comparison of single and double ionization of He for $\lambda = 780$ nm, $t_p = 160$ fs (panel a, expt., [31]); $\lambda = 614$ nm, $t_p = 120$ fs (panel b, expt., [30]); $\lambda = 248$ nm, $t_p = 500$ fs (panel c, expt., [37]). He$^+$ yield: expt., '+'; theory, left hand solid curve. He^{2+} yield: expt., '×'; theory, right hand solid curve. Sequential ionization yield (calculated): dotted curve.

intermediate return times, but only a sequence of 'collisions' (c.f. Fig. 1.4(c)) (each integrated over all intermediate times). Nevertheless, we may note that the time-scale of the back-scattering is of the order of $t_{bs} = \mathrm{r}_0 / \sqrt{2\mathcal{E}_n}$, where r_0 is of the order of the atomic (or ionic) size, which is for the field parameters of present interest much smaller than the first return-time after a half-cycle, $t_{re} = \pi/\omega$. Thus, the much faster 'on the way out' back-scattering of the intermediate Volkov electron, not only 'boosts' the electron energy before the e-2e process, but also reduces the wave-packet spreading effect compared to that expected during the much longer time-scale of the return-time.

For the explicit calculations, to be discussed below, the partial rates $\Gamma^{(e-2e)}$ are calculated from the semi-empirical Lotz formula [33]. The rate of single ionization of the neutral atom and that of the residual He$^+$ ion is obtained from the so-called KFR rate formula [12, 13, 14], corrected approximately for the Coulomb effect in the simple form [34, 21, 22]:

$$\Gamma^{(ATI)} = C\big(Z, E_0, E_B\big)\Gamma^{(KFR)},\qquad(1.10)$$

where the Coulomb correction factor,

$$C\big(Z, E_0, E_B\big) = \big(4E_B/E_0 r_c\big)^{\big(2Z/\sqrt{2E_B}\big)}\qquad(1.11)$$

is obtained from a WKB estimate; $r_c \approx 2/k_B$ is a measure of atomic size in the initial state; Z and E_0 are the residual charge and the peak field strength, respectively.

The ion yields are calculated by using the above formulae for the *basic rates* that appear in the rate equations connecting the various yields. For the purpose of comparison with the experimental data, a laser beam with a Gaussian TEM_{00} mode, and a Gaussian pulse profile [35, 36] having the pulse width, t_p, of a given experiment, is chosen. The final yields are obtained from the solutions of the rate equations at the end of the pulse and integrating them along the longitudinal and transverse directions of the beam.

Fig. 1.5(a-c) shows the results of the full calculations [21, 22] (solid curves) and compares them with the experimental data for He at three wavelengths, $\lambda = 780$ (panel a), 614 (panel b) and 248 nm (panel c), obtained by Walker et al [31], Fittinghoff et al. [30], and Charalambidis et al. [37], respectively. They are seen to agree remarkably well, including the ubiquitous knee-like structure observed at the optical wavelengths. In contrast the calculated sequential yields (dotted curve) deviate greatly from the experimental results by many orders of magnitude over most of the intensity region (c.f. fig. 1.5(a) and (b)). This clearly demonstrates the crucial importance of the NS rates for the observed agreement with the data at the optical wavelengths. For the ultraviolet wavelength (248 nm, see, Fig. 1.5(c)) the prediction of the present theory also agrees with the experimental data and with the *absence* of the knee structure at this wavelength. This implies that, in the intensity domain of the experiments, the NS rates are *not* significant at the ultraviolet wavelength. Indeed, in the latter case the full double ionization yields, calculated including NS rate (solid curve), virtually *coincide* with that calculated from the step-wise process alone (dotted curve). This may be understood from the fact that at the ultraviolet frequency, the ponderomotive energy, $U_p = I/4\omega^2$, too small to effectively

accelerate the intermediate ATI electron to sufficiently high energies, $\mathcal{E}_N$, that is needed for a significant NS contributions via the 'e-2e' process. Moreover, for the ultraviolet frequency the step-wise rate becomes larger (due to the fewer number of photons needed for the ionization of the He^+ ions) than at the optical frequencies. The combined result is the complete domination of the step-wise process and the consequent absence of the knee structure at the ultraviolet wavelength.

1.3.3   Formation of multiple charge-states of complex noble gases

We have extended the NS model for the double ionization to the more challenging problem of non-resonant multiple ionization (n-fold ionization $n \geq 3$) of a complex atom as well [22, 38]. The essential ingredient of this generalization is provided by the diagram given in Fig. 1.6. It shows the general two-step-one mechanism in which the 'active' electron '1' shares its initially absorbed field energy via a 'e-ne' process with the (n − 1) other

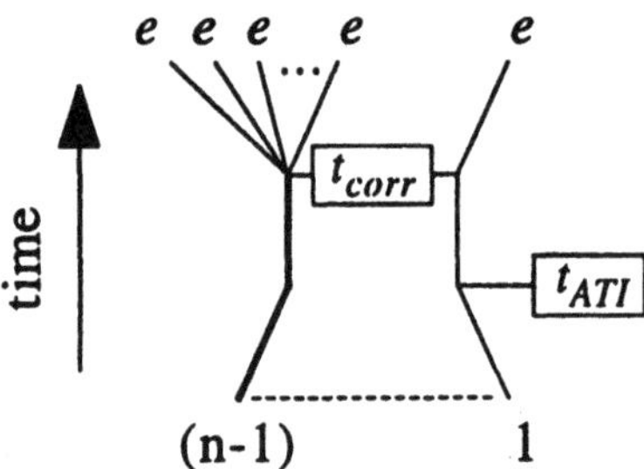

Figure 1.6. Leading Feynman diagram for laser induced non-sequential n-fold ionization of a many-electron atom or ion. The symbols have the same significance as in Fig. 1.4

electrons (n = 2, 3, 4, ...).   Based on the above diagram, a formula (analogous to the case of double ionization (c.f. Eq. 1.9)) can be written down for the NS rate of n-fold ionization of a many-electron atomic species that can be used to analyze the experimental data for non-resonant multiple ionizations of the heavier noble gas atoms.

A detailed discussion of the model formula and comparisons of the calculated results with experimental data is given elsewhere [22, 38]. Here, in Fig. 1.7, we only show the final results of the calculation for Xe upto the six-fold ionization at $\lambda = 800\,\text{nm}$ and $t_p = 200\,\text{fs}$, and compare them with the corresponding experimental data recently obtained by Larochelle et al [39]. As can be seen from the Figure, the theoretical calculations and the experimental yields agree remarkably well with the data (except perhaps for Xe^{3+} at the lowest intensities [40]) as well as reproduce the *multiple knee* structures observed for the higher charge states as well.

To gain a better insight into the above result, we have further calculated the hypothetical yields of the charge states by neglecting specific NS channels. We found for all cases that the neglect of one or more of the NS rates has a drastic effect on the magnitudes of the yields. In the right hand panels, (i) and (ii), we show the results of the detailed calculations for the two highest charge states measured i.e. Xe^{5+} (panel i), and Xe^{6+} (panel ii) and compare them with the experimental data. Note, for example, for Xe^{6+} we show the results of the full calculation (bold solid curve) and four additional curves that are calculated by *neglecting* the following NS channels (thin curves): (a) $0-6$, $1-6$, (b) $0-6$, $1-6$, $2-6$, (c) $0-6$, $1-6$, ..., $3-6$ (d) $0-6$, $1-6$, ..., $4-6$. (In the above notation the integers stand for the charge states involved, e.g. $1-6$ stands for the channel in which an initially singly ionized atom becomes a final six-fold ionized atom.)

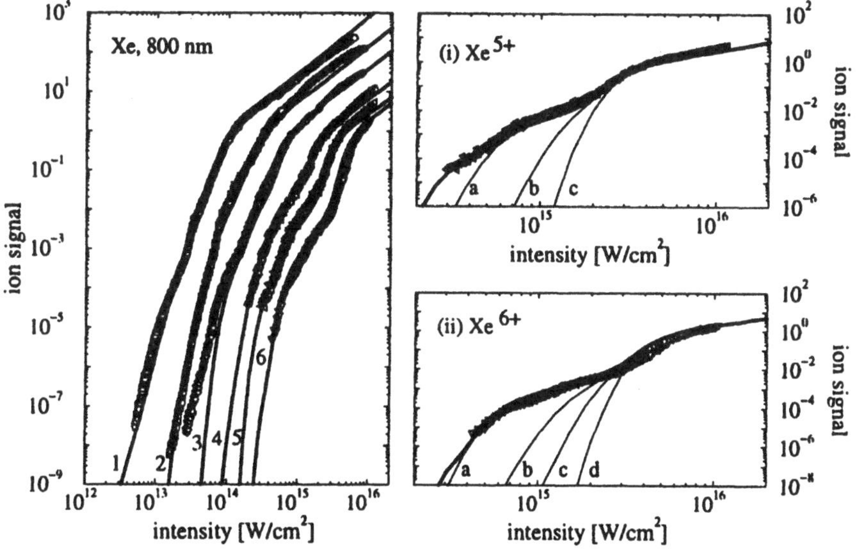

Figure 1.7. Comparison of experiment and theory (solid curves) for *n*-fold ionization (n = 1-6) of Xe. Experimental data for λ = 800 nm, t_p = 200 fs [39].

And similarly for Xe^{5+} in panel (ii). The drastic reduction of yields for neglecting specific NS channels, mentioned above, is now apparent. In general, it is found that only specific NS channels contribute dominantly in different parts of a yield curve. We note from the figures that in general a higher order NS channel become more important at the lower intensities than a lower order channel. For example, we observe in panel (ii) that in the intensity domain of the measurement, the rates for the NS transitions $0-6$ and $1-6$ (c.f. solid curve and thin curve a) make no significant contribution but the channel $2-6$ becomes very significant at the lowest intensities. From this result we may also conclude that the contributions of upto the fourth order NS channels, involving direct escape of upto four electrons, are significantly present in the measured yields of Xe^{6+}. Similar conclusions hold for Xe^{5+} as can be seen from panel (i); (and correspondingly for the other charge states). One finds that the yield curve for a given charge state tends to change its slope in the intensity range between two neighbouring

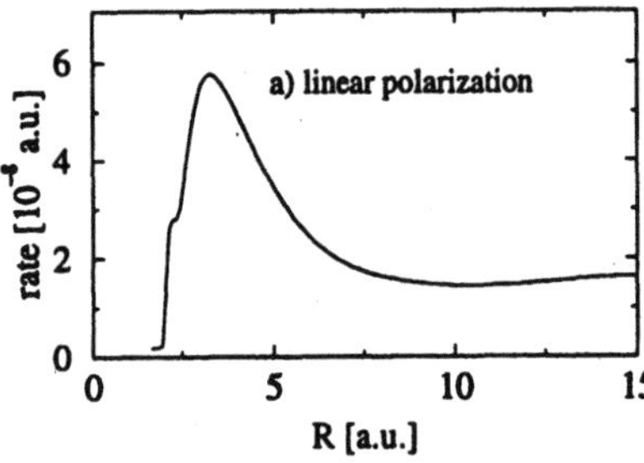
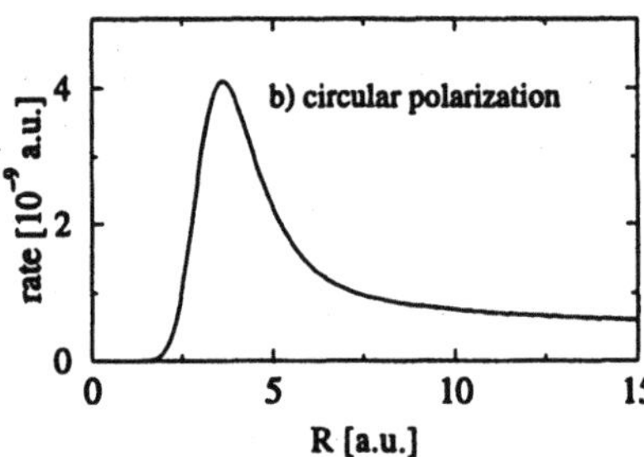

Figure 1.8. Rates of laser induced single ionization of H_2 as a function of intermolecular distance R. A comparison is shown for linear (panel a) and circular (panel b) polarized laser light at $\lambda = 532$ nm and $I = 5 \times 10^{13}$ W/cm². The orientation of the molecular axis is chosen parallel to the polarization axis (panel a) and parallel to the propagation of light (panel b).

dominant channels. This is due to the competition between the neighbouring NS or sequential channels. Note, finally, that a significant knee-like structure covering a broad range of intensity, arises only when one of the two neighbouring channels contributes in the total yield more dominantly, over the intensity range, than the other.

1.3.4 The phenomenon of 'enhanced ionization' of two-electron molecule, hydrogen, and of many-electron molecules, benzene and its cation

Finally, we illustrate the use of the theory for molecular systems and present the results (using the lowest order IMST diagrams) of the total rates of single ionization of the diatomic hydrogen molecule H_2, as well as of the polyatomic benzene molecule and its cation, C_6H_6 and $C_6H_6^+$, respectively. We analyze the dependence of the corresponding rates on the intensity of the laser, for two different polarization, as well as on the separation of the nuclear coordinates. The calculations are performed by generalizing the well-known KFR-formula for atomic ionization [12, 13, 14] including a Coulomb correction [21, 22] to the laser induced ionization of diatomic and polyatomic molecules (e.g. [41, 42]).

The wavefunctions for the neutral molecules and their ions are calculated by Valence-Bond method [43], in the case of the diatomic hydrogen molecule, and the Hückel Molecular Orbital method [43, 44], for the π-electron-system of benzene and its cation.

Fig. 1.8(a, b) shows the results for ionization of H_2 in a linear (panel a) and circular (panel b) polarized laser field at $\lambda = 532$ nm and $I = 5 \times 10^{13}$ W/cm² [42]. The orientation of the axis of the molecule is chosen here as parallel to the laser polarization axis (Fig. 1.8(a)) and parallel to the propagation direction of light (Fig. 1.8(b)), respectively. It can be clearly seen from the Figure that in both cases the ionization rate at the intermediate internuclear distances is strongly enhanced compared to that at the equilibrium distance ($R = 1.64$ a.u.). The maximum of the rate is found to occur at a distance of $R_c \approx 3.5$ a.u. for both the laser polarizations. This result further confirms the existence of the so-called 'enhanced-ionization' phenomenon, that has been discussed by several authors in the past using numerical simulations for H_2^+ and models of H_2 (e.g. [45, 46]).

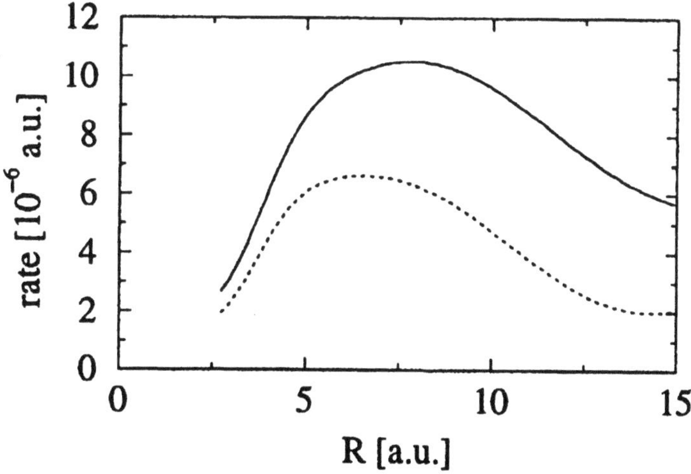

Figure 1.9. Results of calculations of the ionization rates as a function of the ring-radius R for the polyatomic molecule C_6H_6 for two different geometries: a) the laser polarization in the plane of the benzene ring and b) the laser polarization perpendicular to this plane. The parameter of the linearly polarized laser field are $\lambda = 532$ nm and $I = 5 \times 10^{13}$ W/cm^2.

Direct numerical simulations of laser induced ionization, however, is practically unrealizable for the many-electron polyatomic molecules such as benzene and its cation. The S-matrix theory, however, retains its practical usefulness in such cases as well. In fig. 1.9 we present the results of calculations based on the leading single ionization diagram for ionization of C_6H_6, as a function of the radius of the benzene ring. To be specific we have calculated the rates at a wavelength $\lambda = 532$ nm and intensity $I = 5 \times 10^{13}$ W/cm^2. Results are obtained for two different geometries, namely, the laser polarization axis in the plane of the benzene ring (solid curve), and perpendicular to this plane (dotted curve) [42]. In both cases we observe the phenomenon of 'enhanced ionization' which appears in this case as a broad maximum as a function of the ring-radius R, with a peak near $R \approx 7.5$ a.u.

30

Finally, Fig. 1.10(a, b) shows the rate of single ionization for the cation of benzene, $C_6H_6^+$, at the ultraviolet (KrF excimer laser) wavelength $\lambda = 248$ nm, as a function of R. Calculations are made at two intensities $I = 2 \times 10^{14}$ W/cm^2 (panel a) and $I = 2 \times 10^{15}$ W/cm^2 (panel b). Comparing the results in the two panels, one sees that the maximum rate of ionization occurs for both the intensities at about the same critical distance lying between 4.0 – 4.5 a.u. Thus, the critical distance for 'enhanced ionization' in the present case appears to be rather insensitive to the variation of the intensity (here, over an order of magnitude). A similar behaviour in Cl_2 has been observed experimentally [47, 48], recently.

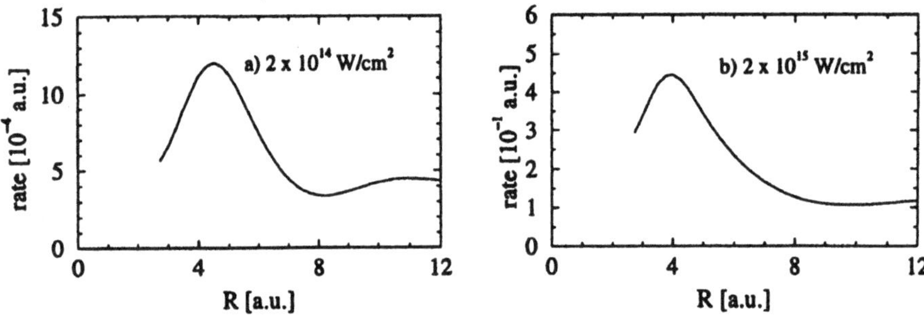

Figure 1.10. Rates of ionization of the cation of benzene, $C_6H_6^+$, at a laser wavelength $\lambda = 248$ nm at two different intensities $I = 1 \times 10^{14}$ W/cm^2 (panel a) and $I = 1 \times 10^{15}$ W/cm^2 (panel b). The polarization axis of the linear polarized laser field is chosen to lie in the plane of the benzene cation.

1.4 CONCLUSIONS

We have given a brief review of the newly developed 'intense-field many-body S-matrix theory' (or IMST), and have demonstrated, with results of concrete calculations, how it can be fruitfully used to analyse a whole class of highly non-linear many-body problems arising from the interaction of intense femtosecond laser pulses with many-electron atomic and molecular systems. Thus, first, the ATI spectrum of He is investigated, which confirms the production of 'hot electron plateaus' due to the rescattering of the ionized electron with the residual core. The results calculated by including a single-rescattering show a much improved agreement with the data compared to the no-scattering results. Second, the double ionization yields of He as a function of the laser intensity at optical and ultraviolet wavelengths are calculated and compared with the experimental data. The results show an excellent agreement between the theory and the experiment. Third, the yields of *multiple* ionization of Xe, leading to the production of high charge states (upto six-fold ionized atoms) are calculated as a function of the laser intensity. ·The results show remarkable agreement with the recently measured experimental data for the same in the field of a femtosecond Ti:Sapphire laser pulse. Fourth, the phenomenon of 'enhanced ionization' of the diatomic molecule H_2 is further confirmed by the present theoretical calculations. Finally, occurrence of the 'enhanced ionization' for the many-electron polyatomic molecules, benzene and its cation, are predicted. The critical radius at which the enhancement for $C_6H_6^+$ occurs is found to be rather insensitive to the change of the laser intensity; this finding is consistent with a similar behaviour in Cl_2, observed experimentally.

ACKNOWLEDGMENTS

We are thankful to Drs. P. Agostini, S. L. Chin, L. F. DiMauro and A. Talebpour for kindly making available to us their published data. We are grateful to Dr. D. Andrae for providing us with the Hartree-Fock wavefunctions for the noble gases and their ions. This work has been partially supported by the *Deutsche Forschungsgemeinschaft* (Bonn) under SPP: 'Wechselwirkung intensiver Laserfelder mit Materie', FA 160/18-1.

REFERENCES

1. A. D. Bandrauk, *Molecules in Laser Fields*, Marcel Dekker, New York (1994).

2. L. F. DiMauro and P. Agostini, Adv. At. Mol. Opt. Phys. **35**, 79 (1995).

3. P. Lambropoulos and H. Walther, Eds. *Multiphoton Processes 1996*, Int. Nat. Conf. Ser. No. 154 (IOP: Bristol) (1997).

4. J. Parker, K. T. Taylor, C. W. Clark and S. Blodgett-Ford, J. Phys. B **29**, L33 (1996).

5. K. T. Taylor, J. S. Parker, D. Dundas, E. Smyth and S. Vivirito, in [3], p. 56 (1997).

6. J. S. Parker, E. Smyth and K. T. Taylor, J. Phys. B **31**, L571 (1998).

7. F. H. M. Faisal and A. Becker, in *Selected Topics on Electron Physics*, D. M. Campbell and H. Kleinpoppen, Eds. (Plenum: New York), p. 397 (1996).

8. F. H. M. Faisal and A. Becker, in [3], p. 118 (1997).

9. A. Becker, *Ph.D. thesis*, Fakultät für Physik, Universität, Bielefeld (1997).

10. M. L. Goldberger ad K. M. Watson, *Collision Theory* (Wiley and Sons: New York), e.g., pp. 186-197 (1964).

11. It can be shown using Schrödinger equation and partial integration that the difference integral

$$-\frac{i}{\hbar}\int_{-\infty}^{t}dt'\left\langle\Phi_f^0(t')\left|V_f(t')-V_i(t')\right|\Phi_i^0(t')\right\rangle=\left\langle\Phi_f(t')\left|\Phi_i(t')\right\rangle\right|_{-\infty}^{t}$$

which does not contribute to the S-matrix.

12. L. V. Keldysh, Zh. Eksp. Teor. Fiz. **47**, 1945 (1964) [Sov. Phys. JETP, **20**, 1307 (1965)].

13. F. H. M. Faisal, J. Phys. B. **6**, L89 (1973).

14. H. R. Reiss, Phys. Rev. A **22**, 1786 (1980).

15. A. Becker and F. H. M. Faisal, Laser Phys. **8**, 69 (1998). We may note that in Eq. (7) of this reference, the momentum k_n has been changed rather *ad hoc* to $k_{n,\,m}$ in order to consistently fit with the on-shell energy $E_{n,\,m}$ of the 'e-2e' rates.

16. B. Sheehy, B. Walker, R. Lafon, W. Widmer, A. Gambhir, L. F. DiMauro, P. Agostini and K. C. Kulander, in [3], p. 98 (1997).

17. G. G. Paulus, W. Becker, W. Nicklich and H. Walther, J. Phys. B **27**, L730 (1994).

18. F. H. M. Faisal and A. Becker, in *Super-Intense Laser-Atom Physics IV*, H. G. Muller and M. V. Federov, Eds. (Kluwer Academic Publishers: Dordrecht), 317 (1996).

19. A. Becker and F. H. M. Faisal, J. Phys. B **29**, L197 (1996).

20. F. H. M. Faisal and A. Becker, Laser Phys. **7**, 684 (1997).

21. A. Becker and F. H. M. Faisal, Phys. Rev. (submitted) (1998).

22. F. H. M. Faisal and A. Becker, Comm. At. Mol. Phys. (in press) (1998).

23. J. H. McGuire, *Electron Correlation Dynamics in Atomic Collisions* (Cambridge University Press) (1997).

24. Note that in the weak field photo-ionization the usual convention for writing the Feynman diagrams, including the TS1 diagram, does not follow a definite direction of time (c.f. e.g. Ref. 22). The structure of the latter diagram is however topologically equivalent to the intense-field TS1 diagram, refer Fig. 4(c), where a definite time direction is followed for a greater intuitive clarity.

25. M. Yu. Kuchiev, Sov. Phys. – JETP Lett., **45**, 2141 (1987).

26. M. Yu. Kuchiev, J. Phys. B **28**, 5093 (1995).

27. M. Yu. Kuchiev, Phys. Lett. A **212**, 77 (1996).

28. P. B. Corkum, Phys. Rev. Lett. **71**, 1994 (1993).

29. T. Brabec, M. Yu. Ivanov and P. B. Corkum, Phys. Rev. A **54**, R2551 (1996).

30. D. N. Fittinghoff, P. R. Bolton, B. Chang and K. C. Kulander, Phys. Rev. Lett. **69**, 2642 (1992).

31. B. Walker, B. Sheehy, L. F. DiMauro, P. Agostini, K. J. Schafer and K. C. Kulander, Phys. Rev. Lett. **73**, 1227 (1994).

32. Thus, for the absorption of n' photons by back-scattering, the dominant order $n' = \alpha_0\left(k_n + k_{n+n'}\right)$, where $k_{n+n'}$ is the back-scattered momentum arising from the incident momentum k_n, and $\alpha_0 = \sqrt{I}/\omega^2$. Using this equation and the identity $k_{n+n'}^2/2 - k_n^2/2 = n'\omega$, one can easily find the energy after the back-scattering, $k_{n+n'}^2/2 = k_n^2/2 + 4k_n\sqrt{U_p} + 8U_p \equiv \mathcal{E}_n$. We may note that $\mathcal{E}_n$ may also be considered as the maximum possible classical energy attainable in the field, which would seem to restrict the applicability of the formula, Eq. (1), for laser parameters outside the so-called 'multiphoton regime'.

33. W. Lotz, Zeit. F. Phys., **216**, 241 (1968).

34. V. P. Krainov, J. Opt. Soc. Am. B **14**, 425 (1997).

35. M. R. Cervenan and N. R. Isenor, Opt. Comm. **13**, 175 (1975).

36. P. W. Milonni and J. H. Eberly, *Lasers* (Wiley and Sons: New York), p. 490 (1988).

37. D. Charalambidis, D. Xenakis, C. J. G. J. Uiterwaal, P. Maragakis, Jian Zhang, H. Schröder, O. Faucher and P. Lambropoulos, J. Phys. B. **30**, 1467 (1997).

38. A. Becker and F. H. M. Faisal, Phys. Rev. (submitted) (1998).

39. S. Larochelle, A. Talebpour and S. L. Chin, J. Phys. B **31**, 1215 (1998).

40. At present this single departure from the trend is not understood; whether it is due to a possible induced resonance (or to other effects) not included in the present model, requires further investigations for its clarity.

41. T. Zuo, A. D. Bandrauk and P. B. Corkum, Chem. Phys. Lett. **259**, 313 (1996).

42. J. Muth, *Diplom thesis*, Fakultät für Physik, Universität Bielefeld, (1997).

43. K. Yates, *Hückel Molecular Orbital Theory* (Academic Press: New York) (1978).

44. E. Hückel, Z. Phys. **60**, 423 (1930).

45. T. Seideman, M. Yu. Ivanov and P. B. Corkum, Phys. Rev. Lett. **75**, 2819 (1995).

46. H. Yu, T. Zuo and A. D. Bandrauk, Phys. Rev. A **54**, 3290 (1996).

47. M. Schmidt, D. Normand and C. Cornaggia, Phys. Rev. A **50** (1994).

48. D. Normand, S. Dobosz, M. Lezius, P. D'Oliveira and M. Schmidt, in [3], p. 287 (1997).

Structure and Properties of Atomic Clusters

Sugata Mukherjee

S. N. Bose National Centre for Basic Sciences,
Block JD, Sector III, Salt Lake City, Calcutta - 700 091, India

A brief review is presented on various properties of small atomic clusters of different elements. Experimental measurements and theoretical explanation of abundance spectra, stability of multiply-charged clusters, magnetic properties, reactivity studies on clusters, and vibrational studies are discussed. A brief account of the *ab-initio* theoretical method based on local density-functional theory and Car-Parrinello molecular dynamics method used for calculating groundstate electronic properties of clusters are presented.

1.   INTRODUCTION

The study of elemental micro-clusters, containing between two to several thousand atoms, provides the possibility to examine the transition from molecule to crystalline solid. The physical and chemical properties of the clusters vary significantly with the number and type of their constituent atoms which can be quite different than in crystalline solid or on solid surfaces. Most of the clusters exhibit some sort of 'magic numbers' which are ascribed to indicate certain abrupt enhancement or decrement of a physical or chemical property (e.g. binding energy, ionization energies, chemical activity etc.) of clusters containing a given number of atoms. These 'magic clusters' can be of significant use in devising novel cluster-assembled nanomaterials. Moreover, clusters of metals, semiconductors, their compounds and alloys were found to exhibit striking structural, magnetic and catalytic properties, not found in the bulk materials. During last few decades great progress has been made to produce, detect and mass-select clusters of a variety of elements and their compounds. For example, it has been possible to produce and detect very weakly bound inert gas clusters [1, 2] as well as clusters of very high boiling-point materials such as carbon [3] and silicon [4] with strong covalent bonding.

In typical experiments, clusters are formed in a supersonic jet expansion [5] of the element vapour from a high pressure chamber into vacuum. The expanding and cooling gas provides thousands of collisions of the atoms before the mean free path between collision increases rapidly with decreasing gas pressure. During this expansion phase clusters of atoms are formed by nucleation and growth [6]. Actually only a small fraction of the atoms form into clusters, the other unbound atoms carry away the heat of condensation of the cluster. This technique works very well for elements whose vapour pressure is high at moderate temperature. Smalley and co-workers [7] have greatly extended the range of elements accessible to cluster formation by introducing nanosecond pulsed lasers to evaporate the material of interest into a buffer gas of helium. Clusters nucleate and grow in the buffer gas which expands and cools. A mass filter is then used to isolate size-selected clusters. Using this novel technique clusters of elements with low vapour pressure (e.g. carbon, silicon, germanium etc.) are produced. Once the clusters are generated, they can be used for further studies.

In the next section we shall discuss about few selected properties of the atomic clusters in detail and in section 3 a brief theoretical account for calculating the groundstate properties of clusters will be discussed.

2. SELECTED PROPERTIES OF CLUSTERS

2.1 Abundance Spectra and Shell Structure

The mass spectrum of the cluster beam gives evidence of the relative abundance of different aggregates present in the beam. In Fig. 1(a) we show the measured abundance spectrum of Na_n-clusters [8, 9] which show a number of features. The spectrum shows peaks for all clusters in the size range 3-70, indicating presence of all clusters in the beam. However, particular prominent peaks were observed at $n = 8, 20, 40, 58, 92$ in addition to a peak at $n = 2$. In Fig. 1(b) the second difference of the energy of sodium clusters, calculated using a model potential is shown, which clearly indicates the most stable clusters having same number of atoms as observed experimentally. Similar mass spectrum was also recorded for potassium clusters [10] as well with prominent peaks at same n values as in sodium clusters. These measurements indicate that Na_n and K_n clusters are particularly stable and therefore more abundant for $n = 8, 20, 40, 58, 92$

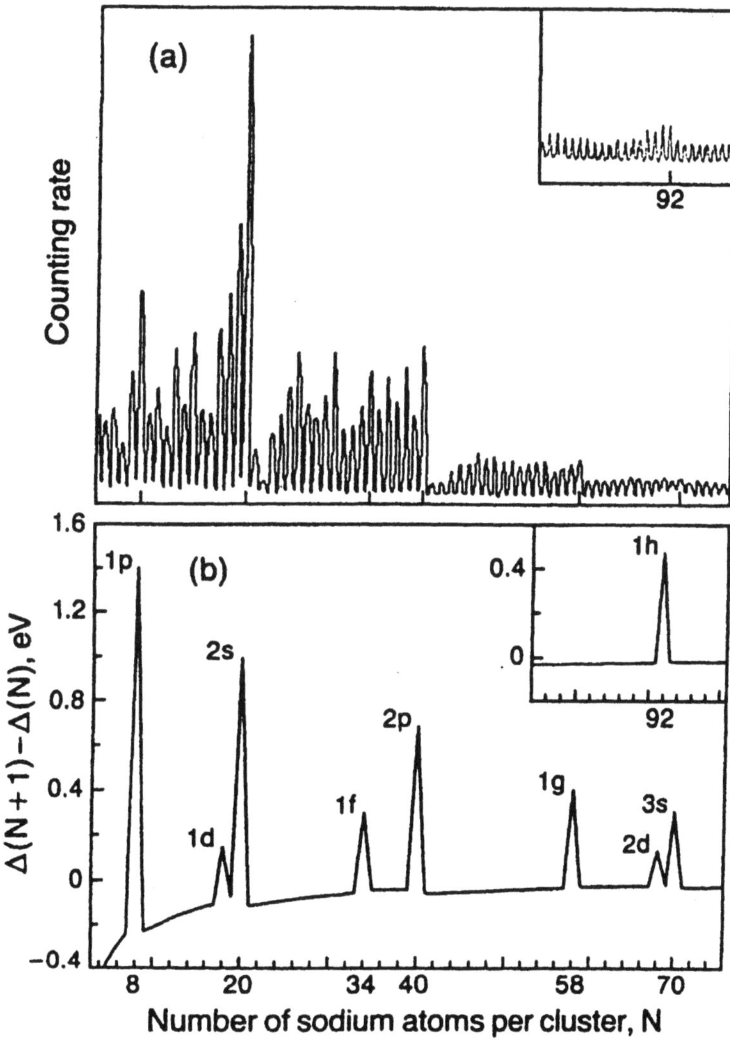

Figure 1. (a) Experimental mass spectrum of sodium clusters as a function of the number of
atoms in the cluster. (b) The second derivative of the energy calculated using
Woods-Saxon type potential. (after Knight et al [8])

which are so called 'magic numbers' for these clusters. The explanation for
the occurrence of such magic numbers was given [8] by assuming the
clusters as spherical droplets of electrons moving in a uniform spherically
symmetric potential due to positive charge background (jellium model). The
total energy of such spherical jellium sphere was calculated [8, 11] using
density functional theory [12] in the local density approximation [13].
When the one-particle Kohn-Sham equations were solved self-consistently
for the jellium sphere, the eigenvalues were found to follow the sequence $1s$,

38

$1p$, $1d$, $2s$, $1f$, $2p$, $1g$, $2d$, $1h$, $3s$, ..., where s, p, d, f, g, ... correspond to angular momentum states $l = 0$, 1, 2, 3, 4, ... with degeneracy $2(2l + 1)$ respectively. Therefore, the shell closing occurs when the total number of valence electrons (or number of atoms in monovalent Na_n and K_n clusters) are $n = 2$, 8, 18, 20, 34, 40, 58, 68, 90, 92... Similar calculations [14] for divalent Mg_n clusters show the occurrence of magic numbers for $n = 4$, 10, 17, 20, 29, 34, 46...; which correspond to the number of electron 8, 20, 34, 40, 58, 68, 92... respectively. However, for trivalent Al_n clusters calculations [14] give magic numbers for $n = 6$, 7, 13, 19, 23, 31... corresponding electron numbers in the clusters being close to shell filling values. It is expected that Al_n^+ should exhibit magic numbers for $n = 3$, 7, 23, 31 etc.

Electronic shell structure was later found to exist in sodium [15] and lithium [16] clusters containing upto few thousands of atoms, with the magic numbers following same sequence as above. Moreover, it was found that the sequence of shell structure follow similar trend as expected in the case of electrons in a spherical cavity describing semi-classical triangular or square orbits, conjectured long ago [17]. The predicted interference effects [18] arising due to slightly different periodicities of the magic numbers were found to exist also in metal clusters, leading to supershell structure in the mass spectrum. Thus, group I metal clusters were found to form a suitable model system for the study of shells with large quantum numbers, where comparison between ordered classical motion and quantum (electronic) shell structure becomes apparent.

Mass spectrum of photoionized clusters, containing upto 20000 atoms, was measured by Martin and coworkers [19]. They found evidence of closed packed polyhedral structure for clusters of varying size range. With increasing size, the clusters grow by adding on layers or shells of atoms on each micro-facets of the polyhedra, retaining the original shape of the cluster. It was found that the clusters of NaCl and other alkali halides tend to grow as tiny cubes with a local atomic arrangement almost identical to that in a crystal. Al and In clusters take the form of octahedra, whereas Na, Ca and Mg clusters were found to form icosahedra. In Fig. 2 we show the mass spectra of large photoionized Na_n clusters as observed by Martin and coworkers [19]. Since closed-shell clusters have high ionization energies [19], minima in the ionization threshold energies occur for such clusters. It is evident from Fig. 2 that the number of atoms in closed-shell clusters of sodium matches closely to that in closed-shell Mackay-icosahedra [20] in the size range between 2000 and 20000 atoms. Similar measurements on Ca_n clusters show clear evidence of icosahedral packing of atoms in the cluster, where addition of shells of atoms on each micro-facet gives rise to secondary maxima observed in the mass spectra. This kind of experiments is ideal to examine the growth of the clusters and so far a large variety of clusters ranging from metal clusters to clusters of C_{60} and metal-coated C_{60} have been studied [21].

The abundance spectra for noble metal cluster ions of Cu_n^+, Ag_n^+ and Au_n^+ has been measured [22]. The cluster ions exhibit electronic shell structure at $n = 3, 9, 21, 35, 41, 59$, with additional peaks at 93, 139 and 199 for Ag_n^+. This data is a clear confirmation of shell structure. At smaller size, the noble metal cluster ions exhibit the odd-even variations in abundances.

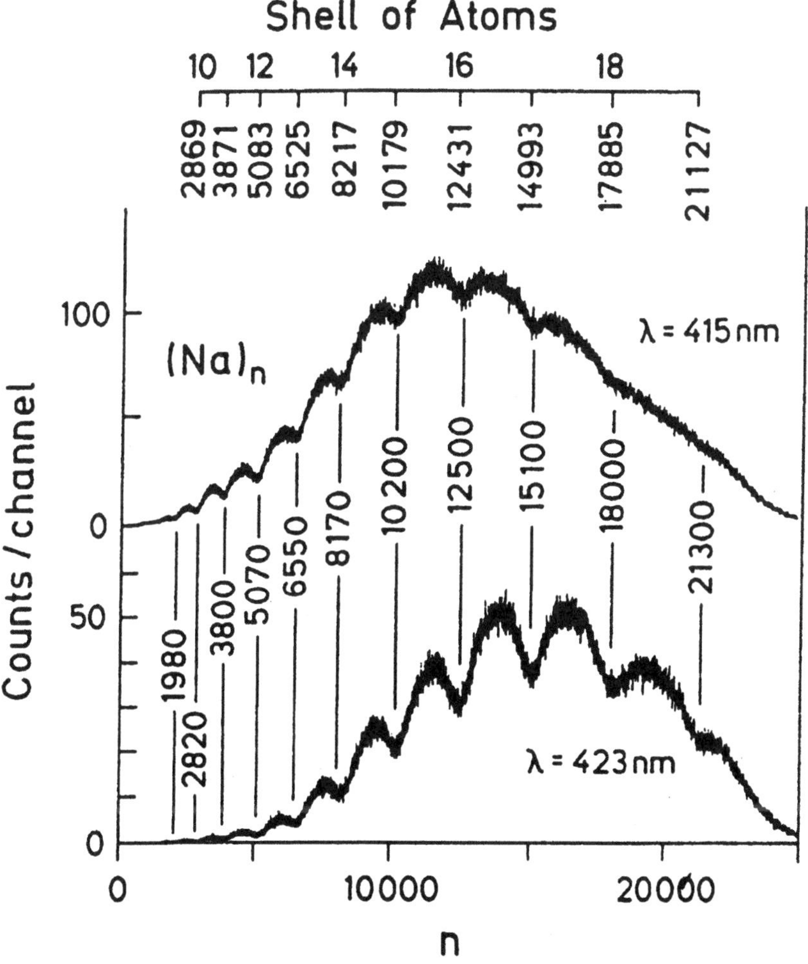

Figure 2. Mass spectra of Na_n clusters photoionized with photons of two different energies. Minima in the ionization threshold occur at values of n corresponding to the icosahedral or cubooctahedral shell closings shown above. (after Martin [19])

The abundance spectra of semiconductor micro-clusters Si_n and Ge_n have been reported. The abundance spectrum of silicon cluster ions indicate [4, 23] that prominent species are Si_6^+ and Si_{10}^+. The photo-fragmentation data [4] indicate that when the initial cluster ion breaks up, the positive charge remains predominantly on the larger fragment and for breakup clusters with n = 7-11, the fragment Si_6^+ is usually prominent. Si_4^+, Si_6^+ and Si_{10}^+ have relatively small total photofragmentation cross section and therefore more stable. The mass spectrum of Ge_n clusters [24] indicate strong peaks for n = 6, 10, 14 and 18. Therefore, both for Si_n and Ge_n clusters most stable species are for n = 6 and 10, which has been verified by various theoretical calculations to be discussed later in this paper.

2.2 Stability of Multiply-charged Clusters

Stability and metastability of small multiply charged clusters and molecules has been studied experimentally [25, 26, 27, 28] in several systems. Stability in multiply-charged clusters is governed by a delicate balance between the repulsive electrostatic interaction and the attractive bonding interaction in the clusters. There exists a critical cluster size, below which the Coulomb repulsion of the charges exceeds the bonding interaction, leading to 'Coulomb explosion' or spontaneous fragmentation of the cluster. Model calculations [29] were performed, by assuming the positive charges localized on the opposite ends of the clusters, to calculate the critical cluster size for doubly and triply charged clusters. This model was later extended [30] to incorporate the effect of screening of the localized charges by the valence electrons of the cluster, and some experimental data [26] on the stability of doubly-charged metal clusters could be explained. However, it was later suggested [31] that the existence of a potential barrier can explain the delicate balance between the cohesive and repulsive energies. Calculations on doubly-charged transition-metal dimers showed [32] that the height of this barrier depends sensitively on the occupation of the d-electron levels. The doubly-charged transition-metal dimers with nearly-filled d-shell undergo a spontaneous fragmentation (their binding energy curves show no barrier), whereas those with half-filled d-shell are metastable and protected against fragmentation by an energy barrier.

Experiments [27] on gold clusters show that the minimum cluster size for Coulomb explosion in Au_m^{2+} is $m = 9$. On the other hand, experiments [28] on Au_2^{2+} could verify the occurrence of Au_2^{2+}. This apparent puzzle on the possible metastability of Au_2^{2+} was explained [33] by calculating the energy of Au_2^{2+} in the excited state, which involves transfer of an electron from highest occupied d-like state to the next unoccupied s-like state. The excitation $d \to s$ causes a loss in binding due to the creation of a d-hole, and at the same time a gain in binding due to occupying a 'bonding' s-state. In effect, this leads to metastable minimum in the energy curve of Au_2^{2+}, which is protected by a small energy barrier [33]. Our result was verified later by relativistic Hartree-Fock calculations performed independently by Pyykkö [34]. Similar metastable minimum in the energy curve was found [35] to occur also in Cu_2^{2+} and Ag_2^{2+}. The transition from metastable Au_m^{2+} to stable counterpart was calculated to occur at $m \sim 6$. Stability of multiply-charged beryllium [36] and magnesium [37] clusters were also calculated using local density-function method. It was found that doubly charged Mg_n-clusters undergo from metastability to stability at $n \sim 9-10$. Be_n^{2+} was found to be stable for $n > 2$.

2.3　Magnetic Properties of Clusters

The magnetic properties of the magnetic metal clusters were investigated in great detail in the recent past. The magnetic moment per atom in a ferromagnetic material depends strongly on the atomic environment. Calculations [38] suggested that the magnetic moment per atom in Fe, Co and Ni, confined to linear chain configuration is larger than those confined on a two dimensional plane and the latter carries a larger moment than the atoms in the bulk. The magnetic moment of the isolated atoms being even larger than in three configurations. The magnetic moment per atom inside isolated ferromagnetic clusters are larger than that in bulk and increases with increasing symmetry of the cluster. However, recent experiments on Fe_n $(60 < n < 250)$ [39] and on Co_n $(50 < n < 260)$ [40] found that the average magnetic moment of these clusters were much smaller than bulk values, increased almost linearly with applied magnetic field for small fields, and depended sensitively on the temperature of the cluster source. The experiments were performed using mass-selected cluster beam, produced in laser vaporization source, and deflected in a Stern-Gerlach gradient magnet. Unlike the atoms, the clusters were found to be deflected only in one direction. The effective magnetic moment was then evaluated from the measured deflection and the applied magnetic field in the gradient magnet. It was explained theoretically [41, 42] that the apparent contradiction between experiment and calculations was essentially due to the effect of temperature induced fluctuations in the cluster in a beam.

Ferromagnetic ordering can be understood in terms of fluctuations with different length scales. Therefore, interesting effects can be expected when the correlation length of the spin system is no longer commensurate with the

42

linear dimension of the particle. In magnetically ordered materials, grains of size typically smaller that 20 nm tend to lose their spontaneous magnetization at temperatures well below Curie temperature. This effect is known as superparamagnetism, since the grain behaves like a giant paramagnet with the magnetic moment fluctuating rapidly under thermal activation. Superparamagnetism at the room temperature is observed only in ultrafine particles where the effect of magnetic anisotropy can be comparable to the thermal energy. The thermal behaviour of the ferromagnetic clusters at a given temperature T is then determined by a competition between the thermal energy $k_B T$ and the magnetic anisotropic energy K. The thermal relaxation time is given by [43],

$$\tau = \tau_0 \exp\left[nK/k_B T_b \right]. \tag{1}$$

Here, τ_0 is the gyromagnetic precession time $\left(\sim 10^{-9} \text{ s}\right)$, n is the number of atoms in the particle, and k_B is the Boltzmann constant. T_b is the so called 'blocking temperature', below which the effect of the lattice anisotropy is dominating, and above which the superparamagnetism is observed.

The average magnetic moment of the cluster was calculated from the mean field solution of a model Hamiltonian containing following three terms to describe the spin system in the external magnetic field.

$$H_{ex} = -J \sum_{\langle ij \rangle} \mathbf{S}_i \cdot \mathbf{S}_j. \tag{2}$$

H_{ex} describes the exchange energy in a Heisenberg model with interacting nearest neighbour spin pairs $\mathbf{S}_i$ and $\mathbf{S}_j$. The magnetic moments $\mu_i = \mu \mathbf{S}_i$ are treated as continuous (classical) vectors. The external field energy is described by,

$$H_h = -\sum_i \boldsymbol{\mu}_i \cdot \mathbf{H}. \tag{3}$$

The lattice anisotropy energy, which fixes the cluster magnetization to an easy axis direction $\mathbf{c}$, is given by,

$$H_{ani} = -K \sum_i (\boldsymbol{\mu}_i \cdot \mathbf{c}/\mu_i)^2. \tag{4}$$

We have calculated the average magnetization component $\langle \mu_z \rangle$ of the cluster along the external magnetic field as a function of the field and observed agreement with the experimental data for both cobalt [41] and iron [44] clusters. In Fig. 3 we show our results for Fe_n clusters compared with the experimental data. The asymmetric Stern-Gerlach deflection profiles can be understood from a mechanism involving the transfer of electronic angular momentum to the cluster rotation or to other degrees of freedom, allowing a depopulation of the high energy magnetic levels [41].

2.4   Chemical Activity of Clusters

Small clusters have a large surface to volume ratio, i.e., relatively large number of atoms are situated at the surface of the clusters. These surface atoms often have a large number of dangling bonds to which other molecules can bind themselves and get dissociated, forming other chemical compounds through a catalytic reaction. In fact, many catalysts used in chemical processes are in the form of highly dispersed small metal clusters [45, 46]. The number of the 'reactive' or surface binding sites may vary drastically with the size and shape of the cluster, thereby making clusters containing a certain number of atoms less reactive (or more reactive) than others. Chemical activity of metal [47] and semiconductor [48, 49] clusters have been measured experimentally for different chemical reactions. The 'magic clusters' in the reactivity data (clusters which were less reactive) was found not to follow the same sequence of magic numbers observed in abundance spectrum discussed before. For example, large silicon clusters containing more than 20 atoms were found to exhibit no magic numbers in the mass spectrum [50], whereas it was found that silicon clusters containing 21, 25, 33, 39 and 45 atoms are much less reactive towards ammonia chemisorption reaction than other silicon clusters in the similar size range [49]. In Fig. 4 the experimental data for the reactivity of silicon clusters has been shown. Further experiments on silicon clusters [51] found that the dissociation energy and the cohesive energy of Si_n-clusters is a smooth and feature-less function of the size for $n \geq 25$. Moreover, well-annealed clusters showed a drastic reduction of the reactivity for ammonia chemisorption which was several orders of magnitude smaller than that on the most stable $(7 \times 7)Si(111)$ surface [52, 48]. Therefore, there is no correlation between the magic clusters as seen in the mass spectrum and those which exihibit reduced reactivity. The reactivity data (Fig. 4) of the silicon clusters have triggered a race to find a suitable structure of Si_n with n = 33, 39 and 45 [53, 54, 55, 56, 57, 58, 35]. For Si_{33} and Si_{45} clusters models were proposed [53] in which a four-fold coordinated 5-atom cluster

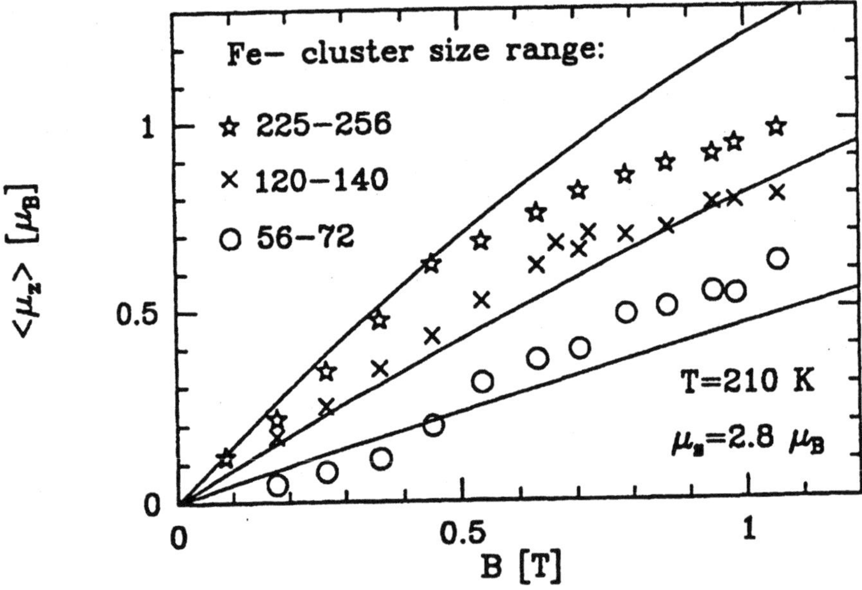

Figure 3. Calculated average magnetic moment of Fe_n clusters (full line) as a function of the applied magnetic field and comparison with experimental data (symbols [40]). (After Jensen et al. [44])

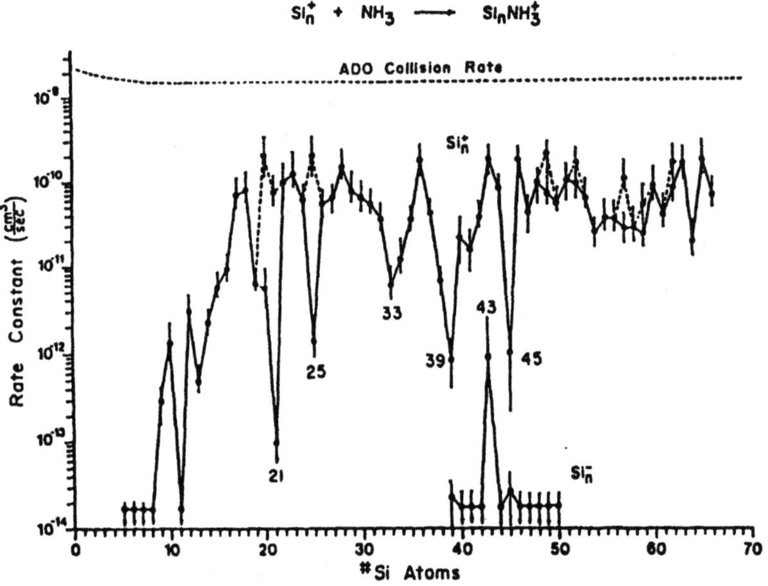

Figure 4. Measured reaction rate of ammonia chemisorption reaction on Si_n cluster ions as a function of the cluster size. (After Alford et al. [49])

was placed inside a fullerene-like cage with 28 and 40 atoms respectively. These structures showed T_d symmetry like in the bulk silicon. Later, calculations on Si_{45} [57] and on Si_{33} [58, 35] showed that above structures were not stable when the atoms were allowed to relax. The later calculations found new novel structures for Si_{45} and Si_{33} which had no central atom, but showed a two-shell structure, energetically more favourable than the previous structures [53]. The Si_{33} had a small 5-atom cage inside the 28-atom fullerene like cage, whereas Si_{45} had a 9-atom cage surrounded by a 36-atom fullerene like cage outside. Both these structures had less number of dangling bonds than previous structures and therefore less reactive. The details about the calculations and the results will be discussed later in this paper.

The experiments on the reactivity of Ni_n and Cu_n clusters for $n = 30\text{-}100$ suggested [47] that the possible structure of the clusters were icosahedron, since the clusters with closed atomic shell (or subshell) exhibited a reduction in the reaction rate with several molecules, e.g. H_2O, NH_3 and N_2. Thus the reactivity data of the clusters are very important for the understanding of the structure of the clusters.

Clusters of bimetallic alloys Cu-Ru and Cu-Os were found to exhibit very interesting miscibility property depending on the size of the cluster. The larger aggregates (bulk) of these alloys are not miscible, Cu being surface segregating in the entire concentration range, whereas in the form of small aggregates these alloys were found to be miscible [59, 46], with the degree of miscibility depending on the size of the cluster. This phenomenon led to the discovery of bimetallic cluster catalysts Cu-Ru and Cu-Os, in which the reaction rate towards ethane hydrogenolysis and dehydrogenation of cyclo-hexane could be controlled by varying the size of the aggregate [46]. The crucial quantity responsible for controlling the chemical activity of these alloy aggregates is the surface segregation of Cu at the surface of the aggregates. Model calculations [60, 61] have been performed on these alloy clusters assuming the cluster shape as icosahedral and cubooctahedral. It was found the strong Cu-segregation is inhibited in alloy clusters due to lower average coordination of the atoms.

2.5   Vibrational Measurements on Clusters

Recently vibrational measurements on size-selected Si_n ($n = 4, 6, 7$) clusters have been performed. The threshold photodetachment or zero electron kinetic energy (ZEKE) measurements [62] on Si_4^- obtained detailed information on the anion and several of the low lying excited states of the neutral Si_4. The observed spectrum was found to be consistent with the calculations by Raghavachari [63] and confirmed the rhombus groundstate structure of Si_4. Later, this result was largely augmented by Raman spectroscopy measurements [64] on size selected Si_n ($n = 4, 6, 7$) clusters deposited on solid nitrogen matrix. They have reported eleven sharp vibration lines for three clusters and comparison of their data with predicted Raman modes by Raghavachari confirmed the structure of the clusters. This is perhaps first ever direct experimental evidence of the compact structures of Si_6 and Si_7 which are very different from microcrystalline forms. This kind of measurements provides the scope of a direct comparison between experiment and theory, from which groundstate geometry of the clusters can be obtained quite accurately.

3. THEORETICAL STUDIES ON CLUSTERS

3.1 *Ab-initio* Total Energy Calculation Method

Most of the *ab-initio* calculations performed by the physicists are based on the density-functional theory (DFT) [12] in the local density approximation (LDA) [13]. LDA calculations have been successfully used in a wide variety of systems ranging from atoms, molecules and clusters to surfaces and bulk materials. The electronic and geometric structure of a system is determined from the quantum mechanical total energy of the system and its minimization with respect to electronic and nuclear degrees of freedom. Normally the nuclei are treated in adiabatic (Born-Oppenheimer) approximation. The density-functional theory allows one to model the electron-electron interactions properly, by mapping the problem of a strongly interacting electron gas onto that of a single particle moving in an effective non-local potential. It was proved by Hohenberg and Kohn [12] that the total energy of the system, including the exchange and correlation, is a unique functional of the density. The minimum value of the total energy functional is the groundstate energy of the system, and the corresponding density is the exact single-particle groundstate density. It was shown by Kohn and Sham [13] that it is possible to replace the many electron problem by an equivalent set of self-consistent one electron equations. The Kohn-Sham total energy functional for a set of doubly occupied electronic states ψ_i with ions at positions $\mathbf{R}_I$ can be written as,

$$E[\{\psi_i\},\{\mathbf{R}_I\}] = 2\sum_i \int \psi_i \left(-\frac{\hbar^2}{2m}\right)\nabla^2\psi_i\,dr + \int V_{ext}(\mathbf{r})\,n(\mathbf{r})\,dr$$

$$+E_{XC}[n(\mathbf{r})] + \frac{e^2}{2}\int \frac{n(\mathbf{r})\,n(\mathbf{r}')}{|\mathbf{r}-\mathbf{r}'|}\,dr\,dr' + \frac{1}{2}\sum_{I\neq J}\frac{Z_I Z_J}{|\mathbf{R}_I-\mathbf{R}_J|}$$

$$(5)$$

where, V_{ext} is the electron-ion potential, $E_{XC}[n(\mathbf{r})]$ is the exchange-correlation functional, Z_I is the charge on I-th ion at position $\mathbf{R}_I$, and $n(\mathbf{r})$ is the electron density given by,

$$n(\mathbf{r}) = 2\sum_i |\psi_i(\mathbf{r})|^2.$$

$$(6)$$

The set of wavefunctions ψ_i that minimize the Kohn-Sham energy functional obey the Kohn-Sham equations,

$$\left[-\frac{\hbar^2}{2m}\nabla^2 + V_{ext}(\mathbf{r}) + V_H(\mathbf{r}) + V_{XC}(\mathbf{r})\right]\psi_i = \varepsilon_i\psi_i.$$

$$(7)$$

Here, ε_i is the eigenvalue of the state i, and V_H is the Hartree potential given by,

$$V_H(\mathbf{r}) = e^2 \int \frac{n(\mathbf{r}')}{|\mathbf{r}-\mathbf{r}'|}\,dr'.$$

$$(8)$$

The exchange-correlation potential is given by the functional derivative,

$$V_{XC}(\mathbf{r}) = \frac{\delta E_{XC}[n(\mathbf{r})]}{\delta n(\mathbf{r})}.$$

$$(9)$$

In the LDA the exchange-correlation energy of an electronic system is written by assuming the exchange-correlation energy per electron $\varepsilon_{XC}(\mathbf{r})$ to be same as the exchange-correlation energy per electron in a homogeneous electron gas that has the same density as the electron gas at $\mathbf{r}$. Therefore,

$$E_{XC}[n(\mathbf{r})] = \int \varepsilon_{XC}(\mathbf{r})\,n(\mathbf{r})dr.$$

$$(10)$$

48

In case of a spin-polarized calculation, this energy should correspond to the density with the same spin polarization as in the actual system. There exists different parameterizations [65] for $\varepsilon_{XC}(\mathbf{r})$, all of which lead to very similar total energy results.

The electron-ion potential is written in terms of pseudopotential,

$$V_{ext}(\mathbf{r}) = \sum_I v_{ps}^I(\mathbf{r} - \mathbf{R}_I),\tag{11}$$

where $v_{ps}^I(\mathbf{r})$ is the pseudopotential for the I-th ion. Several ionic pseudopotentials exist in the literature [66]. In most applications norm-conserving pseudopotentials due to Bachelet et al [67] are used.

3.2   Car-Parrinello Molecular Dynamics

The *ab-initio* molecular dynamics (MD) method developed by Car and Parrinello [68] provides an efficient way for the minimization of the energy functional $E[\{\psi_i\}, \{\mathbf{R}_I\}]$ with respect to electronic $\{\psi_i\}$ and ionic $\{\mathbf{R}_I\}$ degrees of freedom. This method is based on the assumption that the motion of the electrons and ions can be treated in the adiabatic (Born-Oppenheimer) approximation, so that the electrons follow the ions instantaneously as they move and therefore remain very close to the groundstate of the corresponding ionic configuration. In the MD method the time evolution of the electronic and the ionic coordinates are incorporated by introducing an effective Lagrangian,

$$L = \sum_i \mu \langle \dot{\psi}_i | \dot{\psi}_i \rangle + \frac{1}{2} \sum_I M_I \ddot{\mathbf{R}}_I^2 - E[\{\psi_i\}, \{\mathbf{R}_I\}]\tag{12}$$

Here, μ is a fictitious mass associated with the electronic wavefunctions. The equations of motions describing the electron and ion dynamics is given by,

$$\mu \ddot{\psi}_i(\mathbf{r}, t) = -H\psi_i(\mathbf{r}, t) + \sum_j \Lambda_{ij} \psi_j(\mathbf{r}, t),\tag{13}$$

and

$$M_I \ddot{\mathbf{R}}_I = -\frac{\partial E}{\partial \mathbf{R}_I}.\tag{14}$$

The Lagrange multipliers Λ_{ij} arise from the orthonormalization condition of the single particle wavefunctions,

$$\int \psi_i^*(\mathbf{r})\psi_i(\mathbf{r})d\mathbf{r} = \delta_{ij}. \tag{15}$$

It should be noted that the ion dynamics in the MD is real whereas the dynamics of the electronic degrees of freedom is fictitious and serves as a means for solving the Kohn-Sham equations. For a fixed ionic configuration when the acceleration of the orbitals becomes zero, then from Eqn. (15) the eigenvalues the matrix Λ become the eigenvalues of the Kohn-Sham equations.

The force acting on an ion I, is written as,

$$\mathbf{f}_I = -\frac{dE}{d\mathbf{R}_I}. \tag{16}$$

Following the Hellmann-Feynman theorem [69], when each ψ_i is an eigenstate of the Hamiltonian, then the partial derivative $-\left(\partial E/\partial \mathbf{R}_I\right)$ (Eqn. (16)) gives the real physical force on an ion. The equilibrium geometry of the system can be obtained when the forces on each ion vanishes. Thus, the *ab-initio* MD method provides an elegant way to determine groundstate electronic and geometric structure simultaneously. The total energy hyper-surface in the multidimensional configurational space is generally quite complicated, there can be several minima in the energy surface. The problem of finding the equilibrium structure in this situation becomes difficult and can be tackled by using the simulated annealing technique [70]. In this technique, the clusters are heated upto a certain temperature and then allowed to cool slowly so that the system remains close to the BO surface. This procedure has proven very successful in finding new structures which have lower energy than other structures obtained by different method. The calculation for a solid system is performed using a periodic supercell, so that electronic wavefunctions can be written as a sum of plane waves. Supercell method is also applied for clusters with the size of the supercell large enough to avoid any interaction between neighbouring cells. A detail discussion of the pseudopotential plane-wave method can be found in the literature [71, 72]. Apart from the LDA-based calculational methods, there exists several other methods, like Hartree-Fock based methods [73] and quite recently stochastic Quantum Monte-Carlo (QMC) methods [74] have also been applied to clusters.

3.3   Results for Selected Clusters

Results of LDA based calculations on clusters are too numerous to mention. Due to limitation in space, we shall mention only very few results, with some elaboration on the results of silicon clusters.

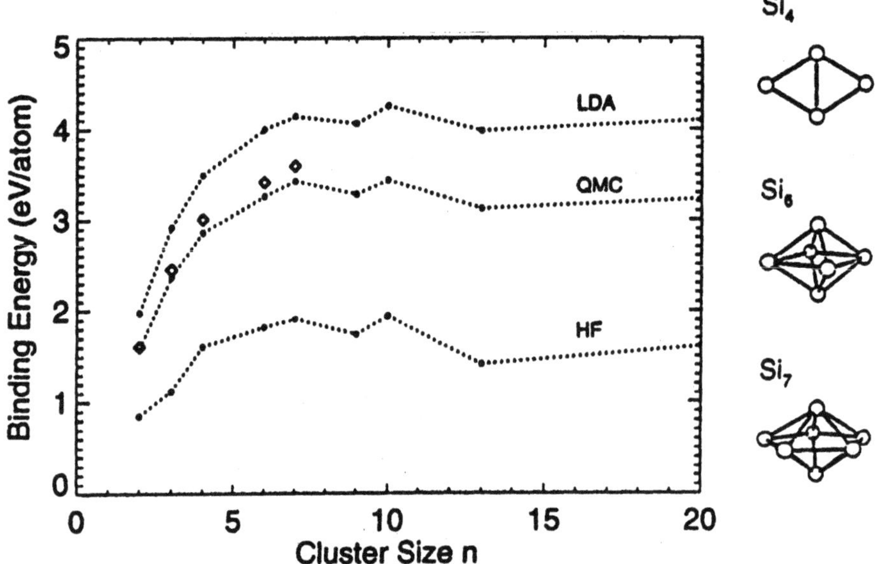

Figure 5. Calculated binding energy of Si$_n$ clusters using LDA, Hartree-Fock (HF) and Quantum Monte-Carlo (QMC) methods, compared to the experimental data (open symbols) [80]. The calculated equilibrium structures of Si$_4$, Si$_6$ and Si$_7$ are shown in the right.

LDA based calculations were performed for a large variety of clusters, e.g. Na, Be, Mg, Al, Sb, Sn, Se, C, B and also for several mixed clusters. A complete discussion with references can be found in the literature [75]. The alkali-metal clusters of Na [76] and mixed clusters Na-K tend to follow the electronic shell closing discussed before. The divalent clusters of Be [78] and Mg [77] show a transition from a weak van der Waals type bonding in dimers to strong covalent metallic-like bonding with increasing size of the clusters. This is expected due to increasing *s-p* hybridization effect with cluster size. Recent calculations on Sr$_n$ ($n = 2 - 20$) clusters [79] showed strong effect of *d*-electrons on the binding energy with increasing size of the clusters. The *d*-electrons were also found to induce icosahedral growth in the Sr$_n$-clusters for $n \geq 13$.

There exist numerous calculations on Si$_n$ clusters using different methods. In Fig. 5 we show the calculated binding energies of Si$_n$ ($n = 2 - $

10, 13, 20) clusters using LDA, Hartree-Fock [63] and Quantum Monte-Carlo [74] methods and compared with the available experimental data [80]. The binding energy trends are similar for all calculations. The LDA overestimates the binding energy by $\sim 15\%$, whereas HF calculation underestimates the binding energy by a large amount. The QMC method agrees well with the experimental data, since it captures most of the correlation effect in the calculation. The structure of the clusters having minimum energy are same for all calculations for $n \leq 7$ and they are completely different than bulk-like cluster fragments. The compact structures obtained for $n = 4, 6, 7$ (Fig. 5) have been confirmed by Raman measurements [64] discussed in section 2.5. Calculations on Si_9 and Si_{10} show that structures like distorted tri-capped trigonal prism and the tetra-capped tetragonal prism respectively, to be the lowest energy structures than other structures. Si_{13} was found to have a structure (with C_{3v} symmetry) having lower energy than icosahedral structure (I_h). Icosahedral structures are favored by Al_{13} clusters and by the doped $Al_{12}X$ (X = B, Ga, C, Si) clusters [81]. Si_{20} was found to have a long prolate shaped structure rather than spherical.

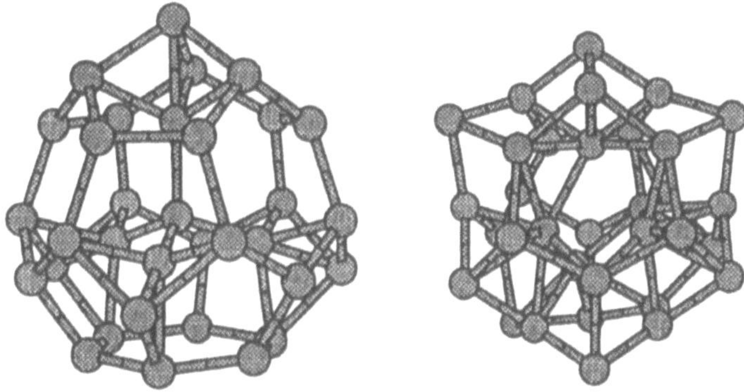

Figure 6. Structures of Si_{33} clusters obtained by *ab-initio* simulations. The geometries referred to as Kax33 (left), and S2 two-shell structure (right) in the text. (After Mukherjee et al. [35])

As mentioned in section 2.4, the structures of Si_{33}, Si_{39} and Si_{45} have been attempted by several groups. Our calculated lowest energy structure of Si_{33} together with that obtained by Kaxiras [53] are shown in Fig. 6. The calculated S2 structure, which has a distorted fullerene-like cage of 28 atoms and a smaller cage of 5 atoms inside, has a lower energy than Kax33 structure. Calculations have been performed using Car-Parrinello type iterative minimization method, as well as using conjugate gradient minimization of the total energy method [71]. Both calculations [58, 35] indicate two-shell geometry (Fig. 6) to be energetically more favourable than Kax33 structure. The fewer number of dangling bonds in S2 structure compared to that in Kax33 structure indicates the former to be less reactive, as observed in the reactivity measurements [49]. An analysis of the total energy surface has indicated [35] that Kax33 structure belongs to a very narrow local minimum in the B-O hyper-surface preserving T_d symmetry, whereas two-shell structure S2 lies near a shallower but energetically more favourable region of the energy surface. Similar two-shell geometries for Si_{45} were found to be energetically more favourable over structures preserving full T_d symmetry [57].

4. CONCLUSIONS

This review is very selective rather than exhaustive. Discussions about ionization energies, polarizabilities, the metal non-metal transitions, optical properties of clusters were not touched upon, due to limited space. However, it should be evident that the experimental and theoretical studies of elemental clusters have emerged as an exciting inter-disciplinary area covering physics, chemistry and materials science. New materials with desired physical and chemical properties can be synthesized in the form of clusters. In the era of fast computers and newly developed efficient simulation techniques, like Car-Parrinello MD method, it is possible to calculate the structure and stability of aggregates containing hundreds of atoms. Improved algorithms which scale linearly with the number of atoms in the system [82] are being developed, so that problems of even greater complexity can be handled in the future.

ACKNOWLEDGMENTS

I would like to thank my former collaborators, P. J. Jensen, K. H. Bennemann, J. L. Morán-López, A. P. Seitsonen and R. M. Nieminen for their contributions. Thanks are also due to P. Jena and V. Kumar for many fruitful discussions.

REFERENCES

1.   O. Echt, K. Sattler and E. Recknagel, Phys. Rev. Lett. **47**, 1121 (1981).

2.   K. Sattler, in *Electronic and Atomic Collisions*, J. Eichler, I. Hertel and N. Solerfoht, ed., Elsevier, Amsterdam (1984).

3.　E. A. Rohlfing, D. M. Cox and A. Kaldor, J. Chem. Phys. **81**, 3322 (1984).

4.　L. A. Bloomfield, R. R. Freeman and W. L. Brown, Phys. Rev. Lett. **54**, 2246 (1985).

5.　J. B. Anderson, in *Molecular Beams and Low Density Gas Dynamics*, P. P. Wegener, ed., Dekker, New York (1974); J. B. Anderson and J. B. Fenn, Phys. Fluids **8**, 780 (1985).

6.　M. S. El-Shall and A. S. Edelstein, in *Nanomaterials: Synthesis, Properties and Applications*, A. S. Edelstein and R. C. Cammarata, ed., Inst. of Physics Publ., Bristol, p. 13 (1996).

7.　T. G. Dietz, M. A. Duncan, D. E. Powers and R. E. Smalley, J. Chem. Phys. 74, 6511 (1981); J. B. Hopkins, P. R. R. Landridge-Smith, M. D. Morse and R. E. Smalley, 78, 1627 (1983).

8.　W. D. Knight, K. Clemenger, W. A. de Heer, W. A. Saunders, M. Y. Chou and M. L. Cohen, Phys. Rev. Lett. **52**, 2141 (1984).

9.　W. A. de Heer, W. D. Knight, M. Y. Chou and M. L. Cohen in *Solid State Physics*, Vol. 40, H. Ehrenreich and D. Turnbull, Ed., Academic, New York (1987) and references cited therein.

10.　W. D. Knight, W. A. de Heer, K. Clemenger and W. A. Saunders, Solid St. Commun. **53**, 445 (1985); W. D. Knight, K. Clemenger, W. A. de Heer and W. A. Saunders, Phys. Rev. B **31**, 2539 (1985).

11.　W. Ekardt, Phys. Rev. B **29**, 1558 (1984).

12.　P. Hohenberg and W. Kohn, Phys. Rev. B **136**, 864 (1964).

13.　W. Kohn and L. J. Sham, Phys. Rev. A **140**, 1133 (1965).

14.　M. Y. Chou and M. L. Cohen, Phys. Lett. A **113**, 420 (1986).

15.　J. Pederson, S. Bjørnholm, J. Borgreen, K. Hansen, T. P. Martin and H. D. Rasmussen, Nature (London) **353**, 733 (1991).

16.　C. Brechignac, P. Cahuzac, M. de Frutos, J. P. Roux and K. Bowen, in *Physics and Chemistry of Finite Systems: From Clusters to Crystals*, Vol I, P. Jena, S. N. Khanna and B. K. Rao, ed., Kluwer Academic, Dordrecht, p. 369 (1992).

17.　R. Balian and C. Bloch, Ann. Phys. (NY) **69**, 76 (1972).

18.　H. Nishioka, K. Hansen and B. R. Mottelson, Phys. Rev. B **42**, 9377 (1990).

19.　T. P. Martin, Phys. Rep. **273**, 199 (1996) and references therein.

20.　A. L. Mackay, Acta Cryst. **15**, 916 (1962).

21. T. P. Martin, S. Frank, N. Malinowski, U. Näher, F. Tast, K. Wirth and U. Zimmermann, in *Clusters and Nanostructured Materials*, P. Jena and S. N. Behera, ed., Nova, Commack (New York), p. 65 (1996).

22. I. Katakuse, I. Ichihara, Y. Fujita, T. Matsuo, T. Sakurai and H. Matsuda, Int. J. Mass. Spectrom. Ion Proc **67**, 229 (1985).

23. Y. Liu, Q. L. Zhang, F. K. Tittel, R. F. Curl and R. E. Smalley, J. Chem. Phys. **85**, 7434 (1986).

24. T. P. Martin and H. Schaber, J. Chem. Phys. **83**, 855 (1985).

25. K. Sattler, J. Mühlbach, O. Echt, P. Pfau and E. Recknagel, Phys. Rev. Lett. **47**, 160 (1981).

26. T. Jentsch, W. Drachsel and J. H. Block, Chem. Phys. Lett. **93**, 144 (1982).

27. W. Schulze, B. Winter and I. Goldenfeld, Phys. Rev. B **38**, 12937 (1988).

28. W. Saunders, Phys. Rev. Lett. **62**, 1037 (1989).

29. D. Tománek, S. Mukherjee and K. H. Bennemann, Phys. Rev. B **28**, 665 (1983).

30. S. Mukherjee and K. H. Bennemann, Surf. Sci. **156**, 580 (1985).

31. B. Delley, J. Phys. C **17**, L551 (1984).

32. F. Liu, M. R. Press, S. N. Khanna and P. Jena, Phys. Rev. Lett. **59**, 2562 (1987).

33. S. Mukherjee, G. M. Pastor and K. H. Bennemann, Phys. Rev. B **42**, 5327 (1990).

34. P. Pyykkö, in *The Effects of Relativity in Atoms, Molecules and in the Solid State*, S. Wilson, I. P. Grant and B. L. Gyorffy, ed., Plenum, New York (1990).

35. S. Mukherjee, A. P. Seitsonen and R. M. Nieminen, in *Frontiers in Materials Modelling and Design*, V. Kumar, S. Sengupta and B. Raj, ed., Springer, Heidelberg, p. 187 (1998).

36. S. N. Khanna, F. Reuse and J. Buttet, Phys. Rev. Lett. **61**, 535 (1988).

37. F. Reuse, M. J. Lopez, S. N. Khanna, V. de Coulon and J. Buttet, in *Physics and Chemistry of Finite Systems: From Clusters to Crystals*, Vol I, P. Jena, S. N. Khanna and B. K. Rao, ed., Kluwer, Dordrecht, p. 241 (1992).

38. F. Liu, M. R. Press, S. N. Khanna and P. Jena, Phys. Rev. B **39**, 6914 (1989).

39. W. A. de Heer, P. Milani and A. Chatelain, Phys. Rev. Lett. **65**, 488 (1990).

40. J. P. Bucher, D. C. Douglass and L. A. Bloomfield, Phys. Rev. Lett. **66**, 3052 (1991).

41. P. J. Jensen, S. Mukherjee and K. H. Bennemann, Zeits. f. Physik D **21**, 349 (1991).

42. S. N. Khanna and S. Linderoth, Phys. Rev. Lett. **67**, 742 (1991).

43. I. S. Jacobs and C. P. Bean, in *Magnetism*, Vol. 3, G. T. Rado and H. Suhl, ed., Academic, New York (1963).

44. P. J. Jensen, S. Mukherjee and K. H. Bennemann, in *Physics and chemistry of Finite Systems: From Clusters to Crystals*, Vol. I, P. Jena, S. N. Khanna and B. K. Rao, ed., Kluwer, Dordrecht, p. 773 (1992).

45. C. R. Helms, in *Interfacial Segregation*, W. C. Johnson and J. M. Blakely, ed., American Soc. of Metals, Ohio, p. 175 (1979).

46. J. H. Sinfelt, Rev. Mod. Phys. **51**, 569 (1979).

47. S. J. Riley, in *Clusters and Nanostructured Materials*, P. Jena and S. N. Behera, ed., Nova, Commack (New York), p. 77 (1996).

48. M. F. Jarrold, Science **252**, 1085 (1991).

49. J. M. Alford, R. T. Laaksonen and R. E. Smalley, J. Chem. Phys. **94**, 2618 (1991).

50. L. R. Anderson, S. Murayama and R. E. Smalley, Chem. Phys. Lett. **176**, 348 (1991).

51. M. F. Jarrold and E. C. Honea, J. Phys. Chem. **95**, 9181 (1991).

52. U. Ray and M. F. Jarrold, J. Chem. Phys. **93**, 5709 (1990).

53. E. Kaxiras, Chem. Phys. Lett. **163**, 323 (1989); Phys. Rev. Lett. **64**, 551 (1990).

54. C. H. Patterson and R. P. Messmer, Phys. Rev. B **42**, 7530 (1990).

55. D. A. Jelski, B. L. Swift, T. Rantala, X. Xia and T. F. George, J. Chem. Phys. **95**, 8552 (1991).

56. M. V. Ramakrishna and J. Pan, J. Chem. Phys. **101**, 8108 (1994).

57. U. Röthlisberger, W. Andreoni and M. Parrinello, Phys. Rev. Lett. **72**, 665 (1994).

58. S. Mukherjee, A. P. Seitsonen and R. M. Nieminen, in *Cluster and Nanostructured Materials*, P. Jena and S. N. Behera, ed., Nova, Commack (New York), p. 165 (1996).

59. J. H. Sinfelt, Y. L. Lam, J. A. Cusumano and A. E. Barnett, J. Catal. **42**, 227 (1976).

60. S. Mukherjee, J. G. Pérez-Ramírez and J. L. Morán-López, in *Physics and Chemistry of Small Clusters*, P. Jena, B. K. Rao and S. N. Khanna, ed., Plenum, (New York), p. 451 (1987).

61. S. Mukherjee and J. L. Morán-López, Surf. Sci. **189**, 1135 (1987).

62. C. C. Arnold and D. M. Neumark, J. Chem. Phys. **99**, 3353 (1993).

63. K. Raghavachari, J. Chem. Phys. **84**, 5672 (1986); K. Raghavachari and C. M. Rohlfing, *ibid.* **89**, 2219 (1988).

64. E. C. Honea, A. Ogura, C. A. Murray, K. Raghavachari, W. O. Sprenger, M. F. Jarrold and W. L. Brown, Nature (London) **366**, 42 (1993).

65. L. Hedin and B. Lundqvist, J. Phys. C **4**, 2064 (1971); S. H. Vosko, L. Wilk and M. Nusair, Can. J. Phys. **58**, 1200 (1980); J. P. Perdew and A. Zunger, Phys. Rev. B **23**, 5048 (1981).

66. D. R. Hamann, M. Schlüter and C. Chiang, Phys. Rev. Lett. **43**, 1494 (1979); G. Kerker, J. Phys. C **13**, L189 (1980).

67. G. B. Bachelet, D. R. Hamann and M. Schlüter, Phys. Rev. B. **26**, 4199 (1982).

68. R. Car and M. Parrinello, Phys. Rev. Lett. **55**, 2471 (1985).

69. H. Hellmann, *Einführung in die Quantenchemie*, Deuticke, Leipzig (1937); R. P. Feynman, Phys. Rev. **56**, 340 (1939).

70. S. Kirkpatrick, C. Gelatt, Jr., and M. Vecchi, Science **220**, 671 (1983).

71. M. C. Payne, M. P. Teter, C. C. Allan, T. A. Arias and J. D. Joannopoulos, Rev. Mod. Phys. **64**, 1045 (1992).

72. R. Stumpf and M. Scheffler, Comp. Phys. Commun. **79**, 447 (1994).

73. See, for example, W. J. Hehre, L. Radom, P. Schleyer and J. A. Pople, *Ab-initio Molecular Orbital Theory*, Wiley (New York) (1986).

74. J. C. Grossman and L. Mitás, Phys. Rev. Lett. **74**, 1323 (1995).

75. V. Kumar, in *Lectures on Methods of Electronic Structure Calculations*, V. Kumar, O. K. Andersen and A. Mookerjee, ed., World Scientific (Singapore), p. 317 (1994).

76. P. Ballone, W. Andreoni, R. Car and M. Parrinello, Europhys. Lett. **8**, 73 (1989); U. Röthlisberger and W. Andreoni, J. Chem. Phys. **94**, 8129 (1991).

77. V. Kumar and R. Car, Phys. Rev. B **44**, 8243 (1991).

78. R. Kawai and J. H. Weare, Phys. Rev. Lett. **65**, 80 (1990).

79. T. Qureshi and V. Kumar, preprint (1998).

80. R. W. Schmude, Q. Ran and K. A. Gingerich, J. Chem. Phys. **99**, 7998 (1993) and to be published.

81. X. G. Gong and V. Kumar, Phys. Rev. Lett. **70**, 2078 (1993).

82. G. Galli and M. Parrinello, Phys. Rev. Lett. **69**, 3547 (1992); X. P. Li, R. W. Nunes and D. Vanderbilt, Phys. Rev. B **47**, 10891 (1993); P.

Ordejon, D. A. Drabold, M. P. Grumbach and R. M. Martin, *ibid.* **48**, 14646 (1993); W. Kohn, Chem. Phys. Lett. **208**, 167 (1993).

Inner Shell Ionization Processes

C. T. Whelan
Department of Applied Mathematics and Theoretical Physics
University of Cambridge, CB3 9EW, UK

INTRODUCTION

In this paper I would like to consider some aspects of electron impact ionization from atomic inner shells. I will be particularly interested in the theoretical understanding of coincidence measurements, i.e. the (e, 2e) process. Fundamentally an (e, 2e) experiment is one where an electron, of well-defined energy and momentum, is fired at a target, ionizes it and the two exiting electrons are detected in coincidence. The energies and positions in space of these electrons are determined by the experiment so in effect all but the spin quantum numbers are then known. We can, therefore, describe it as a kinematically complete experiment; if we could also measure all the spins we would have all the information on a scattering experiment that quantum mechanics will allow. The technique offers both the possibility of a direct determination of the target wavefunction and profound insights into the nature of few body interactions. What information you extract from such an experiment really depends on the kinematics you choose and the target you use. Integrated cross sections can be crude things and you need the full power of a highly differential measurement to tease out the delicacies of the interactions. Indeed often the most intriguing effects turn up in peculiar geometries where the cross sections are small and where a number of relatively subtle few body interactions are at play.

Let us begin by considering the simplest case, the ionization of a Hydrogen atom, in a state n, l, m at non-relativistic energies. Let us assume that we have fast incoming and outgoing electrons and that to a good approximation the ionization is the result of a single impulsive collision. In other words we assume that we can use plane waves for all the continuum electrons then the triple differential cross section can be written:

$$\frac{d^3\sigma}{d\Omega_1\, d\Omega_2\, dE} = \frac{(2\pi)^4\, k_s\, k_f}{4k_0}\left(\mid f+g\mid^2 +3\mid f-g\mid^2\right) \tag{1}$$

where $\mathbf{k}_f$, $\mathbf{k}_s$ are the momenta of the fast and slow exiting electrons and f, g are the direct and exchange scattering amplitudes. If the wave numbers k_f, k_s are sufficiently large and $k_f \gg k_s$, we can neglect g in (1) and

$$f(\mathbf{k}_f, \mathbf{k}_s) \propto \frac{1}{\|\mathbf{q}\|^2} \int d^3 r\, e^{i(\mathbf{q}-\mathbf{k}_s)\cdot\mathbf{r}}\, \psi_{nlm}(\mathbf{r}), \tag{2}$$

where $\mathbf{q}=\mathbf{k}_0-\mathbf{k}_1$, and $\psi_{nlm}(\mathbf{r})$ is the Hydrogenic wave function. We note that the right hand side of (2) is essentially the Fourier transform of the wavefunction of $\psi_{nlm}(\mathbf{r})$ for the momentum $\mathbf{k}_{recoil}$. Now as is well known the momentum space form of the wavefunction, ψ_{nlm}, can be written

$$\hat{\psi}_{nlm}(\mathbf{p}) = F_{nl}(p)Y_{lm}(\hat{\mathbf{p}}) \tag{3}$$

where Y_{lm} is the usual spherical harmonic and F_{nl} is a function of p, the magnitude of $\mathbf{p}$ and the scattering amplitude, (2) is of the form

$$f(\mathbf{k}_1, \mathbf{k}_2) \propto \frac{1}{\|\mathbf{q}\|^2} F_{nl}(k_{recoil})Y_{lm}(\hat{\mathbf{k}}_{recoil}) \tag{4}$$

It is instructive to consider the form of the function F_{nl} for an s state, $1s$ say and a p state, $2p$ say, i.e.,

$$F_{1s}(p) = \frac{2^{5/2}}{\sqrt{\pi}}\frac{1}{(1+p^2)^2}$$

$$F_{2p}(p) = \frac{128}{\sqrt{3\pi}}\frac{1}{(4p^2+1)^3} \tag{5}$$

Clearly $F_{1s}(p)$ has a maximum at $p = 0$ and then decreases, while $F_{2p}(p)$ has a minimum at $p = 0$ and rises to a local maximum at $p_{max} = \sqrt{.05}$ and then decreases monotonically. The condition $k_{recoil} = 0$ defines the Bethe Ridge and the TDCS is zero at this point for a $2p$ state and increases away from it as we vary k_{recoil} until we reach the local maximum, while for the $1s$ state it is monotonically decreasing from $k_{recoil} = 0$. This is the origin of the characteristic double peak structure seen in the TDCS for a p-state in the region of the Bethe Ridge point. Clearly if we move too far from the Bethe Ridge point the TDCS loses this characteristic form, e.g. when we pass beyond p_{max}. This simple analysis illustrates a key point – the TDCS is sensitive to the character of the target wave function. We should also note that if the magnetic sub-levels of the target are not experimentally resolved then the plane wave Born approximation (PWBA), (2), is symmetric about the direction of momentum transfer. In Figure 1 we illustrate the ability of the technique to give us target information, we show the relative TDCS for Ar(2s) and Ar(2p) taken in symmetric non-coplanar geometry, [1], and plotted as a function of k_{recoil}. Plotted are the PWBA results and those of the more sophisticated distorted wave Born approximation. We see that at least in this geometry and kinematics one can use the technique as a relatively good probe of target wavefunctions. Generally speaking, however, the PWBA, is not a particularly good approximation, it takes no account of any interactions of the continuum electrons, this means amongst other things that the final ejected electron wave function is not orthogonal to the initial state, i.e. we have implicitly allowed for spurious autoionization contributions which manifest themselves mathematically in the wrong behaviour of the TDCS as $q \to 0$. In the PWBA the TDCS goes like $1/q^4$ while the actual TDCS goes like $1/q^2$. A much better approximation is the first Born approximation where the direct amplitude is given by

$$ f(\mathbf{k}_s, \mathbf{k}_f) = \left\langle e^{-i\mathbf{k}_f \cdot \mathbf{r}_f} \psi^-(\mathbf{k}_s, \mathbf{r}_s) \left| \frac{1}{|\mathbf{r}_f - \mathbf{r}_s|} \right| e^{i\mathbf{k}_0 \cdot \mathbf{r}_f} \psi_{nlm} \right\rangle \qquad (6) $$

where $\psi^-(\mathbf{k}_s, \mathbf{r}_s)$ is a continuum state of the atom. This is the exact solution when $q \to 0$ but it also exhibits a symmetry about the direction of momentum transfer. In this paper I will not discuss the very interesting advances that have been made in using the (e, 2e) method as a probe of bulk matter, see for example [2], rather I will focus on a few examples of kinematical arrangements where the first Born approximation fails. The natural way to develop this is in terms of 'geometries' – because as I mentioned earlier the way the kinematics are chosen and the apparatus deployed predicates the physics that comes out.

1.1 Coplanar Symmetric-Pochat geometry: the ionization of He ($1s^2$)

In this geometry the two outgoing electrons have equal energies and their momenta $\mathbf{k}_f$, $\mathbf{k}_s$ lie in the same plane as the incident momentum $\mathbf{k}_0$ and make the same angle, θ, with the beam direction. This can be described as a very 'hard' collision, with the incident electron losing over half its initial momentum. If this were a collision between two free particles, conservation of energy and momentum would mean that the two electrons would be detected at 90° to each other, i.e. at $\theta = 45°$. In our case, the target electron is not free but is in a bound state of the Helium atom – thus it has a momentum distribution before the collision and the nucleus may take an active part in the process. Let us assume that ionization occurs as the result of a single electron-electron interaction then, intuitively, one would expect the cross section to be greatest when the angle between the two particles is approximately 90°. Now in this geometry this can happen when $\theta = 45°$ and $\theta = 135°$. In the original experiments, of Pochat et al [3], the TDCSs were measured for Helium targets at impact energies of 100, 150 and 200 eV. The cross sections were found to be peaked around the 45° point and there was a clear indication of a rise at the larger angle. This prompted Whelan and Walters [4] to speculate that a multiple scattering mechanism might be at work. They argued that the large angle structure could be interpreted in terms of the incident electron first elastically back scattering from the atom, essentially the nucleus, and then colliding with the target electron. Motivated by this simple intuitive model, a distorted wave Born approximation calculation was performed, [5].

In this approximation the TDCS is given by

$$\frac{d^3\sigma}{d\Omega_1\, d\Omega_2\, dE} = 2(2\pi)^4 \frac{k_f k_s}{k_0}\left(|f|^2 + |g|^2 - \mathrm{Re}(f^*g)\right) \qquad (7)$$

where

$$f = \langle\, \chi^-(\mathbf{k}_1, \mathbf{r}_1)\, \chi^-(\mathbf{k}_2, \mathbf{r}_2)\,|\frac{1}{r_{12}}|\, \chi^+(\mathbf{k}_0, \mathbf{r}_1)\, \psi_{1s}(r_s)\rangle,$$

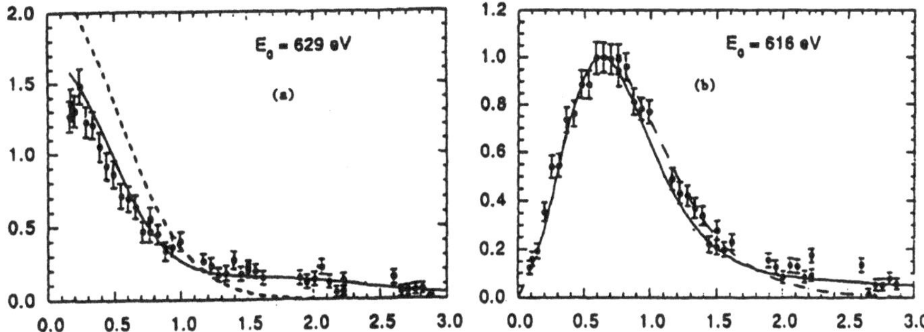

Figure 1. TDCS for the ionization of Ar in non-coplanar symmetric geometry, in (a) we consider the ionization of Ar(2s) at an impact energy of 629 eV, while in (b) it is Ar(2p) and the energy is 616 eV. The experimental data is relative and plotted as a function of k_{recoil} and normalised to the DWBA. Shown are PWBA calculations and DWBA, experiment is from [1]. (We remark that the plane wave impulse as defined in [1] would give an identical curve to the PWBA, in this case).

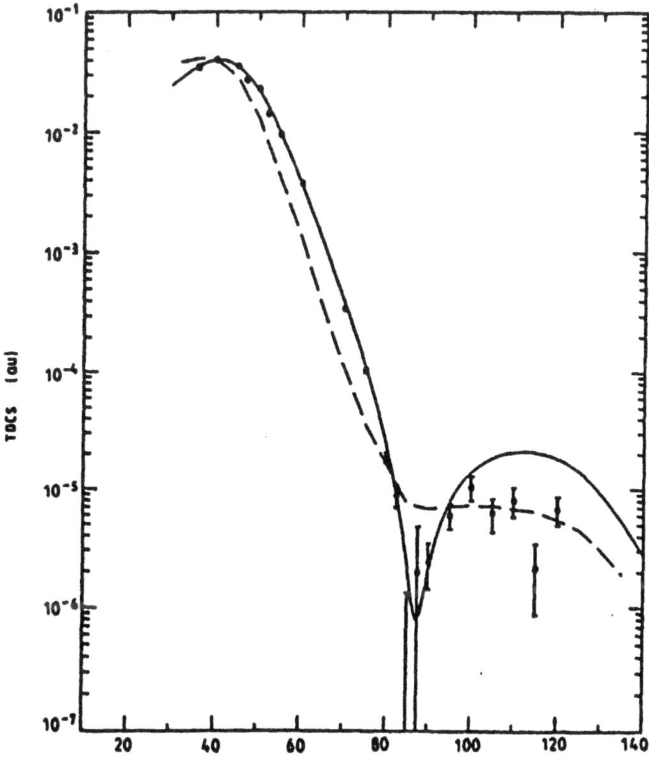

Figure 2. TDCS for the ionization of He in coplanar symmetric geometry. E_0=200 eV. Solid line, DWBA, dashed line 2nd Born calculation of [45] scaled by a factor 0.68; experiment [8].

$$g = \langle\, \chi^-(\mathbf{k}_1, \mathbf{r}_2)\, \chi^-(\mathbf{k}_2, \mathbf{r}_1)\,|\frac{1}{r_{12}}|\, \chi^+(\mathbf{k}_0, \mathbf{r}_1)\, \psi_{1s}(r_s)\,\rangle \tag{8},$$

ψ_{1s} is the $1s$ orbital of *He(1s²)*, taken from the Hartree Fock wave function of [6]; χ^+, χ^- are respectively the distorted waves generated in the static exchange potential of the atom, ion; the final channel distorted waves were orthogonalized to the ground state; f is the direct, g is the exchange scattering amplitudes. In Figures 2 and 3 we show a comparison between these calculations and a number of experiments at impact energies of 500 and 200 eV. Agreement, at these energies, between theory and experiment is good. It is important to understand what the DWBA approximation, as defined in (6) and (7), does and does not contain.

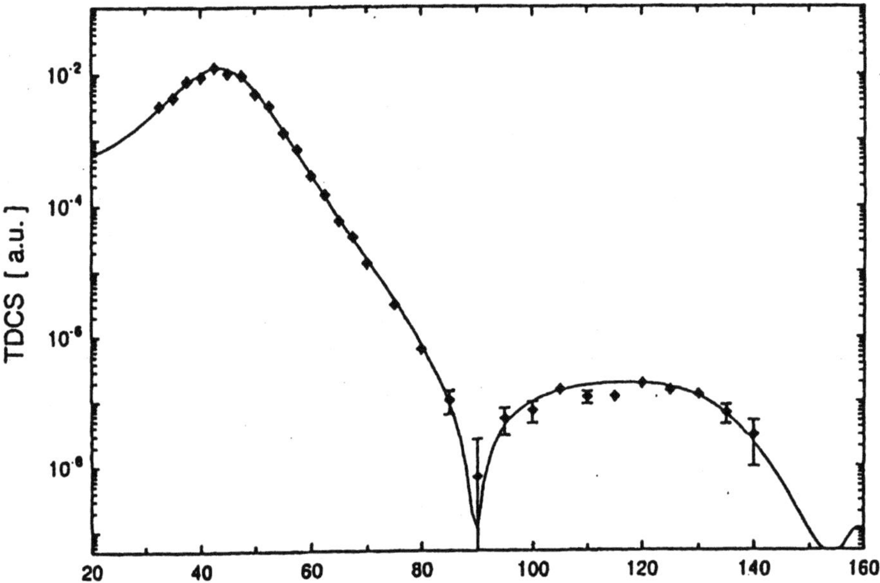

Figure 3. TDCS for the ionization of He in coplanar symmetric geometry. $E_0 = 500$ eV. Solid line, DWBA, experiment, [7].

It includes the elastic scattering of the incident electron to all orders in the static field of the atom, and the same for the outgoing electrons in the field of the ion. The ionizing collision occurs only once, i.e. the interaction between the incident and target electron is treated only to first order and no account of electron-electron repulsion in the final state or polarization in the incident in included. Capture can not take place. Exchange effects manifest themselves both in the indistinguishability of the particles in the final channel, and in the exchange potentials in the elastic channels. In other words we allow for the possibility that the incident electron could exchange with one of the atomic electrons 'before' the ionizing collision and that either of the outgoing electrons could exchange with the remaining 'spectator' electron in the final channel. (In the actual calculations a 'local' exchange approximation is used, for an extensive discussion see [9, 10]). The DWBA is the simplest viable approximation which allows for the mechanism suggested by Whelan and Walters [4] and the good agreement obtained with experiment appears to validate their description. It is further supported by the fact that the large angle peak is observed to be enhanced with respect to the binary as the atom size is increased, [17]; an effect which was also predicted by Whelan and Walters, from their multiple scattering interpretation.

It is worth pausing for a moment to admire Figures 2 and 3 and in particular to note the very sharp dip that theory predicts and experiment finds between the two peaks. It is also worth noting that these results are plotted on a log scale and remarking that the large angle peak is very much smaller than the binary. We will comment on both these features, once again, when we come to look at the relativistic analog of these experiments. There have been a number of experiments for a range of targets in this geometry and for the higher energies and heavier targets agreement with the DWBA was good, see [10] but as the energies were decreased things went badly wrong. Whelan et al [9] suggested that polarization effects needed to be taken into account in the incident channel and post-collisional interactions in the final and were able to come up with a model that had the character of the TDCS right and successfully predicted the shapes of the low energy Hydrogen cross sections in Pochat geometry, [10]. There have been a number of large scale calculations but, in my view, the situation is far from resolved and there is no theoretical calculations about gets both the shape and the absolute size of the TDCS right, for a critical analysis see [11, 12].

1.2 Coplanar Asymmetric – Ehrhardt geometry: the inner shell ionization of Argon

It is instructive to consider the inner shell ionization of Argon and in particular the (e, 2e) studies of Ar(2p), i.e. we are looking at the ionization processes:

$$e^- + Ar(1s^2, 2s^2, 2p^6, 3s^2, 3p^6) \rightarrow 2e^- + Ar(1s^2, 2s^2, 2p^5, 3s^2, 3p^6) \quad (9)$$

The first such measurements were performed at an impact energy of 8 keV, [13, 14], and then by Hink and collaborators at lower energies, in the range of $1.5 - 3$ keV, [15, 16]. In all of these experiments the geometry is coplanar and the TDCS is measured as a function of θ_s of the slow electron for a given fixed angle, θ_f, of the fast. Both angles are given with respect to the beam direction, with θ_f being measured in a clockwise, and θ_s being measured in anti clockwise sense. In a first Born treatment the TDCS will be symmetric about the direction of momentum transfer, $\mathbf{q} = \mathbf{k_0} - \mathbf{k_1}$. This is the binary direction, $-\mathbf{q}$, defines the recoil i.e. the direction of maximum recoil of the ion. It is the lower energy experiments which will concern us here. These have been compared with the Plane Wave Impulse Approximation, and with various forms of the 1st Born [18, 19, 20]. None of these approximations were satisfactory. All were especially poor in predicting the ratio of recoil to binary. A typical example for 2p ionization is shown in Figure 4. Here experiment is compared with a first Born Coulomb wave approximation (FBA-CW), [20, 23].

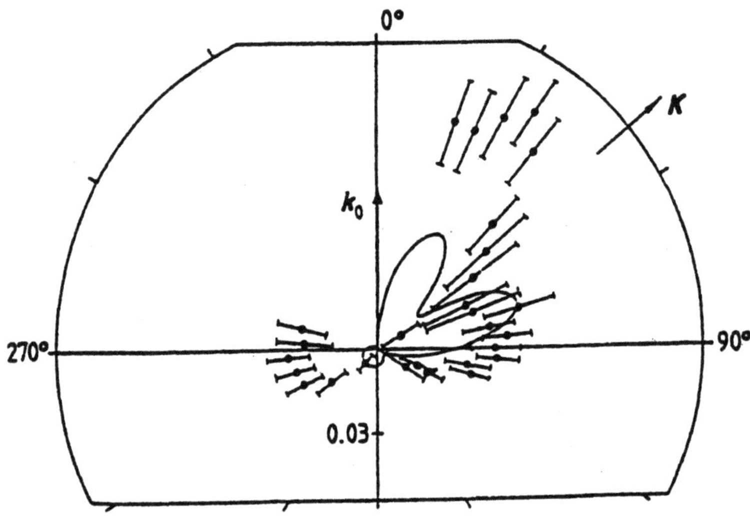

Figure 4. Coplanar TDCS for ionization of Ar(2p) at $E_0 = 1949$ eV, $E_s = 500$ eV, Experiment from Bickert et al [15]; theory FBA-CW.

Since the measurements are relative they have been normalized to theory at the split binary peak. It is clear that agreement is very poor. In the FBA-CW the wavefunctions representing the incident and fast scattered electrons are plane waves, while the slower electron is represented by a Coulomb wave with an adjustable effective charge. Thus this approximation does not allow, in any way, for elastic scattering of the incident electron by the atom prior to ionization nor the effect of the ion on the fast outgoing electron. To some extent, it does include the influence of the ion on the slower outgoing electron through the effective Coulomb potential. It is worth noting that as we would expect from the analysis we have given earlier the FBA-CW predicts a TDCS which is symmetric about the direction of momentum transfer and the binary peak is split exactly as we would expect from a p target. The experiment is distinctly not symmetric about momentum transfer and while it does exhibit a split peak – the ratio of peak heights are different. Zhang et al, [21], argued that one could not neglect the influence of the atomic potential on any of the electrons until much higher energies. This they claimed was especially true for inner shell ionization since one would expect that the ionization process to occur relatively close to the nucleus i.e. in a region where the static potential is at its strongest. Again it was argued that the simplest viable approximation was the DWBA and in [21] results were presented using this approximation. In Figure 5 we show a selection of these results as compared with experiment. Theory and experiment agree that the TDCS is not symmetric about the direction of momentum transfer and there is significant structure in the recoil direction which was absent in the first Born type calculation. Clearly this arises because we have allowed for multiple scattering effects in the elastic channels of the fast electron. Zhang et al explored the effect of elastic scattering by 'switching of' the interaction between the incident electron and the atom, i.e. essentially replacing χ^+ by a plane wave. The result of this calculation is shown in Figure 6 where it is compared with the full DWBA calculation. We see at once that this model calculation has, now, acquired something of the character of the FBA-CW; and that it is the *binary peak* which is most strongly influenced by the distortion effects. Recall that these experiments were relative and all that theory had to compare with was the relative behaviour of the different maxima and minima. Let us emphasize once again that the 1st Born approximation will predict a recoil peak even though the incident and fast electrons do not interact with the nucleus, i.e. are represented by plane waves. This can be understood in terms of the relative motion of the target electron and the ion about their centre of mass prior to the collision. For example Whelan et al, [22], considered the ionization of H in coplanar asymmetric geometry for a range of impact energies and scattering angles, θ_1. They pointed out that in the Hydrogen atom the average value of the momentum of the electron / proton with respect to the centre of mass was one atomic unit and they remarked that when $|k_{recoil}| < 1$ there was a significant recoil peak in the experimental data and when $|k_{recoil}| > 1$ the peak became negligibly small. In heuristic terms an electron coming out in the recoil direction in the second case would correspond to a change in the average momentum which would be characteristic of the proton experiencing a force. Thus the presence of a significant recoil peak in asymmetric geometry does not in itself mean that one should expect strong distortion effects. A pronounced enhancement above the first Born level is a better indication but even that may have other origins for the

Hydrogen case we have already mentioned there is a noticeable enhancement due primarily to post-collisional electron-electron interaction, (2nd Born terms), [23, 24].

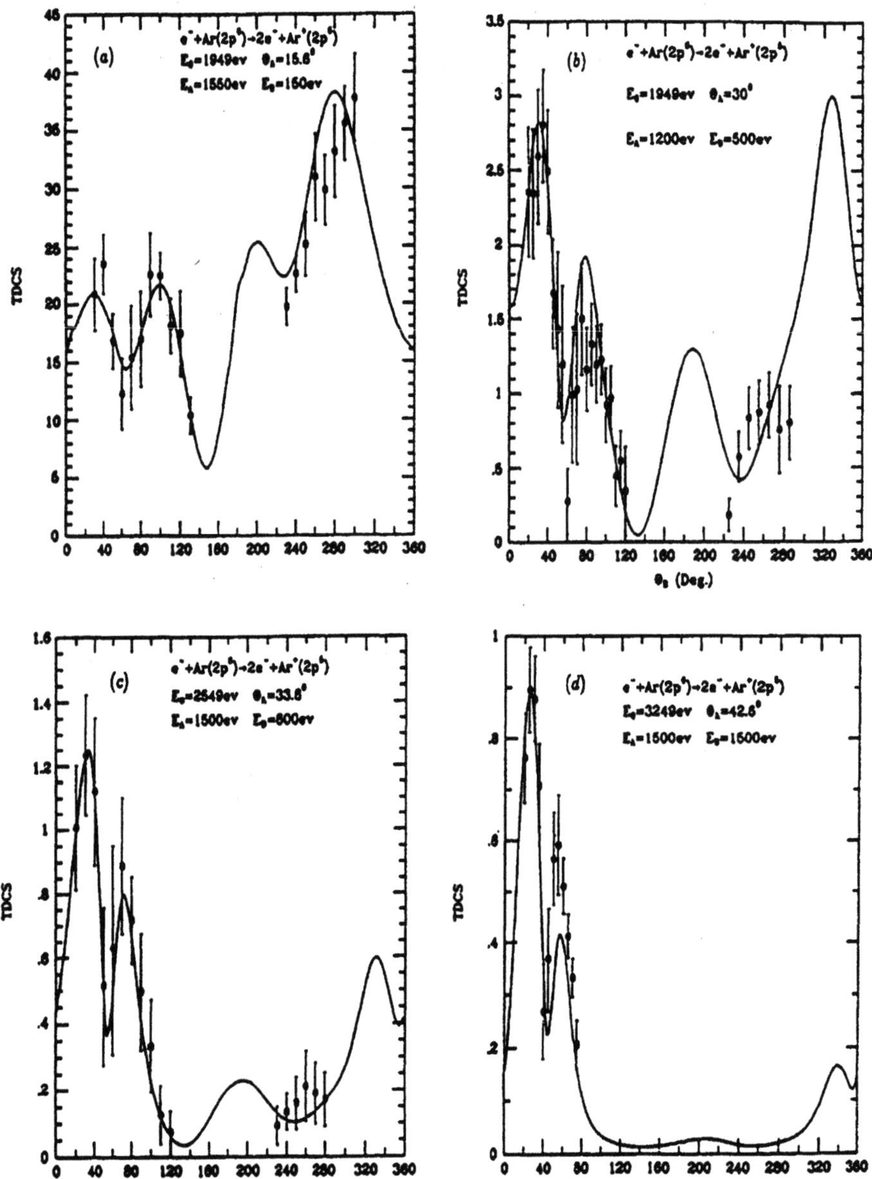

Figure 5. Coplanar TDCS for Ar(2p) in units of 10^{-5} au for various energies and angles. Full curve DWBA. Experiments are from [15] and have been normalized to give best visual fit to the DWBA.

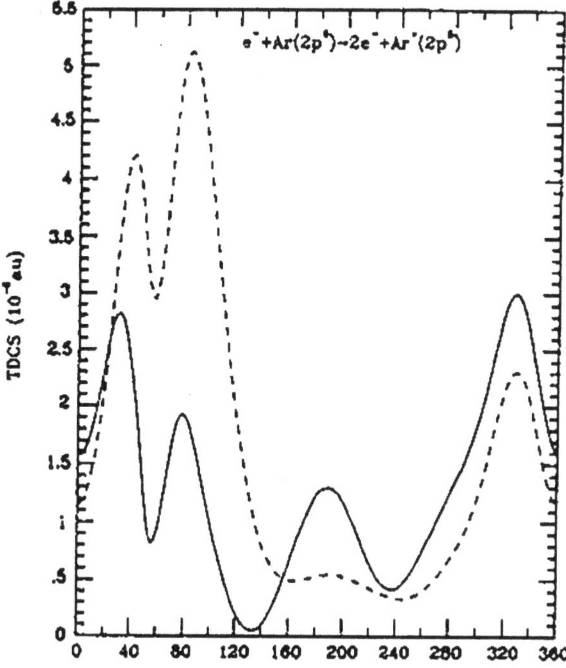

Figure 6. Kinematics are the same as Figure 4, shown is the full DWBA calculation, full curve and model calculation, dashed curve, in the latter the distortion in the incident channel is switched off.

2. Relativistic (e, 2e) processes

There have been (e, 2e) studies at much higher energies than those discussed on the deep inner shells of heavy targets; these began in 1982 with (e, 2e) experiments by Schüle and Nakel [25] at an incident energy of 500 keV on the K shell of silver. The last few years have seen great strides in the area and recently experiments have been performed with transversely polarized electron beams, [26], which represent an important step towards the ideal of a complete quantum mechanical scattering experiment. The description of these processes has necessitated the development of wholly new theoretical and computational methods. This is hardly surprising since the problem is fully relativistic, both the incoming and exiting electrons are fast, the target electron feels the full effect of the field of the highly charged nucleus, it is a Many Body problem with long range Coulomb forces in the final channels. The calculation of the pertinent Triple differential Cross Sections, (TDCS), opens up a whole new area of theoretical study and offers a direct insight into the subtleties of spin dependent, that is to say pure relativistic effects in atomic collision physics.

In the following, atomic units $\left(\hbar = m_e = e = 1\right)$ are used so the numerical value of the vacuum velocity is $c = 137.03604$; the metric tensor is

$$diag(g_{\mu\nu}) = (1, -1, -1, -1),$$

contravariant four vectors are written $x^\mu = (t, \mathbf{x})$ and the summation convention is understood.

2.1   Relativistic Distorted Wave Born Approximation

We have seen that at low energies the DWBA is the simplest possible approximation that we can use to include multiple scattering effects in both the incident and final channels. It works well for a range of geometries where higher order effects – for example polarization of the target in the incident channel or electron-electron repulsion terms in the final are weak. It worked particularly well for the Argon inner shell case. Clearly therefore it is a prime candidate as a theoretical model for the relativistic inner shell problem. Indeed, there are some simplifications at the energies we are working: exchange in the elastic channels is likely to be negligible, final state e-e repulsion will certainly play no role. However, as we mentioned earlier, we are now dealing with a fully relativistic problem. This means that we will have to solve for Dirac spinors rather than Schrödinger wave functions and we will need to include the full QED photon propagator, i.e. including retardation as well as magnetic interactions.

In the earliest attempts to model these processes a number of assumptions were made, which subsequently turned out to be false and it is perhaps valuable to list these here. The simplest approximation one could use is the plane wave Born approximation, (PWBA), in which all the electron wavefunctions are represented by plane waves. At non-relativistic energies this approximation has been used extensively in impulsive experimental arrangements and when it is valid it allows the use of the (e, 2e) method as a means of mapping the momentum distribution of the target electron, see e.g. [27]. A feature of the, low energy, PWBA approximation is that it factorises into two terms, one which is independent of the target wavefunction and the other is in essence the wavefunction of the target in momentum space. In Bell, [28], a relativistic version was proposed in which the TDCS was taken to be the product of the free first order electron-electron (Moeller) cross section and the momentum profile of the bound state. Keller and Whelan, [29], analysed the relativistic plane wave approximation and concluded that in fact it did not factorize and that for any given kinematical arrangement and bound state energy, the 'cross section function' part of the TDCS depended on the spinor structure of the bound state. The results obtained in the relativistic plane wave Born approximation are substantially different from those found in the impulsive treatment of Bell, [28] though both theories share the same, factorized, form in the non-relativistic limit. A number of semi-relativistic variants on the Born approximation were tried. Das and Konar, [30], employed a semi-relativistic Sommerfeld Maue function, [31] for one of the outgoing electrons and Jakubaßa-Amundsen, in a series of calculations, studied the influence of different approximate semi-relativistic scattering and bound state wave functions on the TDCS, [32, 33, 34]. We will discuss the validity of using semi-relativistic wavefunctions below, here we will only comment on another common feature of all these approximations, that is that they only included those spin channels which would contribute at low energies, the other relativistic 'spin flip' channels were not included. In Walters et al [34] it was shown that this was an invalid assumption and that, especially in the symmetric case, they made a very large contribution to the TDCS.

The TDCS for the relativistic (e, 2e) process, where the spins are not resolved, can be written quite generally

$$
\frac{d\sigma}{d\Omega_1 \, d\Omega_2 \, dE_s} = \frac{(2\pi)^2}{c^6} \frac{k_1 k_2}{k_0} E_0 E_f E_s \frac{N_\kappa}{2N_m} \sum_{s_f, s_0, s_s, s_b} \langle k_f, s_f, k_s, s_s \, | \, \hat{S} \, | \, k_0, s_0, \kappa, s_b \rangle^2 \tag{10}
$$

where S is the S matrix operator; $0, f, s$ and b refer to the incoming, the two outgoing and the initially bound electron, respectively, E_0, E_f, E_s and k_0, k_f, k_s are the on shell total energies and momenta of the unbound particles where

$$
E^2 = k^2 c^2 + c^4,
$$

and we are using κ to denote the quantum numbers of the atomic bound states, s are the spin projection operators with respect to the quantization axis, which we take in the beam direction. In the form (3.4) the TDCS is insensitive to spin polarization, we have averaged over the initial spins, s_b, s_0, and summed over the final s_f, s_s. (Hence the factor N_κ/N_m : N_κ is the occupation number of state κ and N_m the number of degenerate states with this set of quantum numbers). Without loss of generality we assume that $E_f \geq E_s$. In Ast et al, [47], a relativistic generalization of the DWBA was introduced. The essential features of this approximation were that as before distorted waves were used to represent the elastic scattering of the incident and exiting electrons in the field of the atom, while the ionizing collision occurred once. The full Dirac equation was solved, numerically, using an effective potential in the elastic channels: thus avoiding semi-relativistic wave functions, all spin flip channels were included and the full photon-propagator employed. One difference was that at low energies, the elastic scattering is solved using a static-exchange potential while at high energies one uses a pure static potential. This is of significance when interpreting spin polarized experiments.

I don't intend to give a discussion of explicitly spin dependent effects here, or ionization from an L shell, or indeed some other very interesting observations (see [35] for a recent review); rather I will just select a few results which particularly appeal to me and hope to illustrate something of the character of deep inner shell ionization with these heavy metal targets.

2.1.1 General Features of the Cross Section

For K shell ionization the TDCS as measured in either coplanar asymmetric or symmetric geometry exhibits only a little structure. One expects to see in the Ehrhardt geometry one peak in the direction of momentum transfer and possibly a second in the recoil direction, and in the Pochat geometry a binary and a large angle peak – associated with elastic scattering from the nucleus. The target wave function has a high degree of symmetry and this symmetry tends to be reflected in the simplicity of the TDCS shape. Thus the major way that relativistic, distortion or other effects manifest themselves is in the *size* of the cross section and hence the significance of *absolute* experiments.

2.1.2 Coplanar Asymmetric – Ehrhardt Geometry

In Figure 7, we compare the experimental results of [36] for a gold target ($E_0 = 500$ keV, $E_1 = 100$ keV, $\theta_f = -15°$) with theoretical results by Das and Konar [30], who use a semi-relativistic Sommerfeld-Maue Coulomb wave function to represent the slow outgoing electron and the Ochkur approximation for the exchange term, and the first Born results of Walters et al [34], where a Darwin-Coulomb wave is used for this electron. Both theories employ plane waves to describe the incoming and fast outgoing electron, and they both over estimate the experiment by about a factor of 4. (It is not clear why the calculations of Das and Konar mispredicts the location of the maximum). Agreement between experiment and the rDWBA is satisfactory. Figure 8 also shows the result of the semi-relativistic Coulomb-Born calculation by Jakubaßa-Amundsen [32, 33]. The results of

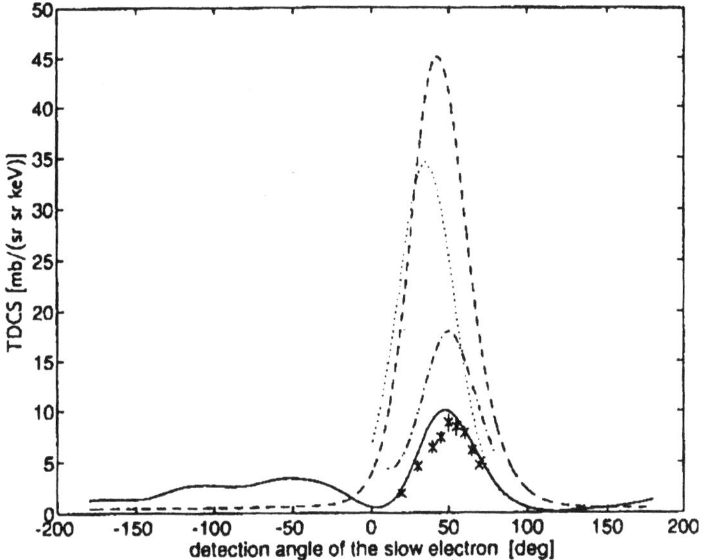

Figure 7. TDCS for the ionization of Gold, $E_0 = 500$ keV, $\theta_f = -15°$; rDWBA (solid line), semi-relativistic Coulomb Born, [32, 33] (dashed dotted) semi-relativistic first Born (long dashed, [34], dotted, [30]), experiment, [36].

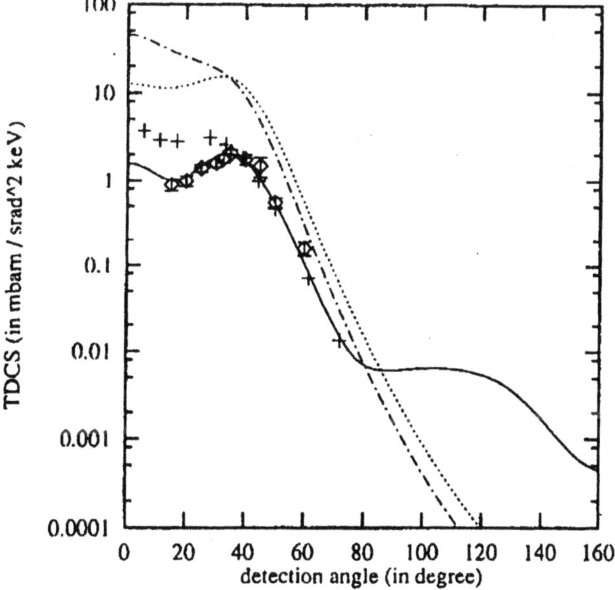

Figure 8. The TDCS for Gold, with $E_0 = 500$ keV in coplanar symmetric geometry, dashed dotted, PWBA, ..., semi-relativistic 1st Born, +'s semi-relativistic Coulomb Born, solid line rDWBA.

this theory, which may be understood as an approximation to the full DWBA, is much closer to the experimental data, but still over predicts them by a factor of 2. Finally we show the rDWBA calculation. A self-consistent relativistic Kohn-Sham LDA potential of a neutral gold atom has been used in all channels. The bound state was described by a relativistic hydrogenic $1s_{1/2}$ wave function. The effects of Slater screening and of orthogonalizing the outgoing distorted waves to the bound state were found to be negligible because the overlap integrals between the bound state and the scattering partial waves are extremely small. Agreement between experiment and the rDWBA is satisfactory. It is worth noting that there is some structure in the recoil region just as in the Argon case.

2.1.3   Coplanar Symmetric Geometry – Pochat Geometry

In the relativistic regime energy sharing collisions have been measured on an absolute scale ([34, 36]). From the theoretical side TDCS have been calculated in plane wave impulse approximation (PWIA) (Fuss et al [37], Bell [28]), semi-classical impulse approximation (Cavalli and Avaldi [38]) and a plane wave Born approximation (PWBA) (Keller and Whelan [29]), a first Born approximation using a semi-relativistic Coulomb function for one outgoing electron ([30, 32, 34]) and in a Coulomb Born approximation, [33]. None of these theoretical approaches could satisfactorily describe the experimentally determined TDCS for gold; either in shape or in magnitude. Walters et al [34] showed that it is important to include all possible spin flip channels in the calculation and their inclusion widens the gap between experimental and theoretical results. It is, therefore, inappropriate to compare our calculations with those which have neglected spin flip and we will not do so. The semi-relativistic Coulomb Born approximation of Jakubaßa-Amundsen [33] while in poor accord with gold does better for silver and copper. In her approximation all unbound electrons are described by non-relativistic Coulomb waves, eigenfunctions of the bare nuclear Coulomb potential, times a plane wave spinor. In Figures 7 and 8 we compare the present rDWBA data (full curve) with the experimental data and other existing calculations for gold at an impact energy of 500 keV.

The full diamonds are the experimental data of Bonfert et al [36] and Walters et al [34], the dashed dotted curve a plane wave Born approximation (Keller and Whelan [29]), the dotted curve represents a semi-relativistic 1st Born type calculation using a Darwin Coulomb function for one outgoing electron (Walters et al [34]) and the crosses are the semi relativistic Coulomb Born calculation of Jakubaßa-Amundsen [33]. We find good agreement between the experimental results and the relativistic DWBA calculation as well as the gross features of their absolute size, the TDCS exhibits some very interesting features that are, as we will see, characteristic of relativistic and distortion effects.

3. The position of the binary peak

In the non-relativistic first Born approximation, as mentioned above, the TDCS is symmetric about the direction of momentum transfer $\mathbf{q} = \mathbf{k}_0 - \mathbf{k}_1$. For a K shell this results in the observation of two peaks, one at $\mathbf{q}$, the binary, and one about $-\mathbf{q}$, the recoil. In the measurements Bonfert et al [36] on gold at 500 keV the binary peak is observed to be shifted away from the direction of momentum transfer. Such a shift has been observed at much lower energies. For example, the experiments of Ehrhardt and co-workers, [40], on H at energies of 250 eV and below show that the symmetry about q being broken with the maxima shifted, the binary peak reduced and the recoil enhanced. The Hydrogen experiments have been analysed by a number of authors [9, 23, 41, 42], and it is now generally accepted that the observed shifts in the binary and recoil peaks is due to post collisional interactions, pci, between the slow ejected and the fast scattered electrons, this interaction is implicitly contained in the 2nd Born term and entirely absent from the standard DWBA calculations. In the case of Bonfert et al, [36], data it is clear that the shift has another origin, outside of the fact that pci effects will be entirely negligible at these energies, there is already a clear indication of the shift in both semi-relativistic and fully relativistic 1st Born calculations, [32, 34, 46]. In order to analyze the shift, Ast et al [39] consider the fully relativistic 1st Born approximation, which is essentially the same as the rDWBA where the distorting potentials for the fast incident and scattered electrons are set to zero and the bound and ejected electrons see an effective potential which is Coulombic with an effective charge $Z_{eff} = Z - 0.3$ [43]. Crucially, it contains the full photon propagator. Now in relativistic physics the electron-electron interaction is mediated by a photon exchange while the non-relativistic interaction can be viewed as an instantaneous interaction between charge densities. In order to understand the role of magnetic and retardation effects Ast et al, [39], considered the propagator corresponding to the Coulomb potential

$$D_{00}^{Coul}(x-y) = -4\pi\,\delta(x_0 - y_0)\lim_{\eta\to 0}\int\frac{d^3k}{k^2+\eta^2}\,e^{i\mathbf{k}\cdot(\mathbf{x}-\mathbf{y})} \qquad (11)$$

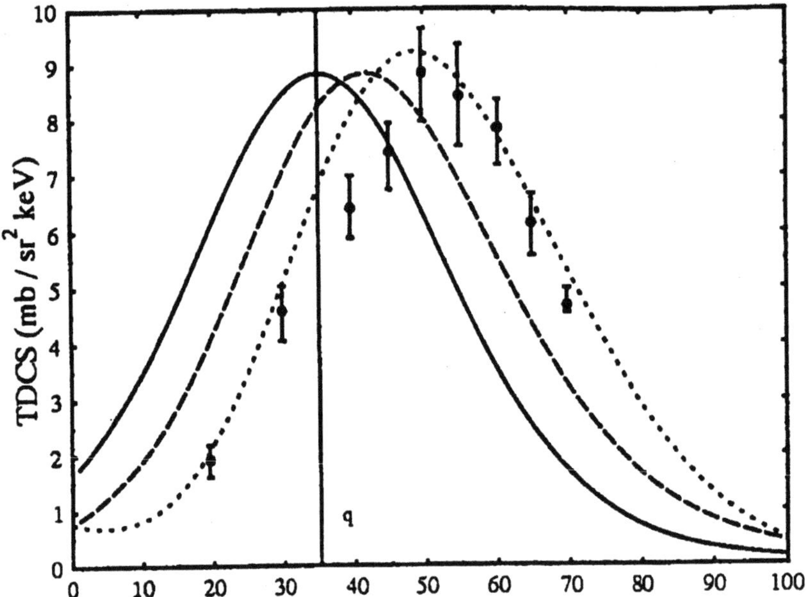

Figure 9. TDCS for gold $E_0 = 500$ keV, $E_s = 100$ keV, $\theta_s = -15°$. Experiment, [36], dashed line, first Born approximation of Walters et al, [34] normalized to the experimental data; solid line, model calculation, (normalized to experiment) dotted line, rDWBA. Note that the rDWBA is not scaled.

and calculated the TDCS for the Au, 500 keV case in asymmetric geometry. We reproduce these results here, Figure 9. The position of the binary peak is insensitive to the use of a semi-relativistic or fully relativistic wavefunction. Clearly the difference

$$\tilde{D}_{\mu\nu}(x-y) = D^0_{\mu\nu} - D^{Coul}_{00}(x-y) \tag{12}$$

represents the relativistic contributions. The spatial components of this propagator $\tilde{D}_{jk}(x-y)$ describe the magnetic interaction between the currents of the two moving charges, the component $\tilde{D}_{00}(x-y)$ incorporates retardation effects. One notes that the position of the peak in the model calculation agrees with the direction of the momentum transfer $(\theta_2 = 35.2°)$. Clearly, the shift obtained in the first Born $(\theta_{2,\,max} = 42°)$ is caused by magnetic and retardation effects. The position of the binary maxima is insensitive to the use of a semi-relativistic wavefunction, [32, 34], we remark that only by doing a full rDWBA calculation can one achieve the full shift to the experimentally observed value, $\theta^{expt}_{2,\,max} = 50°$. Indeed only then as we observed earlier, can we achieve agreement with absolute size of the cross section. Ast et al, [39], performed a similar analysis for equal energy sharing collisions. They chose to plot the results in terms of $\mathbf{k}_{recoil} = \mathbf{k}_0 - \mathbf{k}_1 - \mathbf{k}_2$. Clearly the binary maximum in this geometry, in the non-relativistic 1st Born approximation, corresponds to $\mathbf{k}_{recoil} = 0$. In the relativistic case there is a significant shift away from zero recoil. In contrast to the asymmetric case, however, the semi-relativistic approximations and fully relativistic calculations do not predict the same peak position, Figure 10. For a fuller discussion see [39].

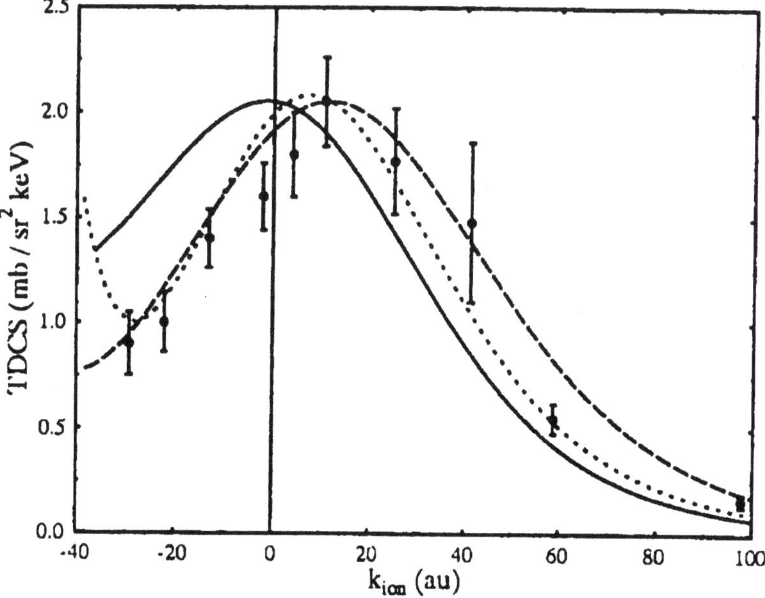

Figure 10. TDCS for the K-shell of gold, E_0 = 500 keV, $E_s = E_f$ = 209.65 keV; symmetric geometry. TDCS plotted as a function of the recoil momentum. Experiment, [34], dashed curve, scaled first Born approximation, [46]; solid line, scaled model calculation; dashed curve, rDWBA.

4.   Spin dependent effects using unpolarized beams on unpolarized targets

In relativistic physics the spin projection quantum number is no longer a good quantum number. In a collision process a flip of the spin of one or both involved particles is always possible. Let us first have a look onto the importance of spin flip processes in the cases so far investigated.

In the relativistic regime experiments have been performed under either highly asymmetric conditions or fully symmetric. In Figure 11 an asymmetric collision is considered. Here, at these impact energies, spin flip processes contribute only a little to the TDCS.

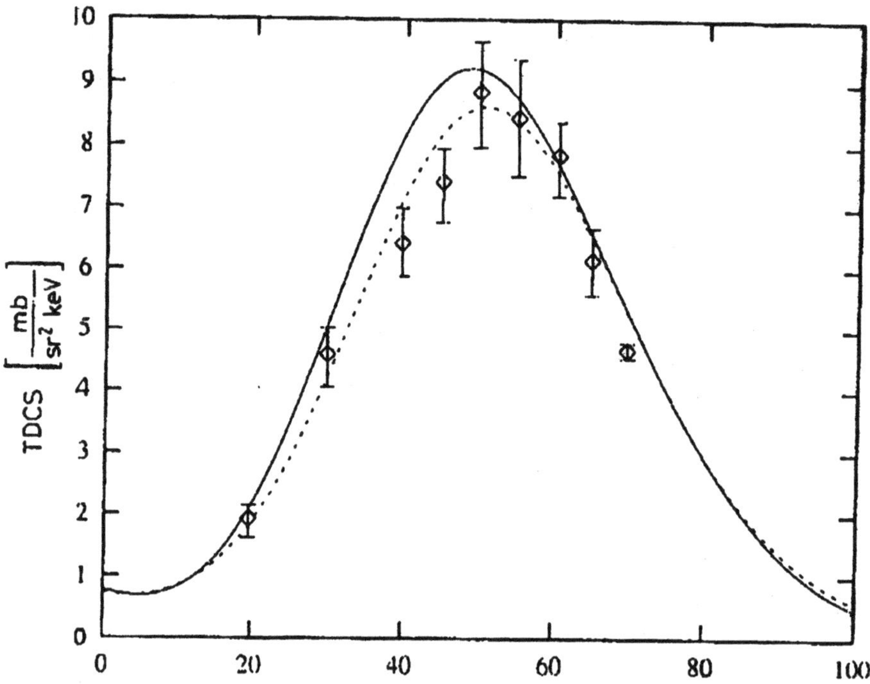

Figure 11. The TDCS obtained in rDWBA are plotted against the detection angle of the slow outgoing electron for a K-shell ionization of gold at an impact energy of 500 keV. The detection angle of the fast outgoing electron (E_1 = 319.3 keV) is fixed at θ_f = -15°, the slow electron carries an energy of 100 keV. The solid curve represents the complete TDCS for all channels and the dashed one the partial TDCS for the non spin flip contributions.

However, the picture changes completely for a fully symmetric collision. If we consider equation (10) there are 16 possible spin channels, only 6 of which would exist in a non-relativistic theory, i.e.

$$(i) \, s_f = s_0 = s_s = s_b,$$
$$(ii) \quad s_f = s_0 = -s_s = -s_b,$$
$$(iii) \quad s_f = -s_0 = -s_s = s_b.$$

where s_i is either +½ or -½. For channel (i) above, both the direct and exchange S matrices are non-spin flip while for (ii) S^{Ex} and (iii) S^{Dir} are spin flip channels. In coplanar symmetric geometry it is clear that channel (i) is zero, i.e. the non spin flip channels with equal spins for both electrons are blocked. Therefore the relative importance of the spin flip channels is enlarged compared to the asymmetric collision. This was first noted by Walters et al, [34], while investigating the semi-relativistic first Born approximation. In Figure 12 the TDCS for gold at an impact energy of 500 keV is plotted against the detection angle θ of one of the outgoing electrons who share energy ($E_1 = E_2 = 209.65\ keV$).

The spin flip channels contribute up to 40% to the TDCS, in the binary region and several orders of magnitude at the minimum point. It is instructive to compare these results with those of Figures 2 and 3. We note that just as in the non-relativistic case the non-spin flip calculation produces a very distinct minimum between the two peaks but the addition of the spin flip terms act to significantly fill up the minimum. Both calculated cross sections are compared with the absolute experimental results, [36]. Whereas the calculated data of the TDCS (solid line) agrees very well with the experimental results, the partial non spin flip TDCS underestimates the experimental results significantly.

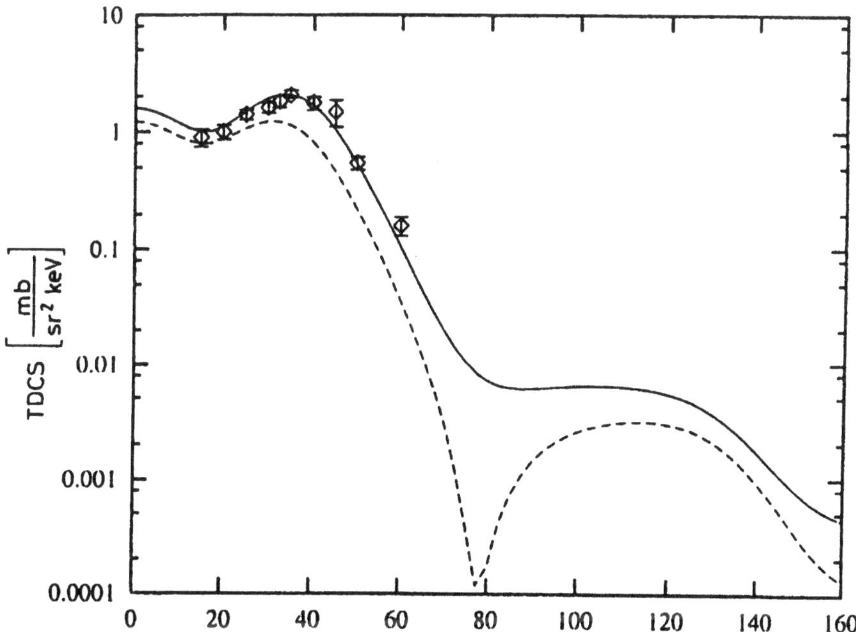

Figure 12. The TDCS in coplanar symmetric geometry for the K shell of gold. $E_0 = 500$ keV. The solid line represents the TDCS obtained in rDWBA including all spin flip contributions and the dashed curve represents the partial TDCS for the non spin flip channels only.

5. Fully relativistic versus semi-relativistic wave functions

In Ancarani et al, [44], a fully relativistic Coulomb Born calculation was performed and compared with the semi-relativistic calculations of Jakubaßa-Amundsen, [33]. It is clear from this work that a large error is introduced by not solving the Dirac equations in the elastic channels.

6. AFTERWORD

In this brief review I have touched on a few (e, 2e) experiments, those which for one reason or another captured my interest and upon which I have worked. I am very conscious that this is a partial review which has left out a number of really exciting new developments. I do hope, however, that I have communicated something of the excitement of the field. Coincidence studies allow one an unparalleled insight into the mechanisms of ionization and the character of the target. The range, variety and sheer novelty of the ever-evolving experimental techniques presents a formidable challenge to theory. The possible geometries are so numerous and the cross sections so small that it is vital that the theorist not only responds to measurements but also takes a role in guiding the on-going experimental programme, but that itself is part of the fun.

ACKNOWLEDGMENTS

I have been very fortunate to be involved in a collaborative project to study relativistic (e, 2e) processes. This joint effort has involved the Queen's University of Belfast, the University of Frankfurt am Main as well as Cambridge. I am grateful to Professor Dreizler, Drs. Ancarani, Ast, Keller and Rasch for very many useful discussions. A special word of thanks is due to Professor Walters with whom I have the privilege of collaborating on a whole range of different (e, 2e) problems. The numerical calculations shown in this work were performed in a number of sites, in particular using the workstation clusters of the Gesellschaft für Schwerionenforschung (GSI), Darmstadt, the Hochschulrechenzentrum Universität Frankfurt and the Hitachi at Cambridge.

REFERENCES

1. A. A. Pinkás, M. A. Coplan, J. H. Moore, S. Jones, D. H. Madison, J. Rasch, C. T. Whelan, R. J. Allan and H. R. J. Walters in *New Directions in Atomic Physics*, edited by C. T. Whelan, R. M. Dreizler, J. H. Macek and H. R. J. Walters, Plenum (1999).

2. G. Stefani in *New Directions in Atomic Physics*, edited by C. T. Whelan, R. M. Dreizler, J. H. Macek and H. R. J. Walters, Plenum (1999).

3. A. Pochat, R. J. Tweed, J. Peresse, C. J. Joachain, B. Piraux and F. W. Byron, Jr., J. Phys. B **16**, L775 (1983).

4. C. T. Whelan and H. R. J. Walters, J. Phys. B. **23**, 2989 (1990).

5. X. Zhang, C. T. Whelan and H. R. J. Walters, J. Phys. B **23**, L509 (1990).

6. E. Clementi and C. Roetti, At. Data and Nucl. Data Tables **14**, 177 (1974).

7. T. Rösel, C. Dupré, J. Röder, A. Duguet, K. Jung, A. Lahamam-Bennani and H. Ehrhardt, J. Phys. B **24**, 3059 (1991).

8. L. Frost, P. Freienstein and M. Wagner, J. Phys. B**23**, L715 (1990).

9. C. T. Whelan, R. J. Allan, H. R. J. Walters, X. Zhang in *(e, 2e) & related processes*, edited by C. T. Whelan, H. R. J. Walters, A. Lahmam-Bennani and H. Ehrhardt, Kluwer, Dordrecht, 1-32 (1993).

10. C. T. Whelan, R. J. Allan, J. Rasch, H. R. J. Walters, X. Zhang, J. Röder, K. Jung and H. Ehrhardt, Phys. Rev. A **50**, 4394 (1994). See also J. Rasch, *Ph.D. thesis*, University of Cambridge (1996).

11. S. P. Lucey, J. Rasch and C. T. Whelan, Proc. Roy. Soc. A, *in press.*

12. J. Rasch and C. T. Whelan in *New Directions in Atomic Physics*, edited by C. T. Whelan, R. M. Dreizler, J. H. Macek and H. R. J. Walters, Plenum (1999).

13. A. Lahmam-Bennani, H. F. Wellenstein, A. Duguet and A. Daod, Phys. Rev. A **30**, 1511 (1984).

14. G. Sefani, L. Avaldi, A. Lahmam-Bennani and A. Duguet, J. Phys. B **19**, 3787 (1986).

15. P. Bickert, W. Hink, C. Dal Cappello and A. Lahmam-Bennani, J. Phys. B **24**, 4603 (1991).

16. P. Bickert, W. Hink and S. Schönberger, *Proc. 17th ICPEAC, Brisbane*, edited I. E. McCarthy, W. R. Mac Gillivray and M. C. Standage (Brisbane: Griffith University) Abstracts, p180 (1991).

82

17. T. Rösel, K. Jung, H. Ehrhardt, X. Zhang, C. T. Whelan and H. R. J. Walters, J. Phys. B **23**, L649 (1990).

18. C. Dal Cappello, C. Tavard, A. Lahmam-Bennani and M. C. Dal Cappello, J. Phys. B **17**, 4557 (1990).

19. A. N. Grum-Grzhimailo, J. Phys. B **18**, L695 (1985).

20. M. J. Brothers and R. A. Bonham, J. Phys. B **19**, 3801 (1986).

21. X. Zhang, C. T. Whelan, H. R. J. Walters, R. J. Allan, P. Bickert, W. Hink and S. Schönberger, J. Phys. B **25**, 4325 (1992).

22. C. T. Whelan, H. R. J. Walters, J. Hanssen and R. M. Dreizler, Aust. J. Phys. **44**, 39 (1993).

23. F. W. Byron, Jr., C. J. Joachain and B. Piraux, J. Phys. B **19**, 1201 (1986).

24. H. R. J. Walters, X. Zhang and C. T. Whelan, in Whelan et al edited, *(e, 2e) & related processes*, Kluwer, Dordrecht, 33-74 (1993).

25. E. Schüle and W. Nakel, J. Phys. B **15**, L639 (1982).

26. H-Th. Prinz, K.-H. Besch and W. Nakel, Phys. Rev. Lett. **74**, 243 (1995).

27. I. E. McCarthy and E. Weigold, Rep. Prog. Phys. **54**, 789 (1991).

28. F. Bell, J. Phys. B **22**, 287 (1989).

29. S. Keller and C. T. Whelan, J. Phys. B **27**, L771 (1994).

30. J. N. Das and A. N. Konar, J. Phys. B **7**, 2417 (1974).

31. M. E. Rose, *Relativistic Electron Theory*, John Wiley, New York (1961).

32. D. H. Jakubaßa-Amundsen, Z. Phys. D **11**, 305 (1989).

33. D. H. Jakubaßa-amundsen, J. Phys. B **25**, 1297 (1992).

34. H. R. J. Walters, H. Ast, C. T. Whelan, R. M. Dreizler, H. Graf, C. D. Schröter, J. Bonfert and W. Nakel, Z. Phys. D **23**, 353 (1992).

35. W. Nakel and C. T. Whelan, Physics Reports, *in press.*

36. J. Bonfert, H. Graf and W. Nakel, J. Phys. B **24**, 1423 (1991).

37. I. Fuss, J. Mitroy and B. M. Spicer, J. Phys. B **15**, 3321 (1982).

38. A. Cavalli and L. Avaldi, Nouovo Cimento Soc. Ital. Fis. D **16**, 1 (1994).

39. H. Ast, S. Keller, R. M. Dreizler, C. T. Whelan, L. U. Ancarani and H. R. J. Walters, J. Phys. B **29**, L585 (1996).

40. H. Ehrhardt, K. Jung, g. Knoth and P. Schlemmer, Z. Phys. D **1**, 3 (1986).

41. H. Klar, A. C. Roy, P. Schlemmer, K. Jung and H. Ehrhardt, J. Phys. B **20**, 821 (1987).

42. E. P. Curran and H. R. J. Walters, J. Phys. B **20**, 337 (1987).

43. J. C. Slater, Phys. Rev. **36**, 57 (1930).

44. L. U. Ancarani, S. Keller, H. Ast, C. T. Whelan, H. R. J. Walters and R. M. Dreizler, J. Phys. B **31**, 845 (1998).

45. F. Mota Furtado and P. F. O'Mahoney, J. Phys. B **22**, 2989 (1989).

46. S. Keller, R. M. Dreizler, L. U. Ancarani, H. R. J. Walters, H. Ast and C. T. Whelan, Z. Phys. D **37**, 191 (1996).

47. H. Ast, S. Keller, C. T. Whelan, H. R. J. Walters and R. M. Dreizler, Phys. Rev. **50**, R1 (1994).

Electron Impact Ionisation: A Progress Report

Hubert Klar

Fakultät für Physik
Albert-Ludwigs-Universität
Hermann Herder Strasse 3, 79104 Freiburg, Germany

We report progress recently achieved for the calculation of multiply differential ionisation cross sections. In particular we emphasise the importance of correlation in the final state. We apply the theory to the ionization of laser-excited targets, and discuss the circular dichroism in (e, 2e) events. Finally we also consider the ionization by spin-polarized electrons. Throughout this paper we use atomic units.

1.   INTRODUCTION

Ionisation of atoms by electron collision has been observed experimentally already more than 100 years ago. The early experiments have measured total cross sections. Only much later experiments with fully determined kinematics became possible [1]. Such a complete experiment needs the coincidence detection of two of the escaping fragments (two electrons, or ion + one electron) including energy analysis. Non-coincidence measurements are not suitable for comparison with theoretical work because calculated probabilities must be averaged over non-observed quantities. Such averages may be critical because one would average also over theoretical errors. Thus a poor theory may yield surprisingly good results. Since the pioneering experiment of Ehrhardt [1] many other coincidence ionisation experiments have been done.

From the theoretical side shortly after the discovery of quantum mechanics the first theory for ionisation by fast electrons was developed [2]. Today Bethe's work is referred to as 'first Born approximation' (hereafter FBA for brevity). The calculated quantity in Bethe's theory is usually called 'generalised oscillator strength', for a review see for instance [3]. The FBA should be applicable for small values of the momentum transfer. Actually calculations based on the FBA are qualitatively in agreement with experimental work. However, full agreement was not achieved. Second order Born corrections have contributed to an amendment [4]. The main shortcoming of the FBA is the total neglect of direct electron-electron interaction in the continuum. The second Born approximation takes this into account, but only approximately and moreover in a rather indirect way. More recently Brauner et al (hereafter BBK for brevity) [5] have applied a correlated 3-body continuum wavefunction to (e, 2e). The wavefunction itself was developed by Garibotti and Miraglia [6] in the context of charge exchange. The essential point is that BBK describes the electron-electron repulsion in the final state. This wavefunction depends explicitly on the electron-electron separation and on the relative electron-electron momentum, see §2 for details.

This progress report summarises the BBK technique, and applications to simple atoms like H, see §2. In §3 we discuss the circular dichroism in (e, 2e) reactions from laser-excited targets. Finally, we consider (e, 2e) from laser-excited atoms by spin-polarised electrons. We use throughout this paper atomic units.

2. SCATTERING THEORY

We treat here the theoretical frame to calculate triply differential cross sections (TDCS) for (e, 2e)-reactions. The TDCS for two escaping electrons (scattered and ionised electron, respectively) is formally given by

$$\frac{d^3\sigma}{d\Omega_a \, d\Omega_b \, dE_b} = (2\pi)^4 \frac{k_a k_b}{k_0} \left| T_{if} \right|^2 \tag{1}$$

where $\vec{k}_0$, $\vec{k}_a$ and $\vec{k}_b$ are the momenta of the incident, scattered and ejected electrons, respectively. The T-matrix element is given by

$$T_{if} = \left\langle \Psi_f^- \left| V \right| \Phi \right\rangle \tag{2}$$

Here the initial state Φ is a product state consisting of a plane wave for the incoming electron and a bound state wave function for the target atom. V is the (non-relativistic) potential between the projectile and the target, and

Ψ_f^- is the final state describing two escaping electrons in the field of an ion; i.e. this should be an exact solution of the *N+1* electron problem.

An exact solution Ψ_f^- of course is not available. The following approximations have been frequently used in the literature.

2.1   Plane Wave Approximation

This simple approximation disregards any interaction and treats both electrons as free particles, i.e.

$$\Psi_f^- = \left(2\pi\right)^{-3} e^{i\left(\vec{k}_a \cdot \vec{r}_a + \vec{k}_b \cdot \vec{r}_b\right)} \tag{3}$$

2.2   Born Approximation

This approximation, also known as Bethe [2] approximation, is applicable at low values of the momentum transfer, i.e. at relatively fast incident electrons, fast scattered and much slower secondary electrons. This approximation treats the fast scattered electron again as a plane wave (free particle) and the secondary electron is represented by an ordinary Coulomb wave $\psi_{\vec{k}_b}\left(\vec{r}_b\right)$,

$$\Psi_f^- = \left(2\pi\right)^{-3/2} e^{i\vec{k}_a \cdot \vec{r}_a} \psi_{\vec{k}_b}\left(\vec{r}_b\right) \tag{4}$$

This implies that the interaction of the slower electron with the ion (often a bare nucleus) is taken into account, but all other interactions in the final state are disregarded.

2.3   BBK

A further improvement of the wavefunction is provided by taking into account both electron-ion interactions as well as the electron-electron repulsion. Exactly this is done by a wavefunction, see also [6, 5], of the following form

88

$$\Psi_f^- = (2\pi)^{-3/2}\, e^{i\left(\vec{k}_a\cdot\vec{r}_a+\vec{k}_b\cdot\vec{r}_b\right)}\; {}_1F_1\left(-i\frac{Z}{k_a},1;-i\left(k_a r_a+\vec{k}_a\cdot\vec{r}_a\right)\right)$$

$$\times\; {}_1F_1\left(-i\frac{Z}{k_b},1;-i\left(k_b r_b+\vec{k}_b\cdot\vec{r}_b\right)\right)\; {}_1F_1\left(i\frac{1}{2\kappa},1;-i\left(\kappa r_{ab}+\vec{\kappa}\cdot\vec{r}_{ab}\right)\right) \tag{5}$$

with $\vec{\kappa}=\frac{1}{2}\left(\vec{k}_a-\vec{k}_b\right)$ and $\vec{r}_{ab}=\vec{r}_a-\vec{r}_b$. The essential property of this wavefunction is that it shows exact asymptotic behaviour in the Redmond [7] limit, i.e. for large particle separations.

2.4   Local Momenta

An even better description of the double continuum state is achieved with help of local space dependent momenta [8]. Actually Alt and Mukhamedzhanov [8] showed the necessity of modified momenta to correctly describe the asymptotic form of the wavefunction in the limiting case of one particle far away from the remaining 2-body subsystem. We present here a slightly different viewpoint and proceed as follows. First we separate off the plane wave factor for the electrons

$$\Psi_f^- = e^{i\left(\vec{k}_>\cdot\vec{r}_> + \vec{k}_<\cdot\vec{r}_<\right)}\,\overline{\Psi} \tag{6}$$

where $\overline{\Psi}$ describes the Coulomb modifications. $r_{>,(<)}$ stands for the larger (smaller) electron-ion separation, and $k_{>,(<)}$ are the corresponding momenta. As in the BBK wavefunction, Eqn. (5), we employ Coulomb waves for 2-body subsystems. For the outer electron located at $r_>$ we use ordinary Coulomb waves for the 2-body subsystem as above,

$$\overline{\Psi}=A\,{}_1F_1\left(-i\frac{Z}{k_>},1;-i\left(k_> r_> + \vec{k}_>\cdot\vec{r}_>\right)\right){}_1F_1\left(i\frac{1}{2\kappa},1;-i\left(\kappa r+\vec{\kappa}\cdot\vec{r}\right)\right)\propto A e^{i\Phi} \tag{7}$$

where the phase Φ being chosen to produce correct asymptotic behaviour,

$$\left(\vec{k}_>\cdot\vec{\nabla}_> + \vec{k}_<\cdot\vec{\nabla}_<\right)\Phi = -\frac{Z}{r_>}+\frac{1}{r_{><}} \tag{8}$$

with the solution

$$\Phi = -\frac{Z}{k_>}\ln\left(k_> r_> + \vec{k}_> \cdot \vec{r}_>\right) + \frac{1}{2\kappa}\ln\left(\kappa r_{><} + \vec{\kappa}\left(\vec{r}_> - \vec{r}_<\right)\right) \qquad (9)$$

and $\vec{\kappa} = \frac{1}{2}\left(\vec{k}_> - \vec{k}_<\right)$. The amplitude in Eqn. (7) describes the motion of the inner electron located at $\vec{r}_<$. For $r_> >> r_<$ the wave equation for $A = A(\vec{r}_<)$ reduces to

$$\left[\Delta_< + 2i\vec{\kappa}_{eff}\cdot\vec{\nabla}_< - \frac{2Z}{r_<}\right]A(\vec{r}_<) = 0 \qquad (10)$$

with the solution

$$A = {}_1F_1\left(-i\frac{Z}{k_{<,\,eff}},1;-i\left(k_{<,\,eff}\,r_< + \vec{k}_{<,\,eff}\cdot\vec{r}_<\right)\right) \qquad (11)$$

where the effective momentum for the inner electron is given by

$$\vec{k}_{<,\,eff} = \vec{k}_< + \vec{\nabla}_<\Phi \qquad (12)$$

and Φ given by Eqn. (9). This momentum modification (12) is identical to the result achieved by [8] to lowest order in $r_</r_>$. $\overline{\Psi}$ in Eqn. (6) has correct asymptotic behaviour also if all three particle separations are large. This is easily seen because each of the confluent hypergeometric functions reduces then to a pure phase factor, and the effective momentum (12) approaches its static value in that limit. Eqn. (7) is still incorrect in the limit of two electrons close together but far away from the nucleus. We investigate therefore this limit now. To this end we introduce Jacobi coordinates $\vec{R} = \frac{1}{2}\left(\vec{r}_< + \vec{r}_>\right)$ and $\vec{r} = \vec{r}_> - \vec{r}_<$. For large values of R and finite values of r we expect a structure of the wavefunction like $\overline{\Psi} = B(\vec{r})e^{i\Lambda}$ where the phase Λ is now defined by the eikonal equation

$$\left(\vec{k}_>\cdot\vec{\nabla} + \vec{k}_<\cdot\vec{\nabla}\right) = -\frac{Z}{r_>} - \frac{Z}{r_<} \qquad (13)$$

90

with the solution

$$\Lambda = -\frac{Z}{k_>}\ln\left(k_> r_> + \vec{k}_> \cdot \vec{r}_>\right) - \frac{Z}{k_<}\ln\left(k_< r_< + \vec{k}_< \cdot \vec{r}_<\right) \qquad (14)$$

For the amplitude B we find then the wave equation ($r << R$)

$$\left[\Delta_r + 2i\vec{\kappa}_{eff} \cdot \vec{\nabla}_r - \frac{1}{r}\right] B(\vec{r}) = 0 \qquad (15)$$

where $\vec{\kappa}_{eff}$ is given by

$$\vec{\kappa}_{eff} = \frac{1}{2}\left(\vec{k}_> - \vec{k}_<\right) + \frac{1}{2}\left(\vec{\nabla}_> - \vec{\nabla}_<\right)\Lambda \qquad (16)$$

We conclude therefore that the wavefunction given by Eqn. (7) should be an accurate solution of the Schrödinger equation provided the relative momentum $\vec{\kappa}$ is replaced by its effective value, see Eqn. (16). We intend to employ this wavefunction for future *(e, 2e)* and *(γ, 2e)* calculations.

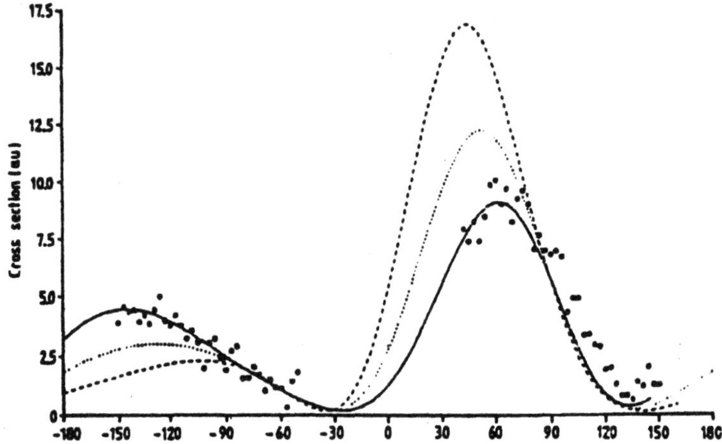

Figure 1. The TDCS for $e^{\pm}$ impact ionisation of atomic hydrogen as a function of the slow electron ejection angle Θ_e. The projectile incident energy is *250 eV* and it is scattered through an angle of 3°. The ionised electron has an energy of *5 eV*. The curves show the TDCS for the electron impact (full curve), positron impact (dotted curve). The Born approximation results are also shown (chain curve). The full circles are experiments made by Ehrhardt [1]; taken from [5].

Although we expect the wavefunction (7) to give better results than (5) we stress that even (5) does quite well. Fig. (1) shows a typical example, the ionisation of H(1s) by electron (positron) impact at an energy of $E_0 = 250$ eV. The scattering angle of the fast electron (positron) is $\Theta = 3°$, the secondary electrons have an energy of $E_b = 5$ eV. The dotted line is the result within the first Born approximation, see Eqn. (4), which is the same for electron and positrons. The solid line corresponds to electron impact and the use of the wavefunction given by (5) whereas the dashed line displays the results for positrons and again the use of (5). The dots are experimental results, see Ref. [5] for more details. Although the kinematic conditions in Fig. 1 well fulfill the validity of the Born approximation it is evident that the BBK result is superior. In particular the huge difference between electrons and positrons comes from long range correlation properly taken into account in the wavefunction given by (5).

3.　　LASER EXCITED TARGETS

3.1　　Orientation and Alignment in (e,2e) Reactions

This subsection is dedicated to the electron impact ionisation of oriented and/or aligned targets, for instance

$$\gamma(polarized) + Na(3s) \rightarrow Na^*(3p;\ \rho)$$

$$Na^*(3p;\ \rho) + e \rightarrow Na^+(^1S) + 2e \tag{17}$$

where ρ stands for the density operator describing the M-state population of the excited $3p$ orbital which depends of course on the initial light polarisation. For the moment we entirely disregard the spin of the electrons. To spin effects we come in the next subsection. It is obvious that an initial circular light polarisation leads to an oriented excited state whereas linearly polarised light creates an aligned excited state. Analogous to the circular dichroism in photo-double-ionisation [9] [10] [11] we discover a final state in which the two continuum electrons must carry the orientation or the alignment. We come now to the question how does the TDCS depend on ρ. At a first glance one might expect a continuous dependence. Actually, however, we will immediately see that only a finite number of cross-section portions really depend on the density matrix ρ. To this end it is useful to rewrite the TDCS for the reaction described by Eqn. (17) in the form

92

$$\frac{d^3\sigma}{d\Omega_a\, d\Omega_b\, d E_b} = C\sum_M \left\langle \Psi_{k_a k_b}\left| V \right| \vec{k}_0\,\Phi_{LM}\right\rangle \rho^L_{MM} \left\langle \vec{k}_0\Phi_{LM}\left| V \right|\Psi_{k_a k_b}\right\rangle \qquad (18)$$

with $C = \left(2\pi\right)^4 \frac{k_a k_b}{k_0}$, see also §2. It is convenient to express the density matrix (above in diagonal form) in terms of state multipoles ρ_{KQ}, see for instance Fano [12],

$$\rho^J_{MM} = \sum_M (-)^{K-J-M}\left\langle J-MJM\,|\,K0\right\rangle \rho_{K0} \qquad (19)$$

This leads to the TDCS decomposition

$$\frac{d^3\sigma}{d\Omega_a\, d\Omega_b\, d E_b} = \sum_{K=0}^{(K)} \rho_{K0}\,\Sigma_0^{(K)} \qquad (20)$$

where the quantities $\Sigma_0^{(K)}$ are given by

$$\Sigma_0^{(K)} = \left\langle \Psi_{k_a k_b}\left| V\, S_0^{(K)}\, V \right|\Psi_{k_a k_b}\right\rangle \qquad (21)$$

with the tensors

$$S_0^{(K)} = C\sum_M (-)^{K+J+M}\left\langle J-MJM\,|\,K0\right\rangle \left| \vec{k}_0\,\Phi_{JM}\right\rangle\left\langle \vec{k}_0\,\Phi_{JM}\right| \qquad (22)$$

From the construction it is obvious that our quantities $\Sigma_0^{(K)}$ play formally the role of orientation and/or alignment parameters as introduced by Fano and Macek in the context of emitted polarised light [13]. Odd values of K describe orientations, whereas even Ks are alignments. For the purpose of lucidity we apply this framework to the Na experiment mentioned above. Disregarding the spin of the electrons for a moment we have simply $L=1$ for p-states, and therefore $K=0, 1, 2$. The corresponding $\Sigma_0^{(K)}$'s read then

$$\Sigma_0^{(0)} = \frac{C}{\sqrt{3}}\left[\left|\left\langle \Psi_{k_a k_b}\right| V \left|\vec{k}_0\Phi_{11}\right\rangle\right|^2 + \left|\left\langle \Psi_{k_a k_b}\right| V \left|\vec{k}_0\Phi_{10}\right\rangle\right|^2 + \left|\left\langle \Psi_{k_a k_b}\right| V \left|\vec{k}_0\Phi_{1-1}\right\rangle\right|^2 \right]$$

$$(23)$$

$$\Sigma_0^{(1)} = \frac{C}{\sqrt{2}} \left[\left| \left\langle \Psi_{k_a k_b} \right| V \left| \vec{k}_0 \Phi_{11} \right\rangle \right|^2 - \left| \left\langle \Psi_{k_a k_b} \right| V \left| \vec{k}_0 \Phi_{1-1} \right\rangle \right|^2 \right] \tag{24}$$

$$\Sigma_0^{(2)} = \frac{C}{\sqrt{6}} \left[\left| \left\langle \Psi_{k_a k_b} \right| V \left| \vec{k}_0 \Phi_{11} \right\rangle \right|^2 - 2 \left| \left\langle \Psi_{k_a k_b} \right| V \left| \vec{k}_0 \Phi_{10} \right\rangle \right|^2 + \left| \left\langle \Psi_{k_a k_b} \right| V \left| \vec{k}_0 \Phi_{1-1} \right\rangle \right|^2 \right] \tag{25}$$

The physical meaning of these three Σ's is the following. $\Sigma_0^{(0)}$ is the TDCS for a statistical M-state population. $\Sigma_0^{(1)}$ is the TDCS difference for right minus left circular polarisation, and $\Sigma_0^{(2)}$ is the cross section difference for unpolarised light minus linearly polarised light. Within the dipole approximation for photoexcitation no other information except for the three Σ's can be obtained. For more details, see [14]. For the purpose of illustration we show a few numerical results we have obtained for *H(2p)* and *Na(3p)* as targets so far. To evaluate the Na-T-matrix element for ionisation

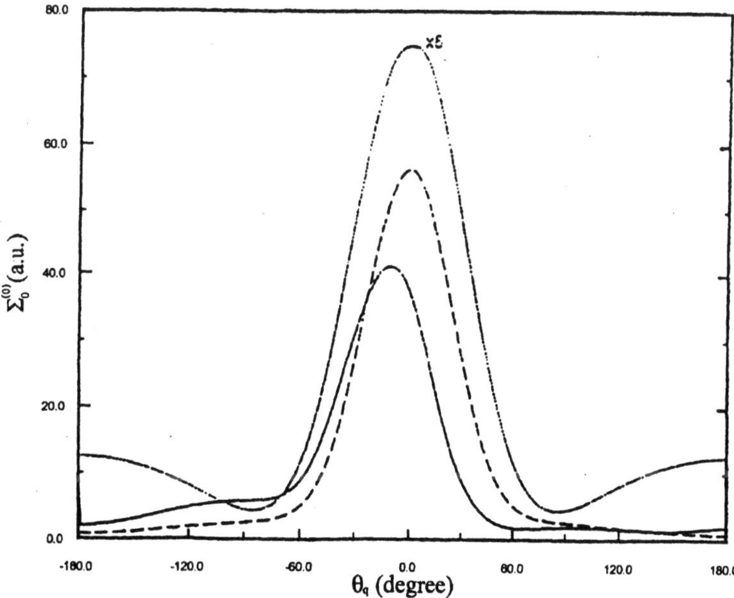

Figure 2. Tensor component $\Sigma_0^{(0)}$ for *H(2p)* as function of the emission angle of the slow electron relative to the direction of the momentum transfer. The impact energy is $E_0 = 150$ eV. Full, dashed and dotted curves present BBK, Born approximation and plane wave approximation, respectively; taken from [14].

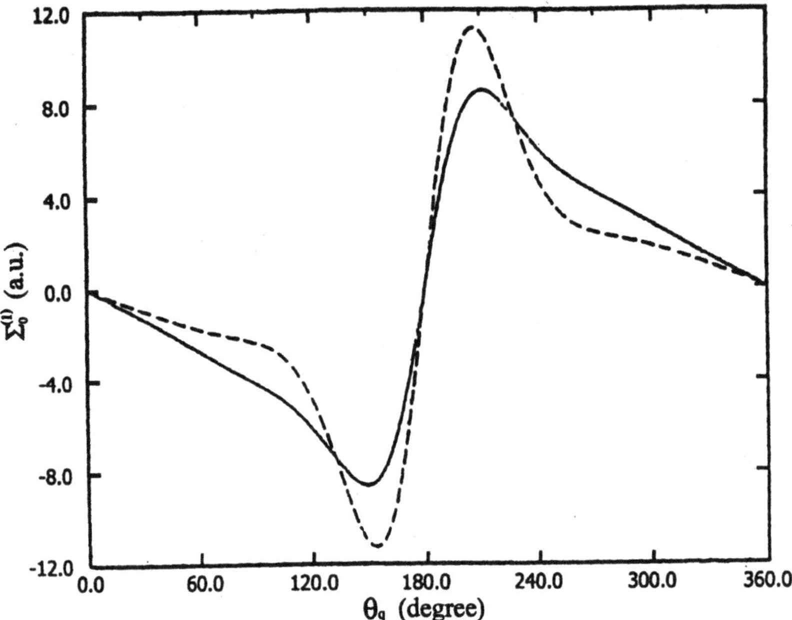

Figure 3. The orientation parameter $\Sigma_0^{(1)}$. The kinematical parameters are the same as in Figure 2, BBK (full curve), Born approximation (dashed curve), taken from [14].

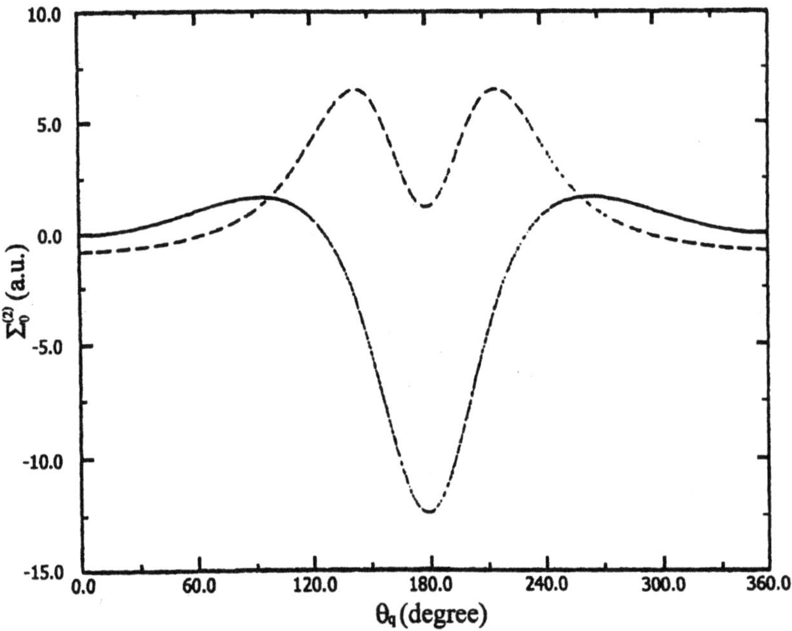

Figure 4. The alignment parameter $\Sigma_0^{(2)}$, notations as in Figure 3; taken from [14].

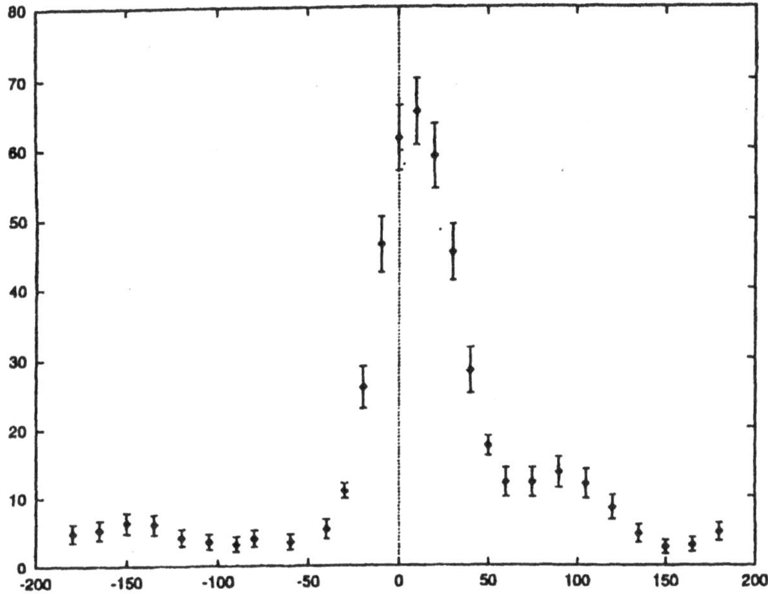

Figure 5. Tensor parameter $\Sigma_0^{(2)}$ for Na, versus emission angle relative to the momentum transfer, incident energy $E_0 = 150$ eV, secondary energy $E_b = 5$ eV, scattering angle 4° taken from [14].

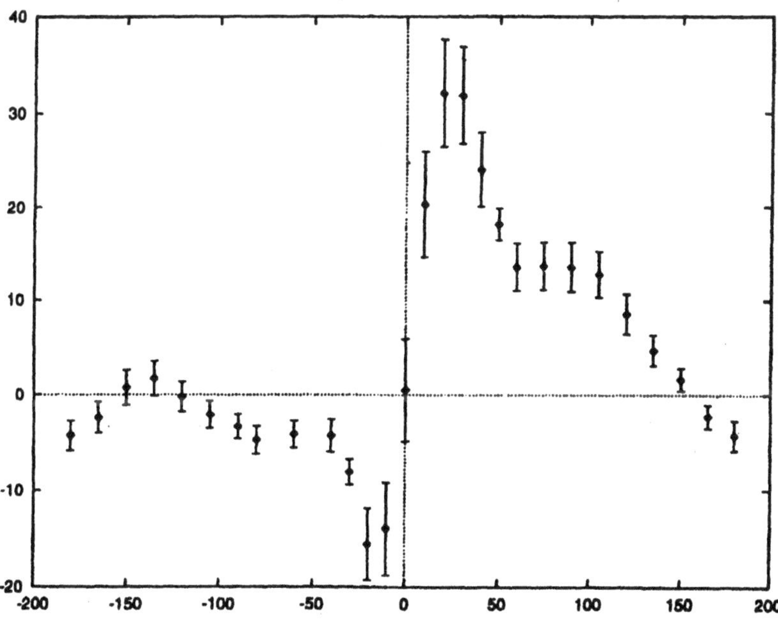

Figure 6. Orientation parameter $\Sigma_0^{(1)}$ for Na, otherwise same as figure 5; taken from [14].

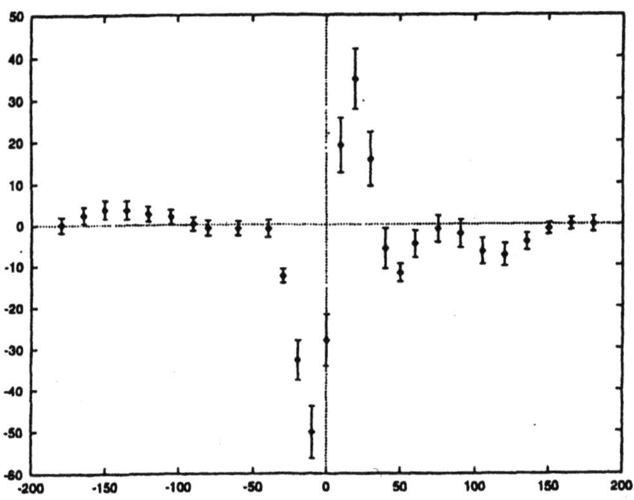

Figure 7 Alignment parameter $\Sigma_0^{(2)}$ for Na, otherwise same as figure 5; taken from [14].

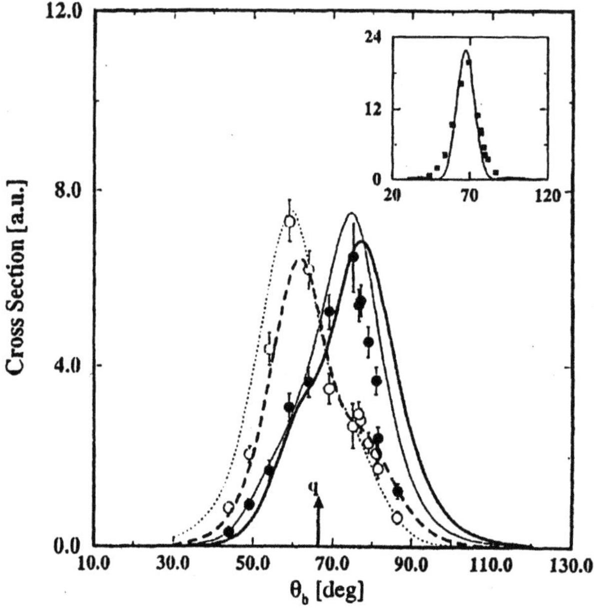

Figure 8 The 150 eV cross section on Na $3^2P_{3/2}$, $m_F = -3$ (filled circles) for $\Theta_a = 20°$, $\Phi_a - \Phi_b = \pi$ and $E_b = 20°$. The corresponding FBA (scaled down by a factor of 2) and BBK cross sections are indicated, respectively. These measurements are normalised to the $m = -1$ BBK cross section peak. The momentum transfer direction is given by $\vec{q}$. The inset shows the ground state Na $3\,^2S_{1/2}$ cross section.

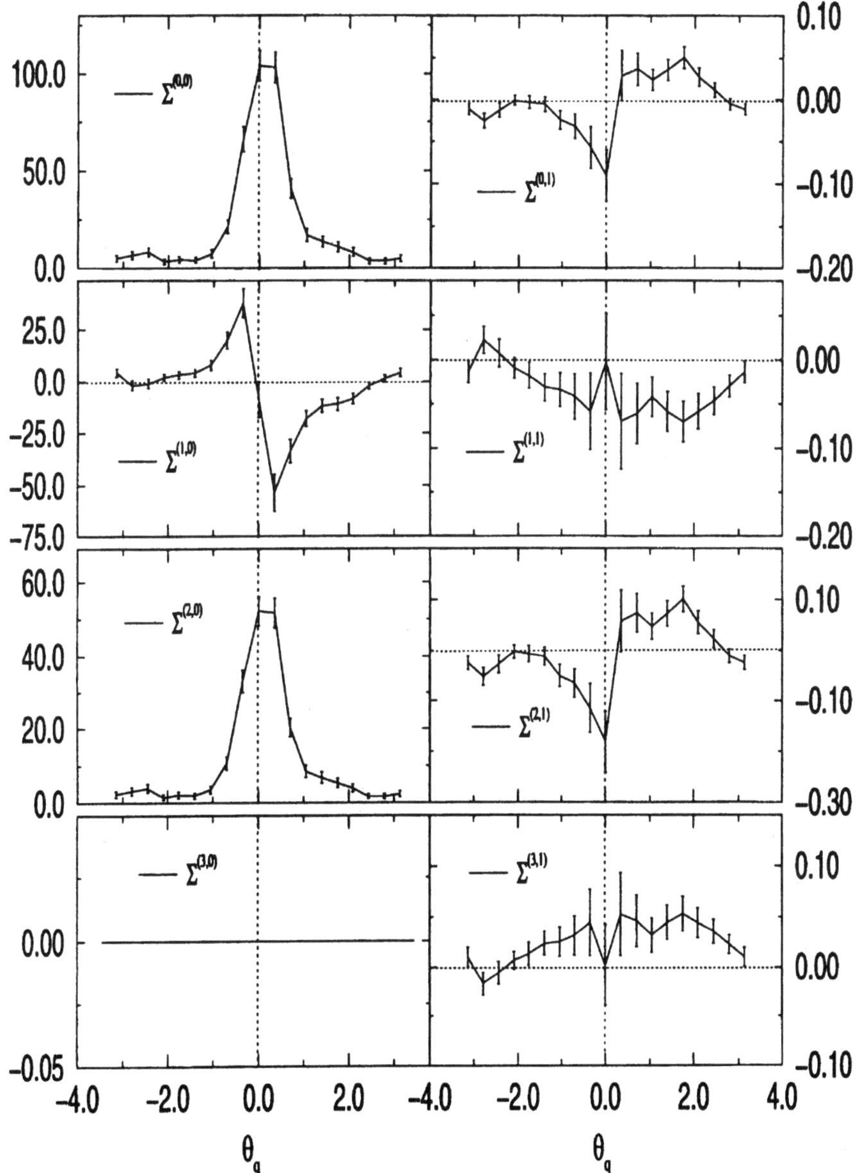

Figure 9 Tensor parameters $\Sigma^{(K\,k)}$ for $Na(p_{3/2})$ with $E_0 = 250$ eV, $E_b = 5$ eV and $\Theta_a = 4°$; taken from [16].

we have used an analytical fit to a Hartree-Fock radial wavefunction for *Na(3p)*, whereas the final state has been approximated by a *BBK*-wavefunction. The latter approximation disregards, of course, any core structure. However, this should not be critical because the *T*-matrix element requires integration over the whole target volume and the contribution from the inner region is depressed by the initial p bound state function. The reason for the H calculation is basically to get a feeling for the structure of the Σ's, a realistic H experiment seems presently too difficult. Fig. (2) shows the result of three calculations for $\Sigma^{(0)}$ at an energy of $E_0 = 150$ eV and $H(2p)$ as target, the secondary electron energy is $E_b = 5$ eV. Plotted is the angular distribution of the ionised electron. It is clearly seen a binary peak near the momentum transfer direction. It is also seen that due to final state correlation the BBK approximation predicts a maximum at a slightly shifted angle. Only the Born approximation has the binary peak exactly at the momentum transfer direction. The plane wave approximation is wrong by a huge factor. Fig. (3) shows the corresponding result for $\Sigma^{(1)}$ except that the incident energy is slightly higher $E = 153.4$ eV. The Born approximation curve (dashed) is antisymmetric around the momentum transfer direction whereas the better BBK approximation (full line) breaks this symmetry. Fig. (4) finally displays $\Sigma^{(2)}$. Here we find a camel back in Born approximation, less pronounced but still present in BBK approximation, see also [14] for more details. For $Na(3p)$ we find similar cross section structures, Figs. (5-7) show the three Σ's analogous to Figures (2-4) for $H(2p)$. The error bars in Figs. (5-7) indicate uncertainties in the Monte Carlo integration evaluating the T-matrix element, see [14] for details. The above theoretical results for $Na(3p)$ are not inconsistent with the experimental observation [15] as is clearly demonstrated in Fig. (8).

3.2   (e, 2e) from laser excited atoms with spin-polarised electrons

We proceed as in §3.1 except we replace Eqn. (18) now by

$$\frac{d^3\sigma}{d\Omega_a\, d\Omega_b\, d E_b} = C \sum_{m_a m_b M m} \left\langle \Psi_{m_a m_b} \right| V \left| \Phi_{JM}\varphi_m \right\rangle \rho^J_{MM}\, \rho^s_{mm} \left\langle \Phi_{JM}\varphi_m \right| V \left| \Psi_{m_a m_b} \right\rangle$$

$$(26)$$

Here the incident electron is now described by a spinor plane wave φ_m with spin projection $m = \pm\frac{1}{2}$, ρ^s_{mm} is the density operator describing the actual experimental spin state of the incident electron, m_a, m_b are the spin quantum numbers of the escaping electrons. All other terms have the same meaning as before. The essential point is now the following. Let us assume we excite for instance the fine structure level $M_J = 3/2$, then we have automatically a spin-polarised target because this situation requires $M_L = 1$ and $M_S = 1/2$. If we hit this target by a spin-polarised projectile we have a fully determined initial state including the spin. A complete experiment would analyse also the spins in the final state, but a coincidence experiment including spin analysis would lead to terribly small cross sections. Therefore we sum over the spin projections m_a, m_b of the final state. We will see below that even this drastically simplified experiment will yield much more information than the previous situation discussed in §3.1. To this end we replace the density operators by state multipoles similar as before,

$$\rho_{MM}^{J} = \sum_K (-)^{K-J-M} \langle J-MJM|K0\rangle \, \rho_{K0} \qquad (27)$$

$$\rho_{mm}^{s} = \sum_k (-)^{k-\frac{1}{2}-m} \langle \tfrac{1}{2}-m\tfrac{1}{2}m|k0\rangle \, \rho_{k0} \qquad (28)$$

Substituting this into Eqn. (26) we arrive at

$$\frac{d^3\sigma}{d\Omega_a \, d\Omega_b \, dE_b} = \sum_{Kk} \rho_{K0}\,\rho_{k0}\,\Sigma^{(Kk)} \qquad (29)$$

which is the generalisation of Eqn. (20), and where the new Σ's are now given by

$$\Sigma^{(K0)} = \frac{C}{\sqrt{2}} \sum_{Mm} (-)^{K-J-M} \langle J-MJM|K0\rangle \sum_{S\mu} \left| \langle \Psi_{S\mu}| V |\Phi_{JM}\varphi_m\rangle \right|^2 \qquad (30)$$

$$\Sigma^{(K1)} = -C\sqrt{2} \sum_{Mm} (-)^{K} - J - M \langle J-MJM|K0\rangle m \sum_{S\mu} \left| \langle \Psi_{S\mu}| V |\Phi_{JM}\varphi_m\rangle \right|^2 \qquad (31)$$

Note that only $k = 0,\, 1$ occur because of $s = \tfrac{1}{2}$. Above we have introduced the total spins $S = 0,\, 1$ for the escaping electrons instead of the individual spins $s_a = s_b = \tfrac{1}{2}$. The $\Sigma^{(K0)}$, Eqn. (30), is proportional to our previous $\Sigma^{(K)}$ because the upper spin index 0 implies averaging over the initial projectile spin polarisation. $\Sigma^{(K1)}$ is novel since it describes effects stemming from the projectile polarisation. Finally we remark that expression (26) reduces to the familiar expression

$$\frac{d^3\sigma}{d\Omega_a\, d\Omega_b\, dE_b} = \frac{C}{2J+1} \sum_{m_a\, m_b\, M} \left| \left\langle \Psi_{m_a m_b} \middle| V \middle| \Phi_{JM}\varphi \right\rangle \right|^2 \qquad (32)$$

for unpolarised electrons and targets where we have used $\rho^J_{MM} = (2J+1)^{-1}$ and $\rho^s_{mm} = \frac{1}{2}$. As an example we have applied this description to the ionisation of $Na\left(3p_{3/2}\right)^*$. Because of $J = 3/2$ we get now $K = 0, 1, 2, 3$ which includes a tensor orientation $(K = 3)$. These parameters we have calculated numerically [16] the results are displayed in Fig. (9) where the kinematical parameters have been $E_0 = 250$ eV, $E_b = 5$ eV and $\Theta_a = 4°$.

ACKNOWLEDGMENTS

Financial support by DFG under contract SFB276 is gratefully acknowledged.

REFERENCES

1. H. Ehrhardt, M. Schulz, T. Tekaat and K. Wilmann, Phys. Rev. Lett. 22, 89 (1969).

2. H. Bethe, Ann. Physik (Leipzig) 5, 325 (1930).

3. M. Inokuti, Rev. Mod. Phys. 43, 297 (1971).

4. H. Ehrhardt, M. Fischer, K. Jung, F. W. Byron, C. J. Joachain and B. Piraux, Phys. Rev. Lett. 48, 1807 (1982).

5. M. Brauner, J. S. Briggs and H. Klar, Journ. Physics B22, 2265 (1989).

6. G. Garibotti and J. E. Miraglia, Phys. Rev. A21, 572 (1980).

7. Redmond (unpublished); see L. Rosenberg, Phys. Rev. D8, 1833 (1964).

8. E. O. Alt and A. M. Mukhamedzhanov, Phys. Rev. A47, 2004 (1993).

9. J. Viefhaus, L. Avaldi, G. Snell, M. Wiedenhöft, R. Hentges, A. Rüdel, F. Schäfer, D. Menke, U. Heinzmann, A. Engelns, H. Klar and U. Becker, Phys. Rev. Lett. 77, 3975 (1996).

10. J. Berakdar, H. Klar, Phys. Rev. Lett. 69, 1175 (1992).

11. J. Berakdar, H. Klar, A. Huetz and P. Selles, J. Phys. B26, 1463 (1993).

12. U. Fano, Rev. Mod. Phys. 29, 76 (1957).

13. U. Fano and J. H. Macek, Rev. Mod. Phys. 45, 553 (1973).

14. J. Berakdar, A. Engelns and H. Klar, Jour. Phys. B29, 1109 (1996).

15. A. Dorn,A. Elliott, J. Lower, E. Weigold, J. Berakdar, A. Engelns and H. Klar, Phys. Rev. Lett. 80, 257 (1998).

16. P. Golecki and H. Klar, J. Phys. (submitted).

Electron Impact Ionisation of Atoms and Molecules

S. P. Khare[*#] and Surekha Tomar[*]

*Physics Department
Chaudhary Charan Singh University, Meerut - 250 004, India

#Inter University Centre for Astronomy and Astrophysics
Pune - 411 007, India

A progress report of the recent investigations carried out by Khare and his associates for the electron impact inner-shell ionisation of atoms and outer-shell ionisation of molecules, in the energy range varying from ionisation threshold to 1 GeV, is presented. The new method of Khare et al [19] is compared with the methods developed by Kim and Rudd [16]. The present ionisation cross sections of the hydrogen atom are compared with those obtained by Kim and Rudd and the experimental data of Saha et al [20].

1.   INTRODUCTION

In the present conference there are more than half a dozen invited talks on the ionisation of the atoms and molecules due to electron, photon and heavy particle impacts.   This indicates the importance of the ionisation process.   We have developed interest in the electron impact ionisation process from the late sixties which is still being continued.   In 1976 a successful formula for the electron impact ionisation cross section of the molecules was proposed by Jain and Khare [1].   This was modified and extended to the dissociative ionisation of the molecules by Khare and Meath [2].   The above methods employed available values of the photoionisation (or the continuum optical oscillator strengths (COOS)) and the Bethe collisional parameter as the inputs to evaluate the ionisation cross sections.

In this talk an attempt is being made to present a progress report of the work done by our group on the ionisation of the inner-shell of the atoms and

Trends in Atomic and Molecular Physics,
Edited by Sud and Upadhyaya. Kluwer Academic/Plenum Publishers, New York, 2000.   103

the outer-shell of the molecules due to electron impact in the last five years. The reason for combining the inner-shell with the outer-shell is due to the fact that we have developed a single method which has given good results for the both types of the ionisation for the impact energy E varying from the ionisation threshold to 1 GeV [3-15].

The recent binary-encounter-dipole (BED) and binary-encounter-Bethe (BEB) models developed by Kim and his associates [16-18] are also discussed. The useful features of the Saksena and Kim models are combined by Khare et al [19] to develop a new method. This new method has been applied to calculate the ionisation cross section of the hydrogen atom. These cross sections are compared with the theoretical cross sections of Kim and Rudd [16] and the experimental data of Saha et al [20].

2. INNER-SHELL IONISATION OF ATOMS

Let us start with the simple approximation of the collision theory namely the plane wave Born approximation (PWBA). This approximation neglects exchange effects as well as the distortion of the wave functions of the projectile and the target. In the PWBA the total cross section for the ionisation of the nl sub-shell of an atom due to an electron of velocity v and energy E is given by [11]

$$\sigma_{nl}^{B1} = \frac{8\pi a_0^2 R}{m v^2} \int_{I_{nl}}^{W_{max}} dw \int_{\ln(K^2 a_0^2)_{min}}^{\ln(K^2 a_0^2)_{max}} \frac{R}{W} \frac{df(w, K^2 a_0^2)}{dw} d\left(\ln(K^2 a_0^2)\right) dw \quad (1)$$

where a_0, R, m and I_{nl} are the Bohr radius, the Rydberg energy, the rest of the electron and the ionisation threshold, respectively. $\hbar K$ is the change in the magnitude of the momentum of the incident electron due to scattering. $\hbar$ is the Planck's constant divided by 2π. w is the energy lost by the incident electron in the ionising collision and $\dfrac{df(w, K^2 a_0^2)}{dw}$ is the continuum generalised oscillator strength (CGOS) for a transition from the bound state $|n|$ > to the continuum. The maximum and the minimum values of $K^2 a_0^2$ at the non-relativistic energies are given by

$$(K a_0^2)_{max, min} = \frac{1}{\sqrt{R}}\left[2E - w \pm 2\sqrt{E(E-w)}\right] \quad (2)$$

and w_{max} is equal to E.

The K-shell ionisation cross section obtained in the PWBA, with the screening parameter given by Slater [21], are shown in Fig. 1.

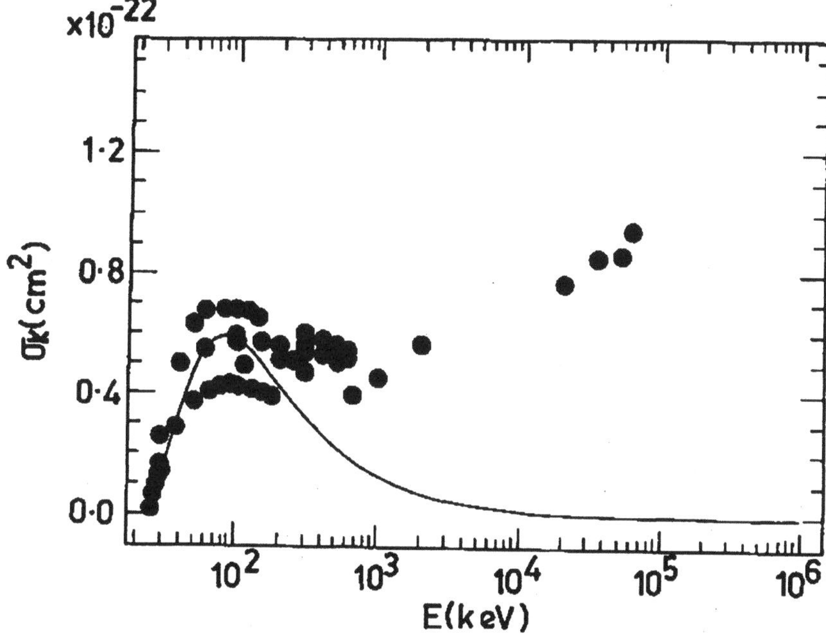

Figure 1. Variation of the K-shell ionisation cross sections of silver atom with the projectile energy E due to electron impact. The curve represents the cross sections obtained in the PWBA and ●·are the experimental cross sections compiled by Long et al [22].

The experimental data compiled by Long et al [22] are also shown for comparison. It is evident from the figure that the agreement between the theory and the experiments is poor. To correct some of the deficiencies of the PWBA exchange, relativistic and Coulomb corrections are introduced. The exchange correction is taken through the Ochkur approximation which multiplies the integrand of (1) by

$$F_{ex}(K^2, k_i^2) = 1 - \frac{K^2}{k_i^2} + \frac{K^4}{k_i^4} \qquad (3)$$

where $\hbar k_i$ is the momentum of the incident electron. With exchange w_{max} reduces to $(E + I_{nl}) / 2$. The Coulomb correction arises due to the fact that while passing through the atomic field the incident electron is accelerated. Hence at the instant of collision the kinetic energy E_c of the projectile is greater than E. Approximately it is given by [11]

106

$$E_c = E + h I_{nl} \tag{4}$$

where h is equal to 4/3 for the K- and L1-sub shells and equal to 8/5 for the L2- and L3- subshells. Due to the relativistic and Coulomb corrections (2) changes to

$$(Ka_0)_{max, min} = \frac{a_0}{\hbar c} \left[\sqrt{E_c(E_c + 2mc^2)} \pm \sqrt{(E_c - w)(E_c - w + 2mc^2)} \right] \tag{5}$$

where c is the velocity of light [11].

Even after the above corrections the modified PWBA is found to underestimate the cross section for E greater than about 0.1 MeV. To understand the reason for the failure of the modified PWBA let us consider the interaction $e^2 / |r_1 - r_2|$ between the incident and the bound electrons, having position vectors r_1 and r_2, respectively. The Fourier transform of the interaction is given by

$$\frac{e^2}{|r_1 - r_2|} = \frac{e^2}{2\pi^2} \int \frac{e^{iK \cdot (r_2 - r_1)}}{K^2} dK \tag{6}$$

and each Fourier component having wave vector K transfers a momentum $\hbar K$ from the incident electron to the bound electron in the direction of K. Hence this interaction is known as the longitudinal interaction. Only this interaction is included in the PWBA. The same momentum is also transferred by a virtual photon emitted by the incident electron and absorbed by the target electron. Since the photon field is perpendicular to K, the above interaction is known as transverse interaction and the ionisation cross section due to this interaction and the ionisation cross section due to this interaction is given by [23].

$$\sigma_t = -\frac{8\pi a_0^2 R}{mv^2} M^2 \{\ln(1 - \beta^2) + \beta^2\} \tag{7}$$

where M^2 is equal to the total dipole matrix squared measured in the units of a_0^2 and β is v/c. M^2 is given by [23]

$$M^2 = \int_{I_{nl}}^{E} \frac{R}{w} \frac{df(w,0)}{dw} dw \tag{8}$$

where $\dfrac{df(w,0)}{dw}$ is the continuum optical oscillator strength (COOS). σ_t is of importance only at ultra high velocities. Furthermore the longitudinal and the transverse interactions are of different parities [23]. Hence the total ionisation cross section in the modified PWBA is

$$\sigma = \sigma_l + \sigma_t \tag{9}$$

where σ_l is given by (1) which includes exchange, Coulomb and relativistic corrections. The ionisation cross sections obtained from (9) by Khare and Wadehra [10] for the silver atom are shown in Fig. 2.

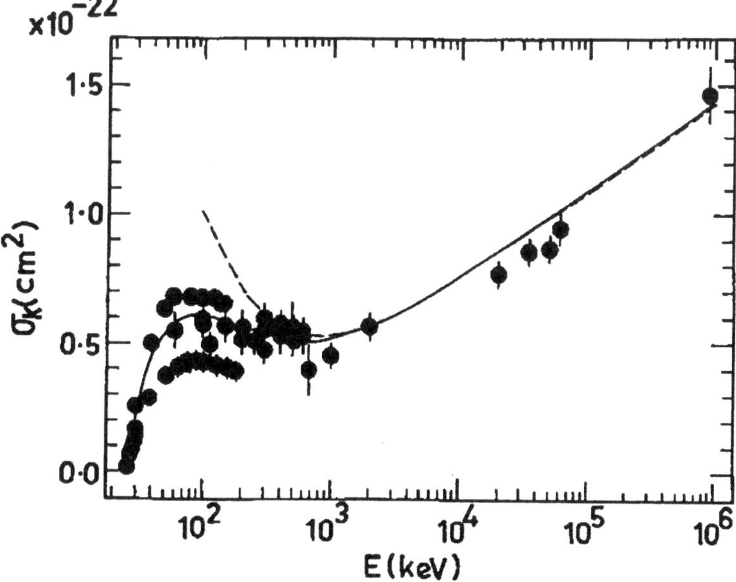

Figure 2. Variation of K-shell ionisation cross section of silver atom with the projectile energy E due to electron impact. The dashed curve represents cross sections obtained in the Born-Bethe approximation. The solid curve is obtained in the modified PWBA which includes exchange, Coulomb and relativistic corrections and also transverse interaction [10].

Considering the spread of the experimental cross sections the agreement between the theory and the experiments is satisfactory over the whole energy range. Khare and Wadhera [11] have calculated σ_K for a number of atoms for E varying from the ionisation threshold to 1 GeV. Practically in all cases the theoretical cross sections are in satisfactory agreement with the experimental data.

For the three sub-shells of L-shell the expressions for the generalised oscillator strengths for the hydrogen atom are given by Holt [24]. The same are converted for the hydrogenic atoms by Khare et al [8, 12] and are utilised by Khare and associates [8, 11, 12] to calculate σ_{L1}, σ_{L2} and σ_{L3} for a number of atoms in the energy range I_{nl} to 1 GeV. In Fig. 3 σ_L ($= \sigma_{L1} + \sigma_{L2} + \sigma_{L3}$) for gold obtained by Khare and his associates [11] are compared with a number of experimental data [25-27]. The agreement between the theory and experiments is good. Khare and Wadehra [11] have obtained ionisation cross sections of the L-subshells for a number of atoms and practically in all cases satisfactory agreement between the theory and experiments is obtained.

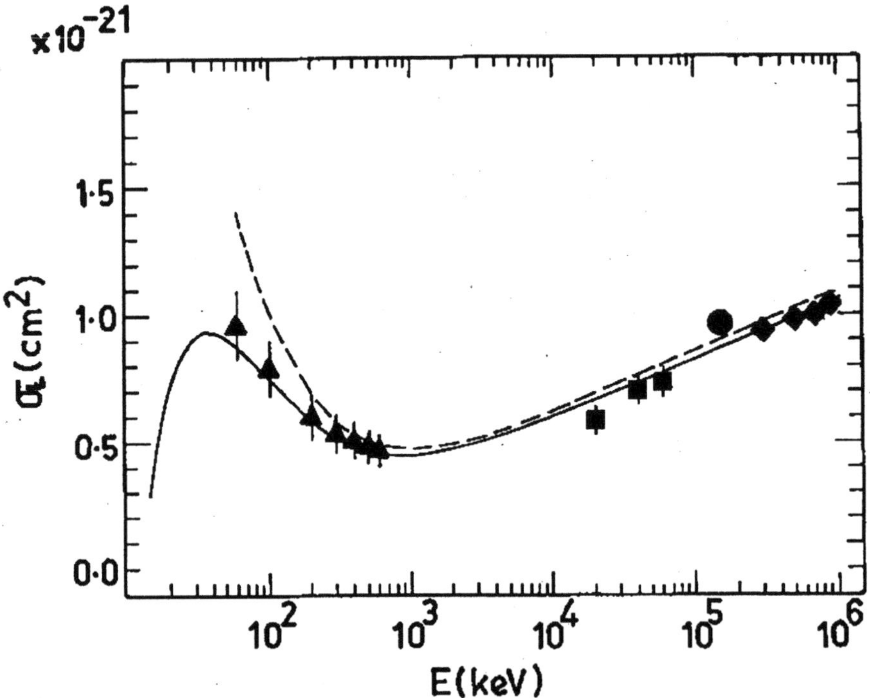

Figure 3. Variation of the L-shell ionisation cross secions of gold atom with the incident electron energy E. The dashed curve depicts cross sections obtained in the Bethe-Born approximation and the solid curve the modified PWBA which includes exchange, Coulomb and relativistic corrections and also the transverse interaction. Experimental data are from refs. [25] (●); [26] (■); and [27] (▲)

Hence it may be concluded that the PWBA with exchange, Coulomb and relativistic corrections and transverse interaction is a successful approximation to calculate K-, L1-, L2- and L3- subshells ionisation cross sections over a very wide energy range.

3.   OUTER-SHELL IONISATION OF MOLECULES

A straightforward extension of the above method to the molecules requires CGOS for these targets. A theoretical evaluation of the CGOS for the multicentred molecules is a difficult task and in most of the cases experimental values of the CGOS are also not available. Hence, Saksena and Kushwaha [13] and Saksena et al [14, 15] used semi-phenomenological relation of Mayol and Salvat [28]. This relation expresses CGOS in terms of COOS and is given by

$$\frac{df(\mathrm{w}, Q)}{d\mathrm{w}} = \frac{df(\mathrm{w}, 0)}{d\mathrm{w}} \, \theta(\mathrm{w}, Q) + h(Q)\, \delta(\mathrm{w}, Q) \tag{10}$$

where θ and δ are the step function and the Dirac delta function, respectively. The recoil energy Q is equal to $K^2 a_0^2 R$ and

$$h(Q) = \int_{I_i}^{Q} \frac{df(\mathrm{w}, 0)}{d\mathrm{w}} d\mathrm{w} \tag{11}$$

I_i is the ionisation threshold of the ith molecular orbital whose ionisation is under consideration. Equation (10) separates the Bethe cross section due to soft collisions (dipole interaction) from the Mott cross section due to hard collisions. According to (10) the soft collisions contribute to σ_i for Q varying from I_l to w. However, for the hard collisions Q is equal to w only. Hence the contribution of the hard collisions are taken only at the Bethe ridge. It describes the collisions between two free electrons.

For E $\geq$ 1 keV the exchange and the Coulomb corrections are negligible. Hence using (10) in (1), integrating over Q and including transverse interaction, we get

$$\sigma = \frac{A}{E'}\left[\int_{I}^{E}\left\{\frac{1}{\mathrm{w}}\frac{df(\mathrm{w}, 0)}{d\mathrm{w}}\ln\left(\frac{\mathrm{w}}{Q_-}\right)+\frac{h(\mathrm{w})}{\mathrm{w}^2}\right\}d\mathrm{w} - \frac{M^2}{R}\{\ln(1-\beta^2)+\beta^2\}\right] \tag{12}$$

Where $A = 4\pi a_0^2 R^2$, $E' = \frac{1}{2}mv^2$ and $Q_- = R\left(K^2 a_0^2\right)_{\min}$. $h(\mathrm{w})$ is also given by (11) but the upper limit of the integrtion changes to w. Saksena et al [15] have employed the theoretical amd the experimental values of

110

$$\frac{df(\mathrm{w}, 0)}{d\mathrm{w}}$$, given by Gallagher et al [29] and Zeiss et al [30] in (12) to

calculate σ for the H_2, N_2, O_2 and H_2O molecules in the energy range 1 keV to 3 MeV. These cross sections in the energy range 1 to 20 keV are shown in Fig. 4 along with a few selected experimental data [31-32].

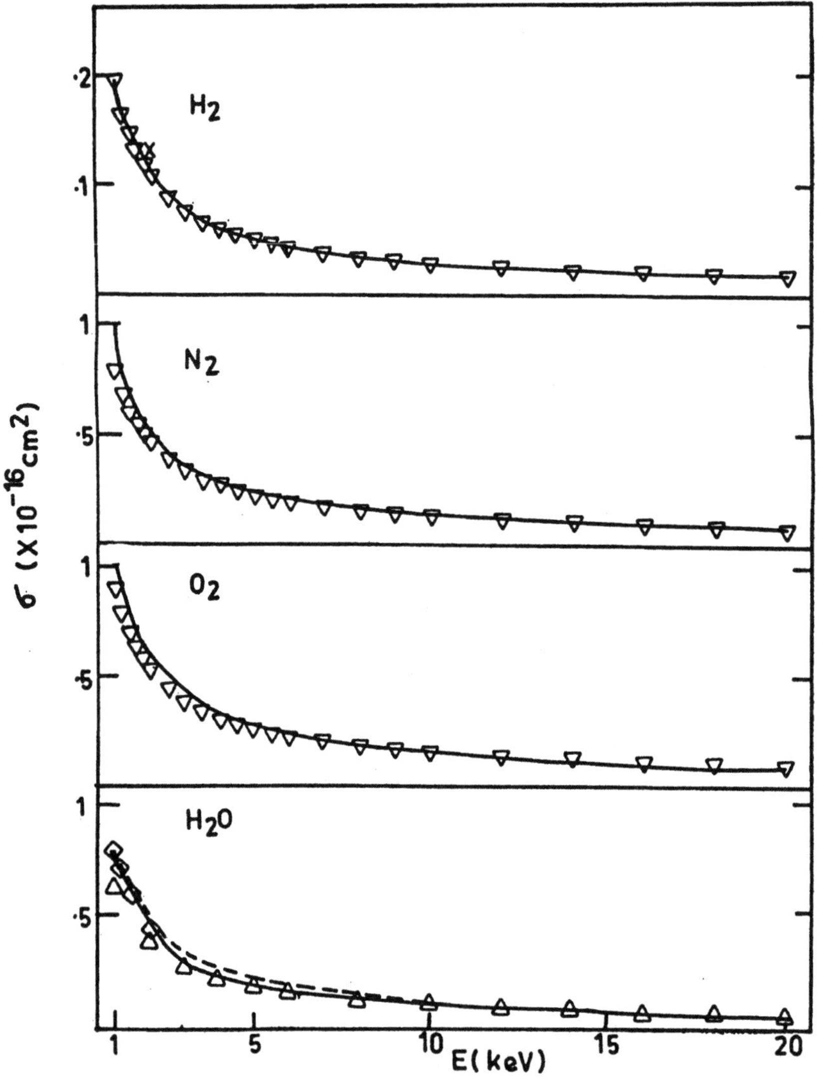

Figure 4. Total ionisation cross sections of H_2, N_2, O_2 and H_2O due to electron impact in the energy range 1 to 20 keV. The solid and dashed curves show the theoretical cross sections of Saksena et al [15] and Kim and Rudd [16], respectively. ∇ and Δ are the experimental cross sections of Schram et al [31] and Shutten et al [32], respectively.

The theoretical cross sections of Kim and Rudd [16] for H_2O are also shown. For all the four molecules the agreement between the theory and the experiment is excellent.

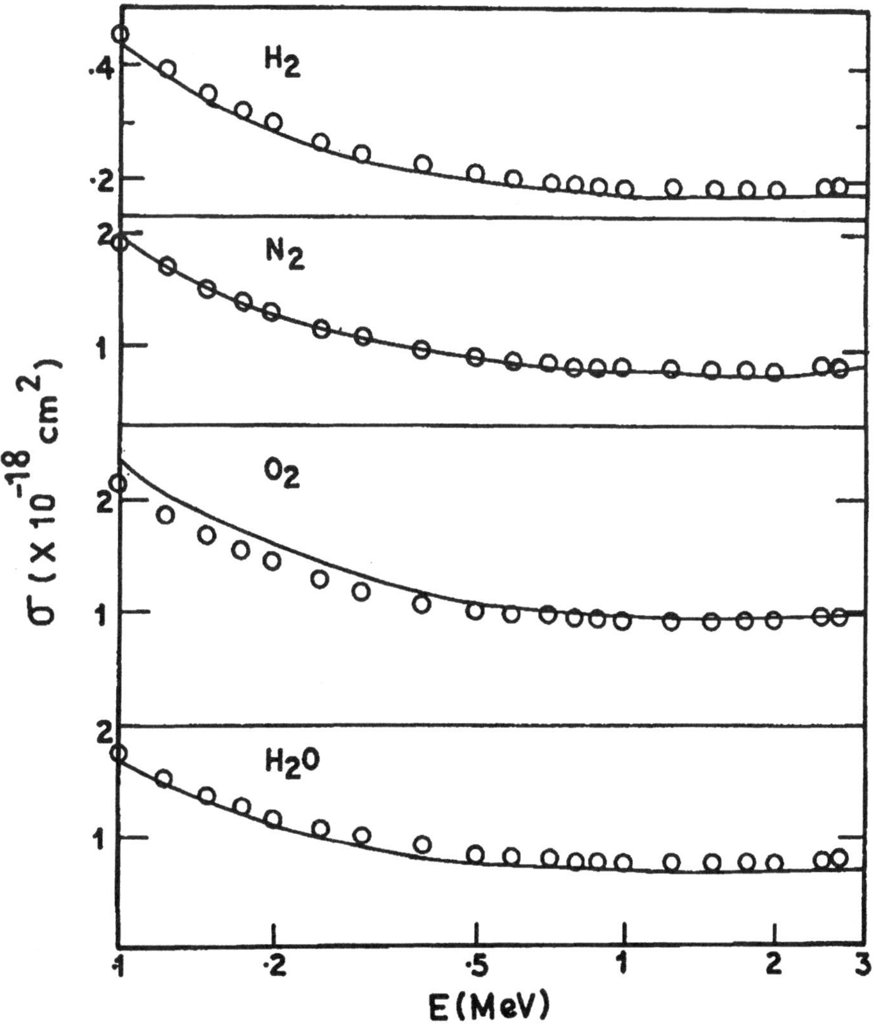

Figure 5. Same as for Fig. 4 but in the energy range 0.1 to 3 MeV. (O) are the experimental cross sections of Reike and Prepejchal [33].

The cross sections obtained by Saksena et al [15] in the energy range 0.1 to 3 MeV are shown in Fig. 5 along with experimental cross sections of Reike and Prepejchal [33]. The agreement between the two sets of the cross section is again very good. Reike and Prepejchal [33] have expressed their cross sections in terms of two parameters M^2 and C_{RP}. The first one is given by (8) and C_{RP}, in the Saksena model, is equal to

$$C_{RP} = \int_I^E \frac{R}{w} \left[\frac{df(w, 0)}{dw} \ln\left(\frac{w}{\beta^2 Q_-} \right) + \frac{h(w)}{w} \right] dw \tag{13}$$

The above equation has been employed by Saksena and Kushwaha [13] to obtain C_{RP} for a number of molecules in the MeV energy region. Their values at 1 MeV for most of the molecules are within 10% of the experimental values of Reike and Prepejchal [33]. These values, as expected, do not change for E > 1 MeV. Hence, it may be concluded that equation (12) is a very successful expression for the calculation of the total ionisation cross section for E > 1 keV.

After having thus established the usefulness of (10) Saksena et al [14] extended their model to low energies. In this energy range (E < 1 keV) the relativistic effect is negligible but exchange becomes important. Saksena et al included exchange through the Ochkur approximation. However it is well known that near the threshold PWBA with exchange overestimates the ionisation cross section. To reduce the overestimation Saksena et al multiplied PWBA-exchange single differential cross section by an adhoc factor $(1 - w/E)$ and obtained for low E

$$\sigma = \sigma_B + \sigma_M \tag{14}$$

where the Bethe cross section σ_B, due to soft collisions (dipole interaction) is given by

$$\sigma_B = \frac{A}{E} \int_I^{\left(\frac{E+I}{2}\right)} \left(\frac{1}{w} - \frac{1}{E} \right) \frac{df(w, 0)}{dw} \left[\ln\left(\frac{w}{Q_-} \right) - \left(\frac{w - Q_-}{E} \right) + \frac{1}{2E^2}\left(w^2 - Q_-^2\right) \right] \tag{15}$$

σ_B, the Mott cross section due to hard (close collisions), is

$$\sigma_M = \frac{A}{E} \int_I^{\left(\frac{E+I}{2}\right)} \left(1 - \frac{w}{E} \right) h(w) \left[\frac{1}{w^2} - \frac{1}{wE} + \frac{1}{E^2} \right] dw \tag{16}$$

Using above equations and the available values of COOS [29, 30] Saksena et al [14] have calculated cross sections for the total ionisation and for the formation of the parent ions of H_2, N_2, O_2, NH_3, H_2O and CO_2 by electron impact for the incident energy E varying from the ionisation threshold to 1 keV. They also obtained dissociative ionisation cross sections of NH_3, H_2O and CO_2 giving rise to NH_3^+, OH^+ and O^+ ions, respectively. Except for H_2 molecule the Saksena model gave the total ionisation cross section in good agreement with the experimental data for $E \geq 300$ eV. For energies lying between about 50 to 300 eV their model is found to slightly overestimate the cross sections. However, at smaller energies ($E \leq 50$ eV) the model underestimates the cross sections for almost all the molecules. This shows that at low values of E, the adhoc factor $(1 - w/E)$ is not adequate and needs modification. For H_2, the cross sections obtained by Saksena et al [14] are found to be lower than the experimental cross sections of Tate and Smith [34], Rapp and Englander Golden [35] and van Wingerden [36] over the whole energy range and tend to follow the experimental data of Schram et al [31, 37]. For the dissociative ionisation their cross sections are also found to be in satisfactory agreement with the experimental data for $E \geq 50$ eV.

Hence it may be concluded that the Saksena model which uses the optical oscillator strengths as the only input gives reliable total ionisation cross section of the molecules due to electron impact over a wide energy range ($E \geq 50$ eV). It also gives dissociative ionisation cross sections with an accuracy of about 20%. However, for $E < 50$ eV, the model requires modification.

4. Binary-encounter-dipole (BED) and Binary-encounter-Bethe (BEB) models

Recently Kim and Rudd [16] have proposed BED and BEB models for molecular ionisation. They neglected exchange in the Bethe part of the ionisation cross section and following binary-encounter theory replaced $\frac{1}{2}mv^2$ by $E + U + I$, where U is the average kinetic energy of the bound electron. Their total ionisation cross section in the binary-encounter-dipole (BED) model is given by

$$\sigma = \sigma_{KBD} + \sigma_{KMD} \tag{17}$$

where the Bethe cross section σ_{KBD} is given by

114

$$\sigma_{KBD} = \frac{A}{E+U+I} \int_I^E \frac{1}{w} \frac{df(w, 0)}{dw} dw \ln\left(\frac{E}{I}\right) \qquad (18)$$

The Mott cross section σ_{KMD} is

$$\sigma_{KMD} = \frac{AN}{I^2}\left(2-\frac{N_i}{N}\right)\frac{1}{(t+u+1)}\left(\frac{t-1}{t}-\frac{\ln t}{t+1}\right) \qquad (19)$$

where $t = E'/I$, $u = U/I$ and

$$N_i = \int_I^\infty \frac{df(w, 0)}{dw} dw \qquad (20)$$

Kim and his associates [16-18] have carried out a number of calculations in the binary-encounter-Bethe (BEB) model. This model is obtained from BED model by taking $N_i = N$ and

$$\frac{df(w, 0)}{dw} = \frac{NI}{w^2} \qquad (21)$$

Hence in the BEB model

$$\sigma = \sigma_{KBB} + \sigma_{KMB} \qquad (22)$$

where

$$\sigma_{KBB} = \frac{AN}{2I^2(t+u+1)}\left(1-\frac{1}{t^2}\right)\ln t \qquad (23)$$

and

$$\sigma_{KMB} = \frac{AN}{I^2(t+u+1)}\left(\frac{t-1}{t}-\frac{\ln t}{t+1}\right) \qquad (24)$$

5.   KHARE, SHARMA AND TOMAR MODEL

Khare et al [19] have incorporated useful features of Kim model with Saksena model to remove the deficiency of the later model at low E. Following Kim and Rudd [16] they replaced $\dfrac{1}{2}mv^2$ by $E'+U+I$ and the factor $(1-w/E)$ by unity. They also dropped exchange from the Bethe part of the cross section. Thus in the BED model the method of Khare et al gives

$$\sigma_{KHBD} = \frac{A}{\left(E'+U+I\right)} \int_I^{E'} \frac{1}{w} \frac{df(w,0)}{dw} \ln\!\left(\frac{w}{Q_-}\right) dw \qquad (25)$$

for the Bethe (soft collision) cross section and the Mott cross section takes the form

$$\sigma_{KHMD} = \frac{A}{\left(E'+U+I\right)} \int_I^{\left(\frac{E'+1}{2}\right)} h(w)\left(\frac{1}{w^2} - \frac{1}{wE'} + \frac{1}{E'^2}\right) dw \qquad (26)$$

Asymptotically (25) goes to

$$\sigma_{KHBD} = \frac{A}{RE'} M^2 \ln\!\left(C_B E'\right) \qquad (27)$$

Where C_B is the Bethe collision praameter. Thus σ_{KHBD} reduces to σ_{KBD} (given by (18)) either by replacing $\ln(w/Q_-)$ in (25) by $\ln(E/I)$ or by replacing C_B in (27) by $(1/I)$. Since the exact value of C_B for the hydrogen atom is $(83.33/I)$ [7] and not $(1/I)$ the cross sections σ_{KHBD} are expected to be greater than σ_{KBD}.

To obtain cross sections in BEB model we take $N_i = N$ and use (21) in (25) and get

$$\sigma_{KHBB} = \frac{ANI}{\left(E'+U+I\right)} \int_I^{E'} \frac{1}{w^3} \ln\!\left(\frac{w}{Q_-}\right) dw \qquad (28)$$

Similarly from (26), we obtain

116

$$\sigma_{KHMB} = \frac{A}{(E'+U+I)} \int_{I}^{\left(\frac{E'+I}{2}\right)} \left(1-\frac{I}{w}\right)\left[\frac{1}{w^2} - \frac{1}{wE'} + \frac{1}{E'^2}\right] dw \qquad (29)$$

Equation (28) with $\ln(w/Q)$ replaced by $\ln(E'/I)$ is identical to (23). Furthermore, (24) is obtained from (29) by replacing E by $E-w-I$ (a similar replacement has been made by Kim and Rudd [16]) and taking $(1-I/w)$ equal to one. Since one is greater than $(1-I/w)$ the cross sections σ_{KHMB} are expected to be less than σ_{KMB}.

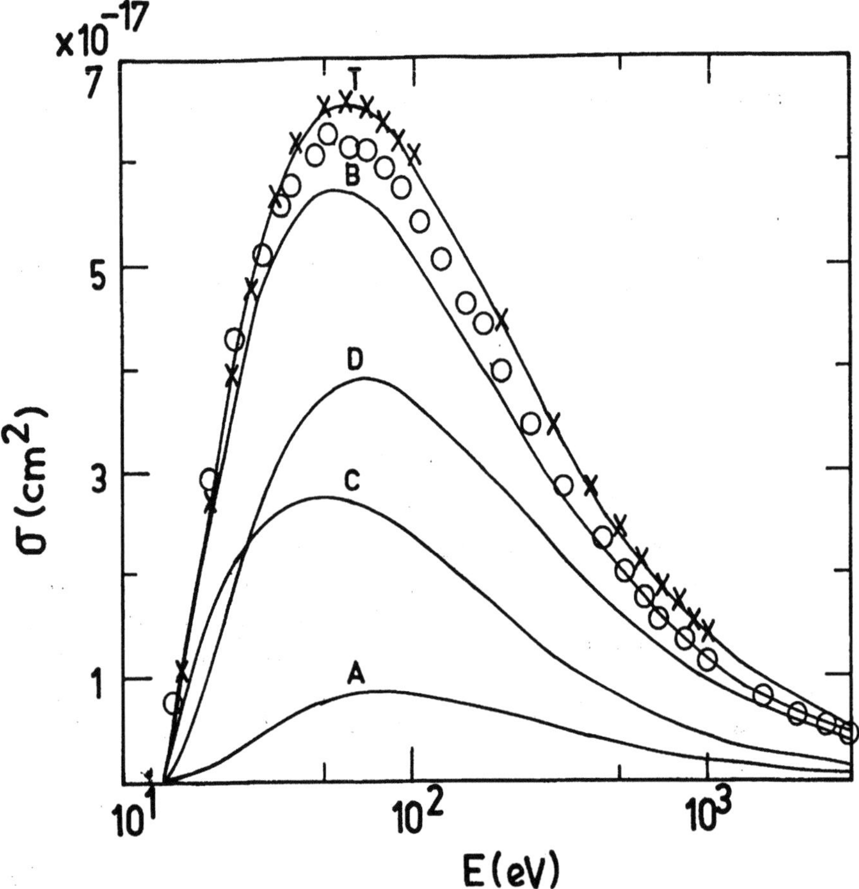

Figure 6. Variation of the ionisation cross sections of hydrogen atom with the incident electron energy E in the binary-encounter-Bethe model. Curves A and B are the present curves obtained from (29) and (28), respectively while the curve T is the sum of the cross sections given by A and B. Curves C and D are obtained from (24) and (23), respectively. The × are the sum of the cross sections given by C and D and O are the experimental cross sections of Saha et al [20].

The above discussion shows that the three cross sections σ_{KBD}, σ_{KBB}, and σ_{KMB} are the approximate forms of the corresponding cross sections obtained by Khare et al [19]. The later authors have not employed asymptotic form of the stopping power cross section given by Bethe. The use of the stopping power cross section, which includes contribution from the discrete excitations also, in a purely ionisation investigation is not desirable.

We have applied the method of Khare et al [19] in the BEB model to the ionisation of the simples atom namely hydrogen atom. In Fig. 6 the present results for σ_{KHMB} and σ_{KHBB} are shown by the curves A and B respectively. Curves C and D represent the cross section σ_{KMB} and σ_{KBB}, respectively. The curve T represents the sum of σ_{KHMB} and σ_{KHBB} while the sum of σ_{KMB} and σ_{KBB} are shown by the crosses. As expected σ_{KHBB} are greater than σ_{KBB} while σ_{KHMB} are less than σ_{KMB}. However the agreement between the curve T and the crosses is remarkable. The reason for this agreement is not yet clear but has been observed for CH_4 molecule also [19]. In the figure the experimental data of Saha et al [20] are also shown. They are about 10% lower than the present theoretical cross sections represented by the curve T.

Finally we conclude that a slight modification of Saksena model [14, 15] has given a method to obtain reliable ionisation cross sections over a wide energy range. It is shown that the expressions obtained by Kim and Rudd [16] for the Bethe and the Mott cross sections can be obtained from the corresponding expressions of Khare et al [19] by making suitable approximations. The application of Khare's method to other molecules is in progress.

ACKNOWLEDGMENTS

Both the authors are grateful to the Council of Scientific and Industrial Research for the financial assistance. One of us (SPK) thanks the University Grants Commission for the financial assistance and the Inter University Centre for Astronomy and Astrophysics, Pune for the hospitality.

REFERENCES

1.   D. K. Jain and S. P. Khare, J. Phys. **B9**, 1429 (1976).

2.   S. P. Khare and W. J. Meath, J. Phys. **B20**, 2101 (1987).

3.   S. P. Khare and S. Prakash, Phys. Rev. **A39**, 6591 (1989).

4.   S. P. Khare and S. Prakash, Phys. Lett. **A140**, 238 (1989).

118

5. S. P. Khare, S. Prakash and W. J. Meath, Int. J. Mass Spectrom. Ion Proc. **88**, 299 (1989).

6. S. P. Khare, Hyp. Inter. **73**, 23 (1992).

7. S. P. Khare, V. Saksena and J. M. Wadehra, Phys. Rev. **A48**, 1209 (1993).

8. S. P. Khare, P. Sinha and J. M. Wadehra, Phys. Lett. **A184**, 204 (1994).

9. S. P. Khare, Hyp. Inter. **89**, 107 (1994).

10. S. P. Khare and J. M. Wadehra, Phys. Lett. **A198**, 212 (1995).

11. S. P. Khare and J. M. Wadehra, Can. J. Phys. **74**, 376 (1996).

12. S. P. Khare, P. Sihna and J. M. Wadehra, Ind. J. Phys. **69B**, 219 (1996).

13. V. Saksena, M. S. Kushwaha, Ind. J. Phys. **B70**, 97 (1996).

14. V. Saksena, M. S. Kushwaha and S. P. Khare, Physica **B233**, 201 (1997).

15. V. Saksena, M. S. Kushwaha and S. P. Khare, Int. J. Mass. Spectrom. Ion Proc. **171**, L1 (1997).

16. Y. K. Kim and M. E. Rudd, Phys. Rev. **A50**, 3954 (1994).

17. W. Hwang, Y. K. Kim and M. E. Rudd, J. Chem. Phys. **104**, 2956 (1996).

18. Y. K. Kim, W. Hwang, N. M. Weinberger, M. A. Ali and M. E. Rudd, J. Chem. Phys. **106**, 1026 (1997).

19. S. P. Khare, M. K. Sharma and S. Tomar, 1999 (to be published).

20. M. B. Saha, D. S. Elliot and H. B. Gilbody, J. Phys. **B20**, 3501 (1987).

21. J. C. Slater, Phys. Rev. **36**, 57 (1930).

22. X. Long, M. Liu, F. Ho and X. Peng, At. Data Nucl. Data Tables **45**, 353 (1990).

23. U. Fano, Ann. Rev. Nucl. Sci. **13**, 1 (1963).

24. A. R. Holt, J. Phys. **B2**, 1209 (1969).

25. H. Ishii, M. Kamiya, K. Sera, S. Morita, H. Tawara, M. Oyamada and T. C. Chu, Phys. Rev. **A15**, 906 (1977).

26. D. H. Hoffmann, C. Brendel, H. Genz, W. Low, S. Muller and A. Richter, Z. Phys. **A293**, 187 (1979).

27. J. Palinkas and B. Schlenk, Z. Phys. **A297**, 29 (1980).

28. R. Mayol and F. Salvat, J. Phys. **B23**, 2117 (1990).

29. L. W. Gallagher, C. E. Brion, J. A. R. Samson, P. W. Langhoff, J. Phys. Chem. Ref. Data **17**, 9 (1988).

30. G. D. Zeiss, Wl J. Meath, J. C. F. MacDonald, D. J. Dawson, Radiat. Res. **63**, 64 (1975); Can. J. Phys. **55**, 2080 (1977).

31. B. L. Schram, F. J. deHeer, M. J. van der Wiel, J. Kistemaker, Physica **31**, 94 (1965).

32. J. Schutten, F. J. deHeer, M. R. Moustafa, A. J. H. Boerboom, J. Kistemaker, J. Chem. Phys. **44**, 3924 (1966).

33. F. F. Rieke, E. Prepejchal, Phys. Rev. **A6**, 1507 (1972).

34. J. T. Tate and P. T. Smith, Phys. Rev. **39**, 270 (1979).

35. D. Rapp and P. Englander-Golden, J. Chem. Phys. **43**, 1464 (1965).

36. B. van Wingerden, R. W. Wagenaar and F. J. deHeer, J. Phys. **B13**, 3481 (1980).

37. B. L. Schram, F. J. deHeer, M. J. van der Wiel and J. Kistemaker, Physica **31**, 94 (1965).

Total Ionization Cross Sections for Electron Scattering from Atomic and Molecular Targets Using Quantum Mechanical Semi-Empirical Approach from Threshold to 400 eV

K. L. Baluja

Department of Physics and Astrophysics
University of Delhi, Delhi - 100 007, India.

Total ionization cross sections (σ_{ion}) by electron impact are presented for a large class of atoms and molecules from the ionization threshold to the intermediate energy region where experimental data are available. A spherical complex optical potential (SCOP), averaged over all molecular orientations, is treated exactly in partial–wave analysis to yield elastic (σ_{el}), absorption (σ_{abs}) and total (σ_t) cross sections. The imaginary part of the SCOP takes into account the loss of flux due to inelastic (mainly the ionization) processes and yield the (σ_{abs}) quantity. We use $\sigma_{ion}(E) = p\,\sigma_{abs}(E)$ to determine the σ_{ion} at an energy E, where p is a scaling factor for the corresponding target. In general, p is less than one and reflects the fraction of ionization cross section in the total inelastic channel. The SCOP term for each target is determined ab initio from target wavefunctions at the Hartree-Fock level. In particular, we present results on the σ_{ion} for He, Ne, Ar, Kr, Xe, H_2, N_2, O_2, F_2, CO, NO, H_2O, H_2S, CO_2, N_2O, NH_3, C_2H_2, CH_4, SiH_4 and CF_4 systems.

1.   INTRODUCTION

The knowledge of electron impact ionization cross sections of atoms and molecules is required in low temperature plasmas, mass spectrometry, gas discharge, lasers, modelling of planetary and cometary atmospheres, radiation physics, etc. In general, the calculations on the ionization cross sections in electron-atom (molecule) scattering are very difficult to perform at the ab initio level because the final state is characterized by two (or more) continuum electrons and the target ion (or ions in case of dissociative ionization); a many-body system interacting mutually through long-range Coulomb forces. Only in the limiting cases, the calculation is feasible for any atomic or molecular systems, for example, at very high impact energies, where the projectile can be treated as a plane wave under the Born approximation, or near the ionization threshold (E_{th}) such as in the Wannier theory [1]. In the non-asymptotic energy region, a large number of classical, semi-classical, empirical, and semi-empirical models have been proposed since 1912 when Thompson [2] first gave his classical formula for electron impact ionization cross section. A full account of various aspects of electron impact ionization of neutral and ionic targets is given in an edited volume by Mark and Dunn [3]. So far the electron impact ionization of atomic, molecular and ionic targets has been most thoroughly investigated through laboratory experiments and a substantial amount of data exists for both single, multiple and total ionization of a broad variety of such targets [4]. In this article we are concerned in the intermediate energy region ($E_{th} - 400$ eV) only.

The total ionization quantity is defined as a sum of all ionization events including dissociative ionization processes in molecules. Very recently, total (σ_{ion}) and partial (σ_{ion}^{n}) ionization cross sections have been measured for several molecules such as the CF_4 [5-6], H_2O, NH_3, CO, CO_2, CH_4 and H_2S [7-10] etc. (this reference list is not complete). Earlier, Rapp and Englander-Golden [11] have performed experiments on the σ_{ion} for a large number of atomic and molecular gases. For rare gases, the experimental σ_{ion} cross sections of Englander-Golden [11] are in excellent agreement with the more recent measurements [12]. Only a few theoretical calculations on the σ_{ion} have been performed on these atoms and molecules by employing semi-empirical formulae based on the Born-Bethe type high-energy theory involving fitting parameters [13-18]. More recently Margreiter et al [15-16] have employed a semi-classical formula based upon the classical binary encounter model, the Born-Bethe theory and the additivity rule in order to determine σ_{ion} and σ_{ion}^{n} for a large number of atomic and molecular systems. The additivity rule assumes that the ionization cross section of a molecule is the sum of cross sections of the constituent atoms; this may be a rather very crude approximation for a realistic situation.

Here we present theoretical results on the σ_{ion} by employing quantum mechanical spherical-complex-optical-potential (SCOP) approach of Jain [19-21], in which the total ionization event is considered in terms of loss of flux from the incoming electron beam. The imaginary part of the SCOP takes into account all possible inelastic channels (of which ionization is the dominant one) and gives the so called absorption or total inelastic cross sections quantity σ_{abs}. In general $\sigma_{abs} = \sigma_{ion} + \sigma_{exc} + \sigma_{diss} + \sigma_{rot} + \sigma_{vib}$; however, at intermediate energies, most of the components (ro-vibrational, σ_{rot} and σ_{vib}, neutral dissociation, σ_{diss} and electronic excitation, σ_{exc}) are in order of 10^{-17} cm$^2 \sim 10^{-18}$ cm^2 in comparison to the dominant σ_{abs} cross section ($\sim 10^{-16}$ cm^2). We use the relation, $\sigma_{ion} = p\sigma_{abs}$ in order to estimate the σ_{ion} for a large number of gases (He, Ne, Ar, Kr, Xe, H_2, N_2, O_2, F_2, CO, NO, H_2O, H_2S, CO_2, N_2O, NH_3, C_2H_2, CH_4, SiH_4, and CF_4). These cross sections will be compared with experimental as well as with previous theoretical data. However in this communication only some results will be shown due to lack of space.

In the next Section II, we provide a summary of our theoretical model, while Section III discusses the results. We use atomic units throughout this paper until otherwise specified.

2.   THEORY

In a proper close-coupling (CC) scheme, the total wave function of the projectile –target system is expanded in terms of a complete states (discrete and continuum) for both the closed and open channels. At intermediate energies, it is impossible to work out such a CC calculation due to a large number of channels. A simple and practical method is to mimic these coupling effects via an effective potential of the projectile-target system. This is the so-called optical potential method [22]. The optical potential, which is generally energy dependent and non-local, becomes complex when open channels are present in the collision event. For a molecular targets, in the present energy region, we can assume that ro-vibrational cross sections are small compared to other elastic and inelastic processes. Thus, for all the present molecular gases, we will ignore an isotropic nature of the optical potential (see Ref. 19 for the validity of this approximation), however, an average over all molecular orientations is performed for the final value of the optical potential. In the SCOP approach [19-21], the effects of target polarization and all possible inelastic open channels (mainly the ionization, dissociation and electronic excitation) can be included via a local complex optical potential, $V_{opt}(r)$. Thus, the scattering equation for the outgoing electron function reduces to,

$$\left[\frac{d^2}{dr^2} - \frac{\ell(\ell+1)}{r^2} + k^2 - 2V_{st}(r) - 2V_{ex}(r) - 2V_{opt}(r)\right] F(r) = 0. \quad (1)$$

124

The optical potential in Equation (1) takes care of closed and open channels. The imaginary part of $V_{opt}(r)$ takes into account the loss of flux due to energetically possible all inelastic channels, while the real part represents the target polarization effects. The optical potential, $V_{opt}(r) = V_{pol}(r) + iV_{abs}(r)$, is very difficult to determine from ab initio methods. Byron and Joachain [23] have discussed various approximations to compute $V_{opt}(r)$ for rare gases from first principles.

In the present work, the static potential, $V_{st}(r)$, is calculated from the unperturbed target wavefunctions Ψ_0 at the Hartree-Fock level, while the $V_{ex}(r)$ term accounts for electron exchange interaction in Equation (1). In the present energy region, a local and real potential model for exchange and polarization effects is adequate. Due to the non-spherical nature of a molecule, the effective potential in Equation (1) ($V_{eff} = V_{st} + V_{ex} + V_{opt}$) is not isotropic. A general expression for $V_{eff}(r)$ for any target can be written in terms of the following multipole expansion around the centre-of mass (COM) of the molecule [24-25],

$$V_{eff}^{p\mu}(\mathbf{r}) = \sum_{\ell h} v_{\ell h}(r)\, X_{\ell h}^{p\mu}(\hat{\mathbf{r}}), \qquad (2)$$

where (pμ) denotes the ground-state symmetry of molecule and the symmetry adapted X functions are defined in terms of real spherical harmonics $S_{\ell m}^{\pm}(\hat{\mathbf{r}})$ [24-25],

$$X_{\ell h}^{(p\mu)}(\hat{\mathbf{r}}) = \sum_{m=0}^{+\ell} b_{\ell h m}^{(p\mu)}\, S_{\ell m}^{q}(\hat{\mathbf{r}}). \qquad (3)$$

The average potential $\left[\frac{1}{4\pi} \int V_{opt}^{p\mu}(\mathbf{r})\,d\hat{\mathbf{r}} \right]$ can be derived from Equation (2) and which comes out to be $v_{pol}(r)/\sqrt{4\pi}$. In order to determine various potential terms, we need the target charge density $\rho(\mathbf{r})$ of a given molecule or atom,

$$\rho(\mathbf{r}) = \int |\Psi_0|^2\, d\mathbf{r}_1 d\mathbf{r}_2 \ldots\ldots\ldots d\mathbf{r}_z = 2\sum_{\alpha} |\phi_{\alpha}(\mathbf{r})|^2, \qquad (4)$$

where Z is the number of electrons in the target, ϕ_i is the ith electronic target orbital and a factor of two appears due to spin integration and α sum being over each doubly occupied orbital. All the four potential terms (V_{st}, V_{ex}, V_{pol} and V_{abs}) are functions of $\rho(\mathbf{r})$. For example,

$$V_{st}(\mathbf{r}) = \int \rho(\mathbf{r_1}) |\mathbf{r} - \mathbf{r_1}|^{-1} d\mathbf{r_1} - \sum_{i=1}^{M} Z_i |\mathbf{r} - \mathbf{R}_i|^{-1}. \qquad (5)$$

The $V_{ex}(r)$ is taken in the Hara free-electron-gas-exchange (HFEGE) model [26] and $V_{pol}(r)$ is calculated in the correlation-polarization (COP) approximation [27]. Thus an accurate evaluation of $\rho(\mathbf{r})$ is important in our SCOP model. For atomic targets, we employed quite accurate charge density [28], while for molecules, we employed various single-centre expansion programs to determine the charge density and various potentials for linear and non-linear molecules [29] (for more details see Ref. 20). In the present high energy region, an exact representation of exchange-polarization correlation is not important; thus the FEGE model for exchange and COP approximation for polarization effects are adequate in the present calculations.

The imaginary part of the optical potential, $V_{abs}(r)$, is the absorption potential which represents approximately the combined effect of all the inelastic channels. An ab initio calculation of absorption potential is still an open problem. Here we employ a semi-empirical absorption potential as discussed by Truhlar and coworkers [30]. The $V_{abs}(r)$, is a function of molecular charge density, incident electron energy and the mean excitation energy, Δ, of the target (see for details reference [30]). After generating the full optical potential of the given electron-molecule (atom) system, we treat it exactly in a partial-wave analysis and evaluate the cross sections.

The (σ_{ion}) cross sections are determined from the corresponding σ_{abs} numbers in a semi-empirical way,

$$\sigma_{ion}(k^2) = p\,\sigma_{abs}(k^2), \qquad (6)$$

where p ($\leq$ 1) is determined arbitrarily in order to make experimental σ_{ion} numbers closer to the theoretical curve. Thus the scaling (Equation (6)) procedure to evaluate σ_{ion} can only be justified with respect to agreement between present σ_{ion} results and the experimental data. We have demonstrated it for a large number of targets (see later). In addition, p is found to be very close to the fraction of ionization cross section relative to total inelastic cross section. Thus, in the absence of any data on the σ_{ion} parameter for any gas, present model can predict an upper limit of σ_{ion} with correct shape near the maximum in the total ionization cross sections.

3. RESULTS AND DISCUSSION

We have shown our results for a number of targets in Figs. 1-7. The value of p (see Equation (6)) is indicated in the figure itself for each target. As mentioned above, the value of p was determined arbitrarily so that the difference between experimental and present σ_{ion} is minimum. For the rare gas atoms, (Figs. 1-2) we show our results for Ne and Kr. Our theoretical curves for σ_{ion} are in good agreement with the experimental data. The maximum discrepancy for Kr occurs at low energies and near the peak. Our best agreement is seen for the Ne atom with p = 0.75.

Fig. 3 shows our results for the N_2 target along with the measurements of Rapp and Englander-Golden [11]. The recent semi-empirical calculations of Margreiter et al [16] are also shown in Fig. 3. We see good agreement between the present theory and the experimental data.

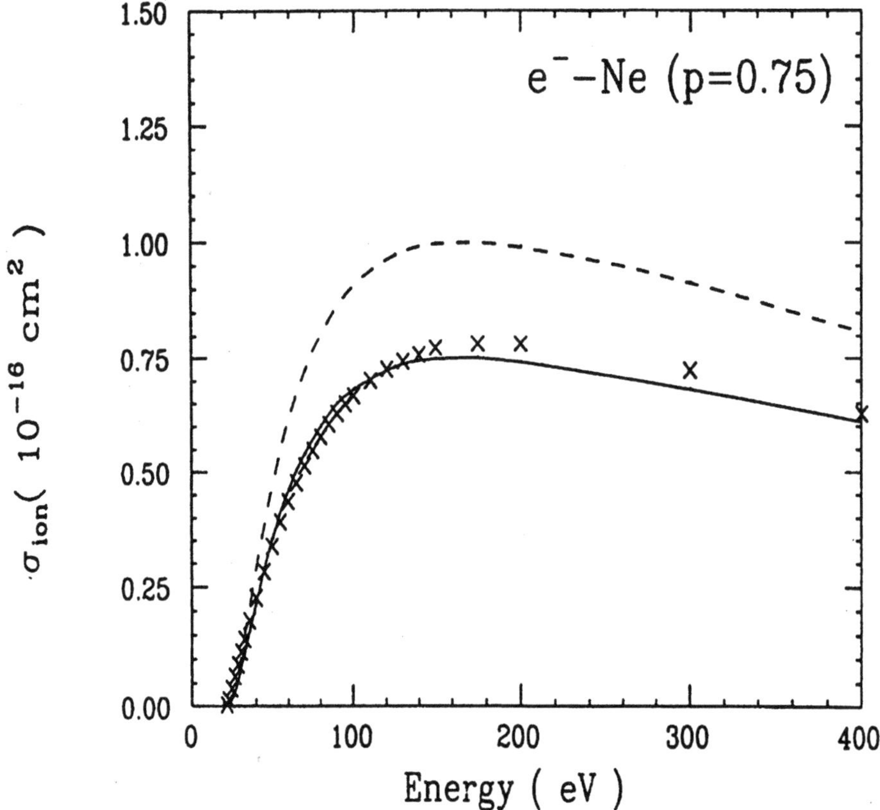

Figure 1. Total ionization (solid curve) and absorption (dashed curve) cross sections for electron-Ne system. The experimental data (crosses) are from Rapp and Englander [11].

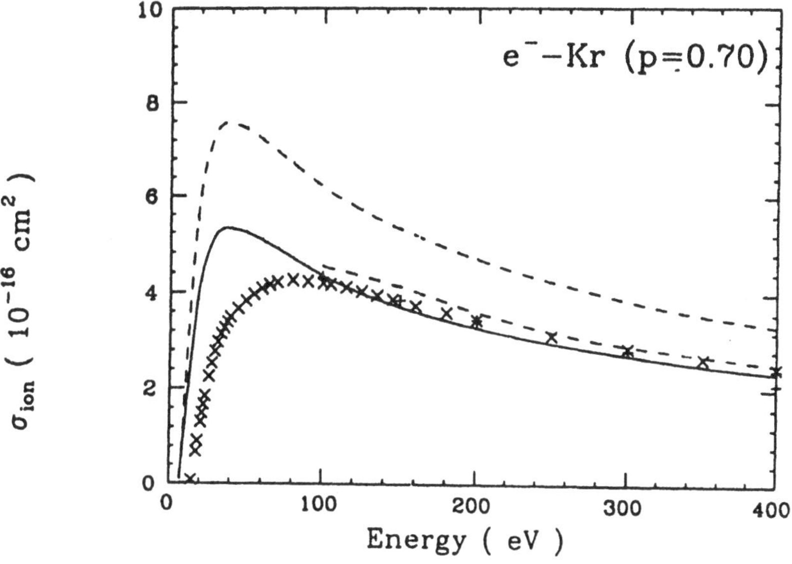

Figure 2. Total ionization (solid curve) and absorption (dashed curve) cross sections for electron-Kr system. The experimental data (crosses) are from Rapp and Englander [11].

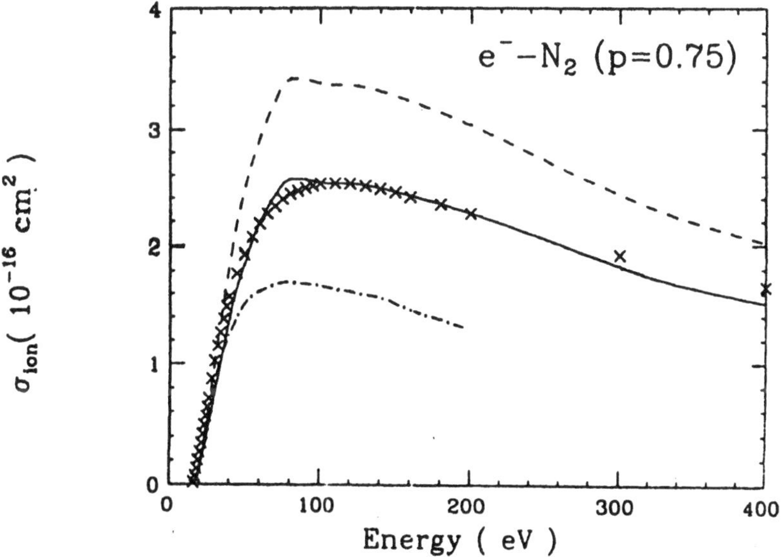

Figure 3. Total ionization (solid curve) and absorption (dashed curve) cross sections for electron-N_2 system. The experimental data (crosses) are from Rapp and Englander [11]. The dotted curves are the theoretical results of Ref. 17.

128

The results for CO are shown in Fig. 4. We see that for CO (Fig. 4) our calculations with p = 0.7 compare very well with experimental data of Rapp and Englander-Golden [11] and Orient and Srivastava [7]. Similarly, the present calculations on the CO_2 molecule with p = 0.6 (see Fig. 5) are in fair agreement with both the measurements [7, 11]. Again the empirical results of Margreiter et al [15] are lower than the present and both the measured values. The σ_{abs} cross sections for the N_2O molecule (with p = 0.6) are shown in Fig. 6 along with the measurements of Rapp and Englander-Golden [11]. Here also, we find a very good agreement between theory and experiment in the present energy region.

In Fig. 7, we compare our C_2H_2 results (with p = 0.7) with other data, both experimental [31] and theoretical [15]. We see a very good agreement between present calculations and other results (both experimental and theoretical) for this molecule.

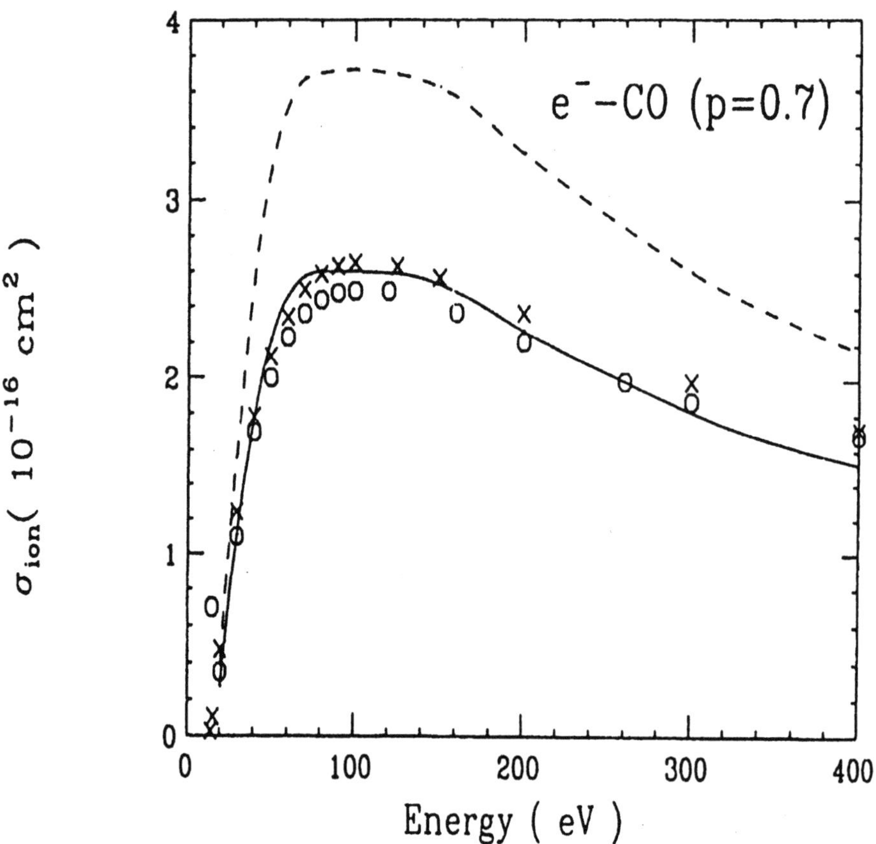

Figure 4. Total ionization (solid curve) and absorption (dashed curve) cross sections for electron-CO system. The experimental data (crosses) are from Rapp and Englander [11]. The open circles are the experimental points of Ref. 7.

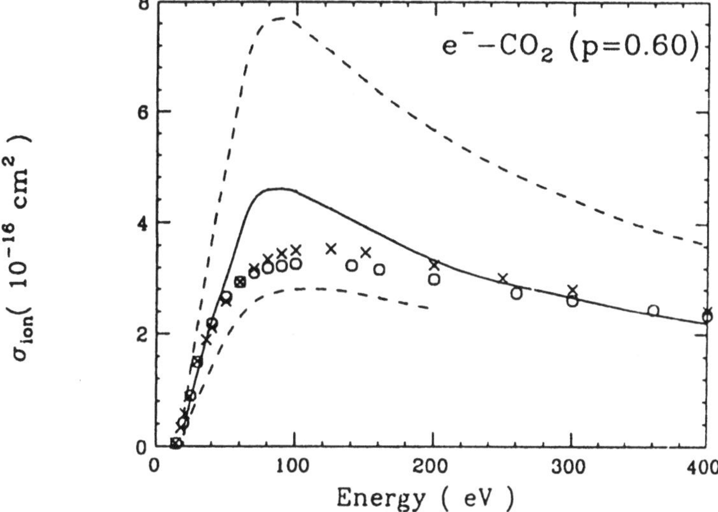

Figure 5. Total ionization (solid curve) and absorption (dashed curve) cross sections for electron-CO_2 system. The experimental data (crosses) are from Rapp and Englander [11].The dotted curve is the calculation of Ref. 15. The open circles are the experimental points of Ref. 7.

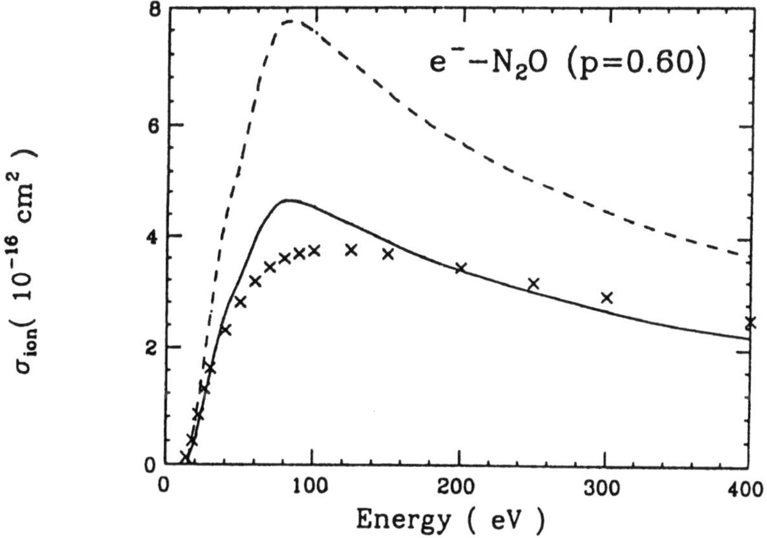

Figure 6. Total ionization (solid curve) and absorption (dashed curve) cross sections for electron-N_2O system. The experimental data (crosses) are from Rapp and Englander [11].

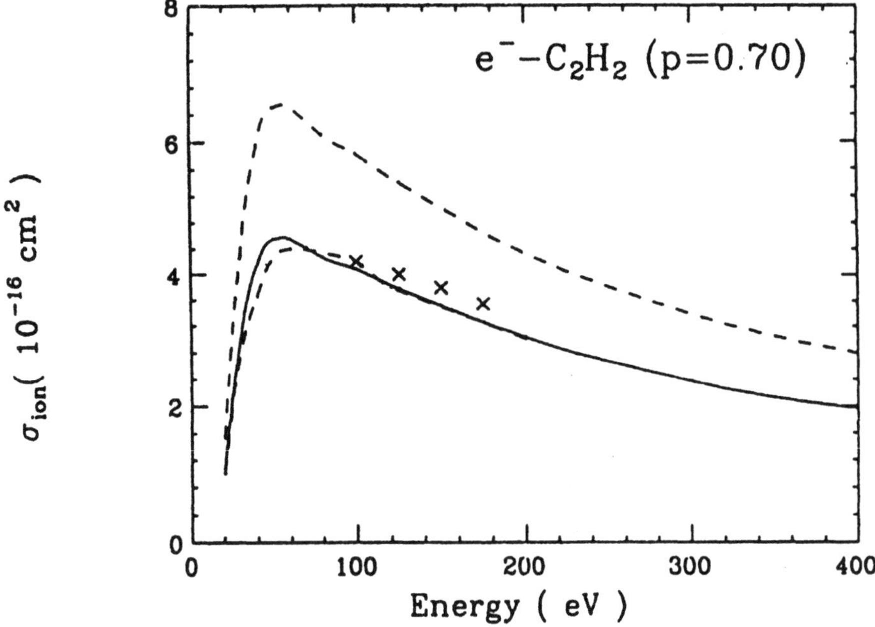

Figure 7. Total ionization (solid curve) and absorption (dashed curve) cross sections for electron-C_2H_2 system. The crosses are the experimental points of Ref. 31.

4. CONCLUDING REMARKS

We have presented quantum mechanical calculations on the σ_{abs} for a large variety of atomic and molecular gases. The σ_{ion} cross sections are obtained from the corresponding total inelastic σ_{abs} quantity by employing Equation (6). For most of the cases, our results compare very well with available experimental data. A most interesting feature of the present theory is that the scaling factor p is less than unity reflecting the physical situation that the ionization process is a fraction of the total inelastic event. In other words, our present theory can predict an upper bound to the ionization cross section, which may be useful for those systems where there is lack of any experimental data.

ACKNOWLEDGMENTS

The author is thankful to Dr. Ashok Jain for many fruitful discussions.

REFERENCES

1. G. H. Wannier Phys. Rev. **90**, 817 (1953).

2. J. J. Thompson, Phil. Mag. **23**, 449 (1912).

3. *"Electron Impact Ionization"*, edited by T. D. Mark and G. H. Dunn (Springer-Verlag, NY 1985)

4. H. Tawara and T. Kato, At. Data Nucl. Data Tables, **36**, 167 (1987).

5. Ce Ma, M. R. Bruce and R. A. Bonham, Phys. Rev. A **44**, 2921 (1991).

6. K. Stephan, H. Deutsch and T. D. Mark, J. Chem. Phys., **83**, 5712 (1985).

7. O. J. Orient and S. K. Srivastava, J. Phys. B **20**, 3923 (1987).

8. S. K. Srivastava and H. P. Nguyen, JPL Publications 87-2 (1987).

9. N. L. Djuric, I. M. Cadez and M. V. Kurepa, Int. J. Mass Spectrom. Ion Process, **83**, R7 (1988).

10. M. V. V. S. Rao, S. K. Srivastava, XVII ICPEAC, book of Abstracts (Brisbane, Australia 10-16 July, 1991) p.254, 257.

11. D. Rapp and P. Englander-Golden, J. Chem. Phys. **43**, 1464 (1965).

12. R. C. Wetzel, F. A. Baiocchi, T. R. Hayes and R. S. Freund, Phys. Rev. A **35**, 559 (1987); T. R. Hayes, R. C. Wetzel and R. S. Freund, Phys. Rev. A **35**, 578 (1987); J. Chem. Phys., **88**, 823 (1988); J. Chem. Phys., **89**, 4035, 4042 (1988)

13. D. K. Jain and S. P. Khare, J. Phys. B**9**, 1429 (1976).

14. S. P. Khare and W. J. Meath, J. Phys. B **20**, 2101 (1987).

15. D. Margreiter, H. Deutsch, M. Schmidt and T. D. Mark, Int. J. Mass Spect. and Ion Proc. **100**, 157 (1990) and references therein.

16. D. Margreiter, H. Deutsch and T. D. Mark, Contrib. Plasma Phys. **30**, 487 (1990).

17. H. Deutsch and T. D. Mark, Int. J. Mass Spect. and Ion Proc. **79**, R1 (1987).

18. H. Deutsch, P. Scheier and T. D. Mark, Int. J. Mass Spect. and Ion Proc. **74**, 81 (1986).

19. A. Jain, Phys. Rev. A **34**, 3707 (1986); J. Chem. Phys. **86**, 1289 (1987).

20. A. Jain, J. Phys. B **22**, 905 (1988)

21. A. Jain and K. L. Baluja, Phys. Rev. A **45**, 202 (1992).

22. C. J. Joachain, *"Quantum Collision Theory"*, (North Holland, NY)

23. F. W. Byron and C. J. Joachain, Jr., Phys. Rev. A **15**, 128 (1977).

24. F. A. Gianturco and D. G. Thompson, Chem. Phys. Lett. **14**, 110 (1976).

25. F. A. Gianturco and A. Jain, Phys. Rep. **143**, 347 (1986).

26. S. Hara, J. Phys. Soc. Japan, **22**, 710 (1967).

27. J. K. O'Connel and N. F. Lane, Phys. Rev. A **27**, 1893 (1983); N. T. Padial and D. W. Norcross, Phys. Rev. A **29**, 1742 (1984); F. A. Gianturco, A. Jain and L. C. Pantano, J. Phys. B **20**, 571 (1987).

28. Froese-Fischer. C, *"The Hartree-Fock Method for Atoms"*, (Wiley, New York, 1977); Jain A, B. Etemadi and K. R. Karim, Physica Scripta **41** 321 (1990); F. Salvat, J. D. Martinez, R. Mayol and J Parellada, Phys. Rev. A **36**, 467 (1987).

29. M. A. Morrison, Comp. Phys. Commun.**21**, 63 (1980); L. A. Collins, D. W. Norcross and G. B. Schmid, Comp. Phys. Commun., **21**, 79 (1980); A. D. McLean and M. Yoshmine, Int. J Quant. Chem. **1S**, 313 (1967); P E Cade and A C Wahl, At. Data. Nuc.data Tables, **13**, 339 (1974); P. E. Cade and W. M. Hup, At. Data and Nuc. Data Tables, **15**, 2 (1975); for polyatomic targets, F. A. Gianturco, D. G. Thompson and A. Jain (private communication)

30. G. Statszewska, D. W. Schwenke, D. Thirumalai and D. G. Truhlar, J. Phys.B **16**, L281(1983); Phys. Rev. A **28**, 2740 (1983); G. Statszewska, D. W. Schwenke, and D. G. Truhlar, J. Chem. Phys. **81**, 335 (1984); Phys. Rev. A **29**, 3078 (1984).

31. A. Gaudin and R. Hagemann, J. Chem. Phys., **64**, 1209 (1967).

Radiative Electron Capture and the Photoionization of Hydrogenlike Ions

Jörg Eichler

Bereich Theoretische Physik,
Hahn-Meitner-Institut Berlin, 14109 Berlin, Germany,
and
Fachbereich Physik,
Freie Universität Berlin, 14195 Berlin, Germany

Radiative electron capture (REC) from a low-Z target into high-Z projectile up to U^{92+} has recently been studied in great detail. This process, in which an almost free electron is transferred from the target to the projectile with the simultaneous emission of a photon plays an important role in plasmas and in the spectroscopy of high-Z one- or few-electron ions. It also provides a means to investigate the inverse reaction, the photoelectric effect, in a regime that is otherwise inaccessible. The theoretical development and some recent results are presented.

1. INTRODUCTION

The photoelectric effect represented the first simple and fundamental confirmation of Planck's quantum hypothesis. Since that time, photoionization has continued to be an object of research. Besides a basic test of quantum theory, the main goal of theoretical treatments has been to assess the contribution of photoionization to the absorption of radiation in matter. In these applications and early experiments, the photon ionizes a neutral target atom, so that, in general, one has to deal with many-electron effects, which render calculations difficult and – in most cases – approximate.

In recent years, it has been possible to approach the pure situation of a one-electron problem from another side, namely by studying the inverse process of radiative recombination (RR), in which a bare nucleus captures a free electron with the simultaneous emission of a photon. Bare nuclei can be produced in a relativistic atomic collision or in an electron-beam ion source (EBIT). In a relativistic ion-atom collision, we consider projectile ions with a velocity comparable to the speed of light, that is with a kinetic energy of 100 MeV/u or more. For sufficiently high energy, all electrons can be stripped, so that one produces bare or almost bare high-Z ions as for example Au^{79+}, Pb^{82+}, or U^{92+}. In these cases, the electronic motion requires a relativistic quantum description by the Dirac equation. On the other hand, the nuclear motion can, to a very good approximation, be described by classical relativistic kinematics.

In practice, no free-electron target of sufficient density is available, so that one uses loosely bound electrons in low-Z target atoms. These electrons can be considered as quasi-free, which is a very good approximation for high-energy high-Z projectiles. Experimentally, radiative electron capture (REC) can be identified by detecting the down-charged projectile in coincidence with a x-ray photon.

2.   EXPERIMENTAL METHODS

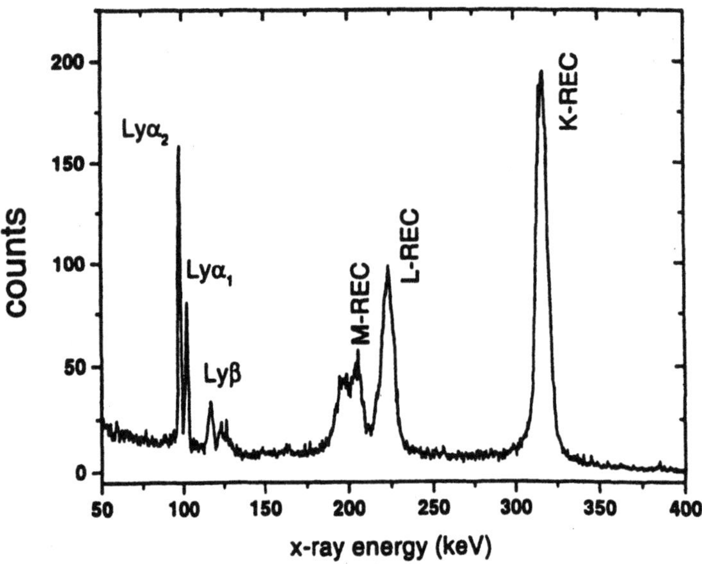

Figure 1. Typical x-ray spectrum registered in coincidence with the particle counter for 358 MeV/u U^{92+} projectiles interacting with N_2 gas targets. The spectra (laboratory frame) were measured at the observation angle 132° and are not corrected for detection efficiency.

The most recent technical development in the field of relativistic heavy-ion physics consists in relativistic heavy-ion storage rings, for example the ESR at the GSI in Darmstadt. This electron cooler ring provides the basis for atomic collision experiments dealing with beams of unprecedented quality [1, 2]. Here, the interaction of intense beams of cooled high-Z ions with low-density gaseous matter can be studied without any beam collimation, guaranteeing almost completely background-free experimental conditions. In order to demonstrate the clean conditions at the ESR jet target we show in Fig. 1 an x-ray spectrum associated with electron capture measured for 358 MeV/u $U^{92+} \rightarrow N_2$ collisions at an observation angle of 132° [3]. Besides the Lyman groundstate transitions, the most prominent features observed in the spectrum are due to radiative electron capture into the ground- and into the excited projectile states.

3.   THEORETICAL TREATMENT OF RADIATIVE ELECTRON CAPTURE

For free electrons, nonradiative capture is excluded by energy and momentum conservation. If, however, the electron transfer is accompanied by the emission of electromagnetic radiation, the emitted photon acts as a third body carrying away energy and momentum released by the formation of the final bound state.

In the following, we assume that (i) the projectile charge is large compared to the target charge, $Z_P >> Z_T$ and (ii) the projectile velocity is large compared to the electron velocity in the target atom, $v >> v_i$. Under these conditions, the target electron can be considered as quasi-free and its momentum spread within the target can be disregarded. Radiative electron capture (REC) is then equivalent to radiative recombination (RR) in the moving projectile frame. This process, in turn, is the inverse of photoionization.

3.1   Approximate approach to photoionization and radiative recombination

We start with the nonrelativistic Born approximation for photoionization. This assumption implies for hydrogenlike ions with charge Z the conditions

136

$$\hbar\omega \;\ll\; m_e c^2$$

$$\alpha Z \;\ll\; 1 \tag{1}$$

$$\varphi_f \;=\; e^{i\,\mathbf{k}_f \cdot \mathbf{r}}.$$

Here $\hbar\omega$ is the energy of the incident photon, m_e is the electron mass, $\alpha \approx 1/137$ the fine-structure constant, φ_f represents the wave function of the emitted electron with the wave vector $\mathbf{k}_f$. If $\mathbf{p}_{op} = \frac{\hbar}{i}\nabla$ is the momentum operator, $\hat{\mathbf{u}}$ the polarization vector of the photon and $\mathbf{k}$ its wave vector, the matrix element for the transition from the initial bound-state wave function φ_i to the final state is given by

$$\begin{aligned}
M_{fi} &= \int \varphi_f^{*}(\mathbf{r})\,(\mathbf{p}_{op}\cdot\hat{\mathbf{u}})\,e^{i\,\mathbf{k}\cdot\mathbf{r}}\,\varphi_i(\mathbf{r})\,d^3r \\
&= (\mathbf{k}_f\cdot\hat{\mathbf{u}})\int e^{i\,\mathbf{q}\cdot\mathbf{r}}\,\varphi_i(\mathbf{r})\,d^3r,
\end{aligned} \tag{2}$$

where $\mathbf{q} = \mathbf{k} - \mathbf{k}_f$. This is the Fourier transform of the initial state. Now, if $\mathbf{k} \parallel \hat{\mathbf{e}}_z$ and $\hat{\mathbf{u}} = \hat{\mathbf{e}}_x$, then $\mathbf{k}_f\cdot\hat{\mathbf{u}} = k_f \sin\theta \cos\varphi$ with θ being the angle between the electron and the photon direction and φ the azimuthal angle. Inserting the Fourier transform of the initial 1s-state [4] and averaging over the photon polarization, we obtain for the differential cross section

$$\frac{d\sigma_{\text{ph}}}{d\Omega} \propto |M_{fi}|^2 \propto \frac{\sin^2\theta}{(1-\beta\cos\theta)^4}, \tag{3}$$

where $\beta = v/c$. If retardation is neglected, $\exp(i\,\mathbf{k}\cdot\mathbf{r}) \to 1$, we have the dipole approximation, $\mathbf{q} \to \mathbf{k}_f$ and

$$\frac{d\sigma_{\text{ph}}^{\text{dip}}}{d\Omega} \propto \sin^2\theta. \tag{4}$$

Radiative recombination is the inverse reaction in the projectile system. We then have to replace $\theta \to \pi - \theta$ and apply the Lorentz transformation to the laboratory frame, i.e.

$$\cos\theta = \frac{\cos\theta_{\text{lab}} - \beta}{1 - \beta\cos\theta_{\text{lab}}}$$

$$\frac{d\Omega}{d\Omega_{\text{lab}}} = \frac{1}{\gamma^2\left(1 - \beta\cos\theta_{\text{lab}}\right)^2} \tag{5}$$

with $\gamma = 1/(1-\beta^2)^{1/2}$. Inserting into the differential cross section (3) including retardation, one obtains the result

$$\frac{d\sigma_{RR}}{d\Omega_{lab}} \propto \sin^2\theta \tag{6}$$

This peculiar effect of cancellation between the effects of retardation and Lorentz transformation in the nonrelativistic limit has been first observed by Spindler et al [5]. It is remarkable that this behaviour is still approximately valid for rather high projectile energies and high charge numbers, see, e.g., Fig. 2.

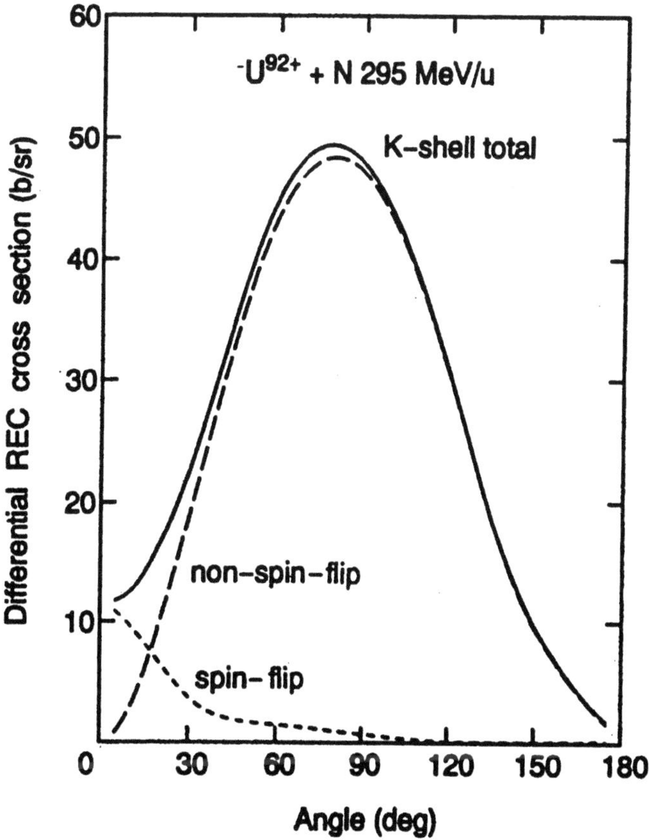

Figure 2. Calculated differential K-shell cross sections for 295 MeV/u U^{92+} on N as a function of the laboratory angle. Spin-flip and non-spin-flip contributions are shown separately. Approximate Hartree-Fock wave functions [11] have been used for the target atom. All partial waves for the continuum electron with Dirac quantum number $|\kappa| \leq 10$ have been taken into account. From [8].

138

3.2 Exact treatment of photoionization and radiative recombination

In a rigorous relativistic description [6], one has to adopt the Dirac theory for the electron subject to a Coulomb potential in the initial bound and final continuum state. Assuming that the electron spin is not detected and the incoming photons (whose momenta define the quantization axis) are unpolarized, one obtains the differential cross section for a single electron [7, 8]

$$\frac{d\sigma_{ph}}{d\Omega} = \frac{\alpha m_e c^2}{8\hbar\omega} \frac{\lambda_c^2}{2j_i+1} \sum_{\mu_i}\sum_{m_s}\sum_{\lambda} \left|M_{p,i}(m_s, \lambda, \mu_i)\right|^2, \tag{7}$$

where we have averaged over the $(2j_i + 1)$ angular momentum projections μ_i in the initial bound state and have summed over the spin components $m_s = \pm\frac{1}{2}$ of the emitted electron. Furthermore, we have averaged over the circular polarizations $\lambda = \pm 1$ of the incoming photon.

Adopting relativistic units $\hbar = c = m_e = 1$, the transition matrix element can be written as

$$M_{p,i}(m_s, \lambda, \mu_i) = \int \psi_{p, m_s}^+(\mathbf{r})(\alpha \cdot \hat{\mathbf{u}}_\lambda) e^{i\mathbf{k}\cdot\mathbf{r}} \psi_{j_i, \mu_i}(\mathbf{r}) d^3 r, \tag{8}$$

where $\psi_{j_i, \mu_i}(\mathbf{r})$ and $\psi_{p, m_s}(\mathbf{r})$ are exact bound and continuum Coulomb Dirac wave functions, the latter being given by a partial-wave expansion [6]. Eq. (7) represents the rigorous relativistic cross section for photoionization of a hydrogenlike atom to first order in the photon interaction. The effect of retardation, that is, of all multipole orders are included. The formulas have to be evaluated numerically [8, 9, 10].

In the following denoting coordinates with respect to the projectile frame with a prime, the cross section $\sigma_{RR}(E', \theta')$ for radiative recombination is related by the principle of detailed balance to the cross section $\sigma_{ph}(\omega', \theta')$ of its inverse reaction, namely the photoelectric effect. By multiplying $\sigma_{RR}(E', \theta')$ with the phase space ratio k'^2/p'^2, we replace the phase space factor of the outgoing electron by that of the emitted photon. For the transition of an electron into a specific substate of the final shell, we hence obtain

$$\frac{d^2\sigma_{RR}(E',\theta')}{dE'\,d\Omega'_{ph}} = \frac{\left(\gamma - 1 + |\varepsilon_i|/m_e c^2\right)^2}{\gamma^2 - 1}\,\frac{d^2\sigma_{ph}(E',\theta')}{dE'\,d\Omega'_{el}} \tag{9}$$

In applications to relativistic ion-atom collisions, the z-direction is usually defined as the direction of the projectile (or photon) momentum. This is opposite to the direction of the electron (or photon) momentum as seen from the projectile. Hence for REC, the angle θ' of the photoelectric effect or of the radiative recombination has to be replaced by $\pi - \theta'$, or $\cos\theta'$ is replaced by $-\cos\theta'$.

As a final step, one has to transform all relevant quantities from the projectile into the laboratory system using Eq. (5).

3.3   Radiative electron capture

A loosely bound target electron may be considered as approximately free in a high-energy collision. In this limit, REC is identical with radiative recombination (RR), in which an electron, initially moving with the velocity $-\mathbf{v}$ towards the projectile, is captured into a bound state with the simultaneous emission of a photon of energy $\hbar\omega'$ and wave number $\mathbf{k}'$.

If the cross section σ_{RR} for radiative recombination is known, see Sec. 3.2, it is natural to refer all momenta to the projectile frame. If the target electron has the momentum $\mathbf{q}$ with respect to the target nucleus, its momentum in the projectile frame $\mathbf{q}'$ is obtained by a Lorentz transformation. Assuming that the momentum $\gamma m_e v$ of an electron travelling with the relative speed of the target towards the projectile is large compared to the electron momentum q in a low-Z target atom, one may use the impulse approximation [12] to write the double-differential REC cross section in the projectile frame (primed coordinates) as

$$\frac{d^2\sigma_{REC}}{d\Omega'\,d(\hbar\omega')} = \int d^3q\,\frac{d\sigma_{RR}(q')}{d\Omega'}\left|\widetilde{\varphi}_i(\mathbf{q})\right|^2 \delta\!\left(\hbar\omega' - |\varepsilon'_f| + \gamma|\varepsilon_i| - T_e + \gamma v q_z\right),$$

$$\tag{10}$$

where $\widetilde{\varphi}_i(\mathbf{q})$ is the Fourier transform of the initial electronic target wave function. The delta function expresses the energy conservation between the final electronic energy $E'_f = m_e c^2 - |\varepsilon'_f|$ and the photon energy $\hbar\omega'$ in the projectile frame on the one hand and the initial electronic energy $E'_i = \gamma\left(E_i - v q_z\right)$ (also in the projectile frame) on the other hand. Here $|\varepsilon_i|$ and $|\varepsilon'_f|$ are the binding energies of initial target and final projectile states.

Furthermore, $T_e = m_e c^2 (\gamma - 1)$ is the kinetic energy of an electron travelling with the same speed as the projectile.

The initial momentum distribution of the target electron centred around $\mathbf{q} \approx 0$ gives rise to a peak in the double-differential cross section (10) around

$$\hbar \omega_0' = \gamma E_i - E_f' = |\varepsilon_f'| - \gamma |\varepsilon_i| + T_e. \tag{11}$$

The Doppler broadening described by the integral over the electronic momentum $\mathbf{q}$ is approximately related to the Compton profile of the initial state. The required momentum wave function can be obtained from a hydrogenic model or from a Hartree-Fock calculation [8, 11].

Specifically, within the nonrelativistic dipole approximation of Stobbe [13], an explicit expression for the RR cross section can be given. In the nonrelativistic limit, the factor in Eq. (9) is $(\hbar \omega'/m_e c^2)^2 (1/\beta^2 \gamma^2) = (\hbar \omega'/2 m_e c^2)(1 + \nu^2)$, so that the Stobbe cross section for radiative electron capture into an empty K-shell $(j_i = \tfrac{1}{2})$ as

$$\sigma_{RR}^{Stobbe} = \frac{2^8 \pi^2 \alpha}{3} \lambdabar_c^2 \left(\frac{\nu^3}{1+\nu^2} \right)^2 \frac{e^{-4\nu \arctan(1/\nu)}}{1 - e^{-2\pi \nu}}. \tag{12}$$

The constants in front of the ν-dependent terms make up a factor of 9164.7 barn.

The rough behaviour of the cross section for radiative electron capture from a neutral target into the projectile for large values of γ is given by

$$\sigma_{fi}^{REC} \propto Z_P^5 Z_T \frac{1}{\gamma}, \tag{13}$$

where the factor Z_P^5/γ arises from the capture process, and the factor Z_T simply counts the number of electrons available in the neutral target atom.

4. DIFFERENTIAL CROSS SECTIONS AND COMPARISON WITH EXPERIMENT

Angular distributions are particularly sensitive to the REC reaction mechanism and to details of atomic wave functions. We should remark here that for high projectile energies, high-Z projectiles and low-Z target atoms, the momentum distribution within the target is negligible and amounts to a correction of the order of a percent [8, 10]. It is therefore indeed justified to identify radiative electron capture with radiative recombination.

A particularly clear-cut example for angular distributions is provided by REC into the K shell of a relativistic high-Z ion. Figure 2 shows the theoretically predicted [8] differential cross section for radiative electron capture from a nitrogen target into the K shell of bare U^{92+} ions at an energy of 295 MeV/u. We observe two features: (i) The angular distribution behaves roughly like a $\sin^2\theta$ distribution as predicted by a nonrelativistic theory including retardation, i.e. all multipole orders, see Eq. (5). In fact, for lower projectile energies and smaller charge numbers, the distributions are even closer to a $\sin^2\theta$ form. (ii) At forward angles, the cross section comes about only magnetic spin-flip transitions, since the conservation of angular

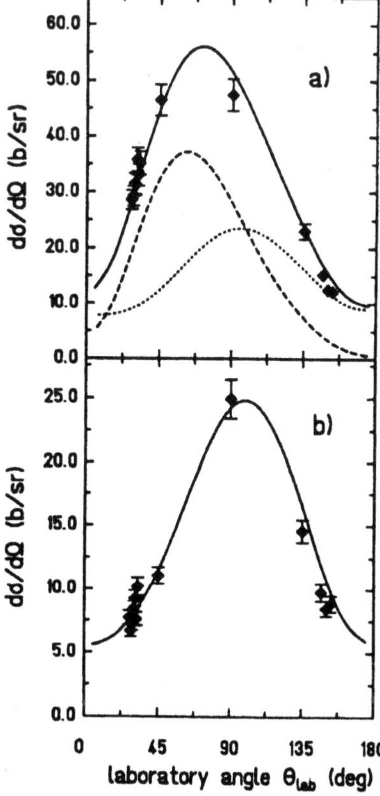

Figure 3. Experimental angular distribution of *L*-REC radiation (diamonds) for (a) REC into the $2s_{1/2}$ and $2p_{1/2}$ states and (b) REC into the $2s_{3/2}$ level. The solid line shows the result of relativistic exact calculations. Theoretical results for capture into the $2s_{1/2}$ state [dashed line in (a)] and for capture into the $2p_{1/2}$ state [dotted line in (a)] are shown separately. From [14].

momentum rigorously forbids the emission of electrons at angles of 0° or 180° without spin-flip. Recently, measurements near 0° performed at the GSI Darmstadt [15] have confirmed the predicted angular distribution similar to Fig. 2. In this way, the electron spin and magnetic transitions manifest themselves in a particularly clear-cut manner. For high-Z ions, the large fine structure splitting provides direct experimental access to the investigation of the angular distributions associated with REC into the j-sublevels of the L-shell. A very first experimental study of the angular distributions for REC into the L-shell sublevels was conducted at the GSI Fragment Separator for 89 MeV/u $U^{90+} \to C$ collisions [14]. The results obtained within this experiment are depicted in Fig. 3. In the figure the measured angular distributions for REC into the $j = 1/2$ and $j = 3/2$ levels of the L-shell of He-like uranium are plotted in comparison with the predictions of the exact relativistic theory [8]. The data for capture into the $j = 1/2$ –levels show a considerable bending of the angular distribution into the forward direction whereas the distribution for capture into the pure p-state ($j = 3/2$) exhibit a slight enhancement at backward angles.

The nonrelativistic Born approximation, see Eq. (3), as well as the rigorous treatment show that for high velocities $\beta \to 1$, the peak of the distribution is increasingly shifted towards forward angles owing to the effect of retardation. This means that the detailed structure is compressed into a narrow forward cone. On the other hand, the Lorentz transformation to the laboratory system decompresses the peak so that details become more visible. This is illustrated in Fig. 4, which contains two pairs of identical cases, i.e. with the same photon energy, viewed as photoionization and, alternatively, as radiative recombination [10]. This demonstrates that REC (or RR) is the most practical way to study the photoelectric effect at high energies [16].

With the increasing accuracy in measuring angular distributions and the rigorous relativistic treatment of radiative recombination developed recently [8], it may become possible one day to identify radiative corrections to the cross sections. The effects will increase with increasing collision energy and increasing projectile charge. The theory of QED corrections to the photoeffect in high-Z systems is presently being studied [17]. It has been found that contributions from vacuum polarization amount to about 0.5% for 300 MeV/u U^{92+} projectiles and to about 1.5% for 3.0 GeV/u at forward angles. The self-energy correction is more difficult to calculate and is presently being considered.

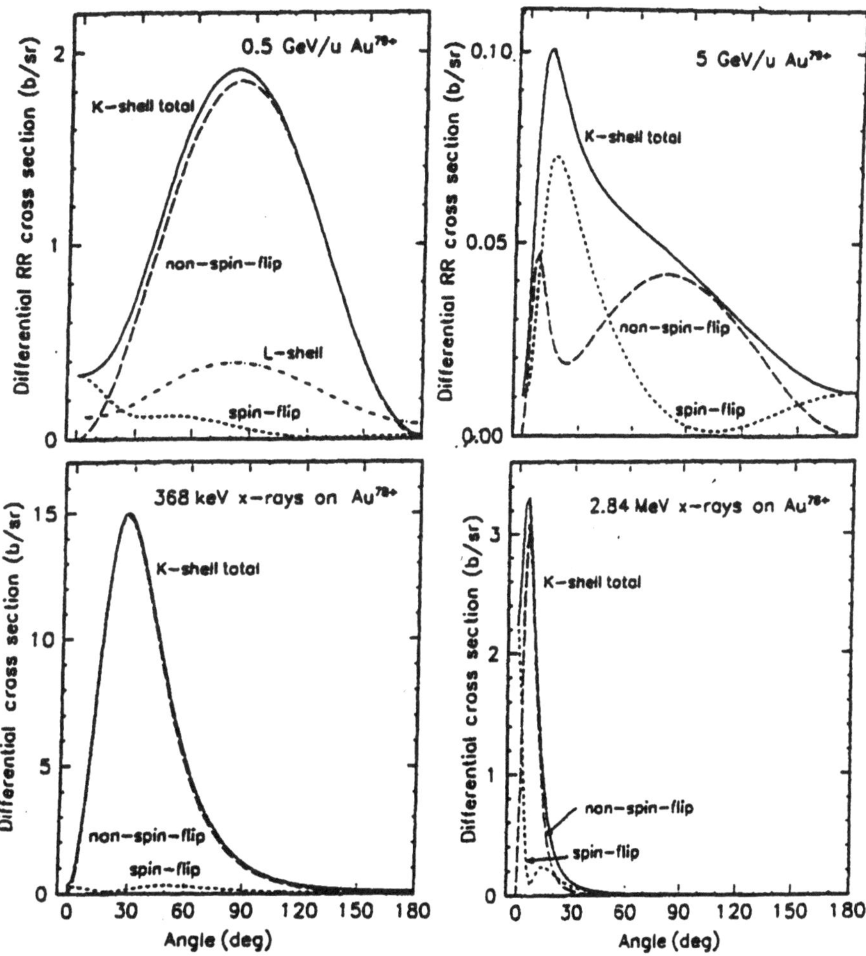

Figure 4. Differential cross sections for radiative recombination (upper row) for bare Au projectiles with energies of 0.5 and 5.0 GeV/u as a function of the laboratory angle. For comparison, the lower row shows the corresponding cross sections for the photoelectric effect. Adapted from [10].

5. SUMMARY AND OUTLOOK

The possibility to accelerate high-Z ions to relativistic velocities and to strip them of all or almost all of their electrons, has opened up a new field of atomic physics. One has gained access to the spectroscopy of heavy one-electron and few-electron ions up to uranium and hence to new tests of quantum electrodynamics in strong fields.

Certainly the best understood process is radiative electron capture, which is closely connected with radiative recombination and the photoelectric effect. For this reaction, calculated differential and total cross sections are in very good agreement with experimental results for K, L, and M shell capture. It has been shown that at high photon energies, radiative recombination is the most practical way to study the dynamics of the photoeffect in a regime that is not accessible otherwise. The decompression of the angular distribution allows for a detailed study of the angular dependence. At small forward angles it is possible to identify magnetic spin-flip transitions in a unique fashion by comparing experimental results with the theoretically predicted behaviour of the angular distribution. It is furthermore expected that concrete predictions for the effects of QED corrections to the photoeffect and to radiative recombination will be available in the near future.

ACKNOWLEDGMENTS

This work would not have been possible without the intense and delightful collaboration with A. Ichihara and T. Shirai on the theoretical side, with Th. Stöhlker on the experimental side and with Th. Beier, V. M. Shabaev, V. A. Yerokhin and G. Soff within the field of radiative corrections. The author is also indebted to the Japan Atomic Energy Research Institute for continuing support and hospitality and to the St. Petersburg State University for the hospitality during a fruitful visit.

REFERENCES

1. F. Bosch, Nucl. Instr. and Meth. B **23**, 190 (1987).

2. F. Bosch, Nucl. Instr. and Meth. A **314**, 1992 (1992).

3. Th. Stöhlker, Phys. Scripta **T73**, 29 (1997).

4. H. A. Bethe and E. E. Salpeter, *Quantum Mechanics of One- and Two-Electron Atoms*, Academic, New York (1957).

5. E. Spindler, H.-D. Betz and F. Bell, Phys. Rev. Lett. **42**, 832 (1979).

6. For a monograph on the subject, see, e.g., J. Eichler and W. E. Meyerhof, *Relativistic Atomic Collisions*, Academic Press, San Diego (1995).

7. R. H. Pratt, A. Ron and H. K. Tseng, Rev. Mod. Phys. **45**, 273 (1973).

8. A. Ichihara, T. Shirai and J. Eichler, Phys. Rev. A **49**, 1875 (1994).

9.   J. Eichler, A. Ichihara and T. Shirai, Phys. Rev. A **51**, 3027 (1995).

10. A. Ichihara, T. Shirai and J. Eichler, Phys. Rev. A **54**, 4954 (1996).

11. E. Clementi and C. Roetti, At. Data Nucl. Data Tables **14**, 177 (1974).

12. M. Kleber and D. H. Jakubassa, Nucl. Phys. A **252**, 152 (1975).

13. M. Stobbe, Ann. Phys. (Leipzig) **7**, 661 (1930).

14. Th. Stöhlker, H. Geissel, H. Irnich, T. Kandler, C. Kozhuarov, P. H. Mokler, G. Münzenberg, F. Nickel, C. Scheidenberger,T. Suzuki, M. Kucharski, A. Warczak, P. Rymuza, Z. Stachura, A. Kriessbach, D. Dauvergne, B. Dunford, J. Eichler, A. Ichihara and T. Shirai, Phys. Rev. Lett. **73**, 3520 (1994).

15. Th. Stöhlker, T. Ludziejewski, F. Bosch, R. W. Dunford, C. Kozhuharov, P. H. Mokler, H. F. Beyer, O. Brinzanescu, F. Franzke, J. Eichler, A. Griegal, S. Hagmann, A. Ichihara, A. Krämer, D. Liesen, H. Reich, P. Rymuza, Z. Stachura, M. Steck, P. Swiat and A. Warczak, Phys. Rev. Lett., submitted.

16. Th. Stölker, P. H. Mokler, C. Kozhuharov and A. Warczak, Comments At. Mol. Phys. **33**, 271 (1997).

17. Th. Beier, A. N. Artemyev, J. Eichler, V. M. Shabaev and V. A. Yerokhin, Physica Scripta, in press.

Electron Excitation of Autoionizing States in Alkalis

Rajesh Srivastava

Department of Physics
University of Roorkee, Roorkee - 247 667, India

Improved distorted wave calculations are performed for the electron impact excitation of the lowest lying autoionizing states $np^5 (n+1) s^2$ ^2P in alkali atoms. The results for total excitation cross sections and alignment parameters are examined in the light of recent experimental measurements for $2p^5 3s^2$ ^2P and $3p^5 4s^2$ ^2P states in Na and K from the Freiburg group (Germany).

1.   INTRODUCTION

We consider here the electron impact excitation of the lowest-lying doublet of autoionizing states $np^5 (n+1)s^2$, $^2P_{3/2,\ 1/2}$ of the alkali atoms *viz.* Na (n = 2), K (n = 3), Rb (n = 4) and Cs (n = 5) and their subsequent decay to the continuum. In order to understand the excitation take the following two processes i.e. electron impact direct ionization and ionization through the excitation autoionization of these levels.

$$e_i^- + A\,[np^6 (n+1)s\ ^2S_{1/2}] \rightarrow e_s^- + A^+[np^6\ ^1S_0] + e_e^- \qquad (1a)$$

$$e_s^- + A^*\,[np^5 (n+1)\,s^2\ ^2P_{1/2},\ ^2P_{3/2}] \qquad (1b)$$

Trends in Atomic and Molecular Physics,
Edited by Sud and Upadhyaya. Kluwer Academic/Plenum Publishers, New York, 2000.   147

148

Where, e_i, e_s and e_e represent respectively the incident, scattered and ejected electrons and A can be any alkali atom *viz.* Na (n = 2), K (n = 3), Rb (n = 4) and Cs (n = 5). Here, the first process (1a) gives only the contribution from direct ionization of the alkali atoms while in the second process (1b) ionization takes place as a two step process i.e. first the alkali atom is excited to either $np^5(n+1)s^2\ ^2P_{1/2}$ or $np^5(n+1)s^2\ ^2P_{3/2}$ state which then in second step autoionizes by ejection of an electron leading to ionization. In the first step, the excitation of $^2P_{3/2}$ autoionizing state leads to its alignment. Through an experiment the alignment of this excited $^2P_{3/2}$ state can be determined via the non-isotropic emission of autoionizing electrons from the decay of this state i.e. in the second step. The angular anisotropy is given by [1, 2]

$$I_{3/2}(\theta) = I_{3/2}\left[1 + \alpha_2\ A_{20}(E)\,P_2(\cos\theta)\right] \qquad (2)$$

$A_{20}(E)$ and α_2 are the alignment parameter and the decay parameter of the $^2P_{3/2}$ state, respectively. E is the incident electron energy and θ, is the angle of the ejected electron from the direction of the incident electron beam. $P_2(\cos\theta)$ is the second order Legendre polynomial. The isotropic part $I_{3/2}$ is proportional to the total excitation cross section of the $^2P_{3/2}$ state.

The alignment parameter $A_{20}(E)$, of the excited $^2P_{3/2}$ state can be related to the electron impact total excitation cross section $\sigma(JM_J)$ of the magnetic substates $|M_J|$ of the autoionizing $^2P_{3/2}$ state with J = 3/2 occurred through first step of the process (1b) and is given by

$$A_{20}(E) = \frac{\sigma\left(\frac{3}{2}\ \frac{3}{2}\right) - \sigma\left(\frac{3}{2}\ \frac{1}{2}\right)}{\sigma\left(\frac{3}{2}\ \frac{3}{2}\right) + \sigma\left(\frac{3}{2}\ \frac{1}{2}\right)} = \frac{\sigma\,(np1) - \sigma\,(np0)}{\sigma\,(np0) + 2\sigma\,(np1)} \qquad (3)$$

The second equality of equation (3) is for excitation of the unresolved fine-structure ^2P state with magnetic quantum number $m_l = 0, \pm1$ and is given in terms of the total excitation cross section $\sigma(npm_l)$ for promotion of an npm_l electron to an $(n+1)s$ orbital [2,3]. The factor α_2 is determined only by the decay process i.e. second step of the process (1b) and is in general dependent on the decay amplitudes of the different ejected autoionizing electron partial waves. However, in the case of (1b) only one autoionizing electron wave is emitted which reduces α_2 to a pure geometrical factor [1-4] and we have $\alpha_2 = -1$.

Further, the relation (2) is only valid if the effect of interference between direct (process (1a)) and autoionizing transition (process (1b)) in the entire process (1) can be neglected. This is true in the present case [5] and the excitation of the autoionizing states may thus be treated like excitation of a discrete state. Also, the electron impact excitation of the lowest autoionizing $np^5(n+1)s^2$ ^2P states from the ground $np^6(n+1)s$ ^2S state in different alkalis viz. Na (n = 2), K (n = 3), Rb (n = 4) and Cs (n = 5) can be expressed in essence as excitation of an inner shell electron from the orbital $np \rightarrow (n+1)s$ and we can treat alkalis to behave like one electron system as is considered in most previous calculations [3, 4, 6].

All the earlier theoretical and experimental efforts on the present study have been very well described and analysed by Pangantiwar and Srivastava (PS) [3] and in the recent experimental papers of Matterstock *et al* [4] and Feuerstein *et al* [6]. In their experiment Matterstock *et al* [4] measured for K, the angular distribution of the ejected autoionized electron from the $^2P_{3/2}$ states and obtained its alignment $A_{20}(E)$ and the ratio of the cross sections of the $^2P_{3/2}$ and $^2P_{1/2}$ states. In the later experiment from the same group Feuerstein *et al* [6] reported the measurement of the relative excitation cross sections of the unresolved lowest-lying autoionizing state $2p^5 3s^2$ ^2P of sodium for electron impact energy lying in the range from threshold excitation to 1.5 keV. They then normalized the relative excitation cross section data with their theoretically calculated non-relativistic plane wave Born approximation results at the high energy of 1.5 keV. In the papers [4, 6] they have also very briefly presented their new non-relativistic distorted wave approximation (DWA) calculations for K and Na alkali atoms and have given detailed comparisons of their calculations with their measurements and the DWA calculations of PS and others. They concluded that there is still need for a suitable better theoretical calculation to compare with their experiments. Here we report our calculation in which earlier errors are avoided in order to be able to compare with experiment in better manner. We present here our results for the total excitation cross sections and alignment parameters for the lowest doublet autoionizing levels in all alkalis. We described the excitation of the autoionizing levels in the framework of · both the non-relativistic distorted wave and relativistic distorted wave (RDW) approximation theories.

The non-relativistic distorted wave calculations of PS had used few approximations which could have been avoided. These were in the treatment of exchange in their calculation e.g. in evaluating the exchange T-matrix they used the Ochkur-Bonham approximation [7] (which may be valid for very high energies only) as well as they did not consider the effect of exchange in obtaining the distorted waves. Further for the choice of distortion potential they used the Mott and Massey version [8] where initial channel distortion potential is described by the ground state of the atom and the final channel distortion potential is described by excited state of the atom [8]. On the other hand the recent calculations mostly use now the distortion in both the initial and final channels to be described by the excited state of the atom. In addition, PS used in their calculation for the initial np and final (n + 1)s orbitals the wavefunctions obtained for the ground state wavefunction of the alkali atom rather than taking the wavefunction for the initial np orbital from the ground state and final (n + 1)s orbital from the excited state wavefunction of the atom. It would therefore be now desirable when the different new reliable experiments [4, 6] are coming up to test the authenticity of these simplifications taken by PS and also to improve agreement with experiment to perform better calculations and compare with them. Therefore, in the present work we again take up the calculation in non-relativistic distorted wave approximation theory where the different approximations / simplifications taken by PS mentioned above have been avoided and (or) further tested. For this purpose we carried out calculations for all the lowest autoionizing levels in the alkali atoms using a distorted wave method as described in the paper of Verma and Srivastava [9]. In this method one can see that the treatment of exchange for the T-matrix does not use the Ochkur-Bonham approximation and also the distorted waves are obtained using exchange potential given by Furness and McCarthy [10] which has been widely used and can be said to be quite suitable and successful in the intermediate energy region. We choose the distortion potential to be described by the excited state of the atom as is current practice. In addition the initial np and final (n + 1)s orbitals representing the ground and excited autoionizing levels have been obtained in both the ways i.e. when both the orbitals are obtained from the ground state wavefunction (referred as GG) as well as when the initial orbital is obtained from the ground state wavefunction and the final orbital is obtained from the excited state wavefunction (referred as GE). These different wavefunctions are obtained by using the computer code of Froese Fisher [11]. Though in the later choice problem of non-orthogonality of the inner core orbitals of the ground and excited state wavefunctions is there but then this will provide us further test of the suitability of the former choice of the orbital wavefunctions as adopted by PS.

In addition to non-relativistic calculation presented here another parallel set of calculation for all the autoionizing states in different alkalis have been performed using RDW method which is described in detail elsewhere [12] and here only the results will be compared. In fact, the advantage of using the RDW method is that one can obtain the results for both the fine-structure states of the autoionizing levels $^2P_{1/2}$ and $^2P_{3/2}$ separately without any type of approximation. Further, the wavefunctions used therein to describe the ground and the two excited autoionizing doublet levels are obtained in an average level calculation from the multi-configurational Dirac-Fock (MCDF) program of Grant et al [13]. In fact, the ground $np^6\,^1S_0$ and the two excited states $np^5(n+1)s^2\,^2P_{1/2}$ and $^2P_{3/2}$ are represented by Dirac-Fock orbitals and the wavefunctions have been obtained simultaneously such that they are represented through same inner orbitals. This avoids any normalisation and the orthogonality problems of the different orbital wavefunctions among themselves as well as between the ground and excited state wavefunctions as faced in most earlier non-relativistic calculations (see Feuerstein *et* al [6]). Also in the RDW method the exchange effects in the T-matrix and in solving the distorted electron waves (which is obtained by solving Dirac equations in the static potential of excited state of the atom) are taken exactly by taking proper antisymmetrisation into account. Consequently, we believe our RDW calculation to be quite reliable which can provide meaningful comparison to our non-relativistic DWA calculation presented here. The numerical details of both the non-relativistic and relativistic theories used here are given in ref. [14].

2. RESULTS AND DISCUSSION

In Figures 1 and 2 respectively, we have compared first our non-relativistic distorted wave calculations for cross section of the unresolved autoionizing ^{2}P state and $A_{20}(E)$ parameter with the available experimental results [4, 6, 15-18]. In each figure we present for our both the GG and GE type DWA calculations with the three different kind of results viz. (i) normal distorted wave calculation with exchange (say, DWE) (ii) distorted wave calculation with only direct T-matrix i.e. without exchange T-matrix (say, DWD) and, (iii) distorted wave results with distortion potential set to zero in both the channels i.e. Born approximation (BA). The available [6, 18] experimental cross section results only for Na are also included but have been renormalized at 1.5 keV to the RDW results [12] which we consider to be the best and discuss more about this aspect later. It should be added here that in the light of the non-relativistic distorted wave calculation presented here, the DWE calculation of PS are in fact similar to the GG type while the plane wave and distorted wave calculations of Matterstock *et al* [4] and Feuerstein *et al* [6] are similar to the GE type calculation. Therefore, for sake of clarity and meaningful presentation we do not compare our results with any of these earlier theoretical calculations which have the undesirable feature pertaining to the wavefunctions used in them as we discussed earlier. However, one could easily compare these results with ours qualitatively.

From Figure 1, we find for each GG or GE type DWA calculation and for all alkali atoms that the behaviour of the three different kind of results i.e. DWE, DWD and BA when compared among themselves is somewhat similar. For example, we can see distinctly that these three calculations

152

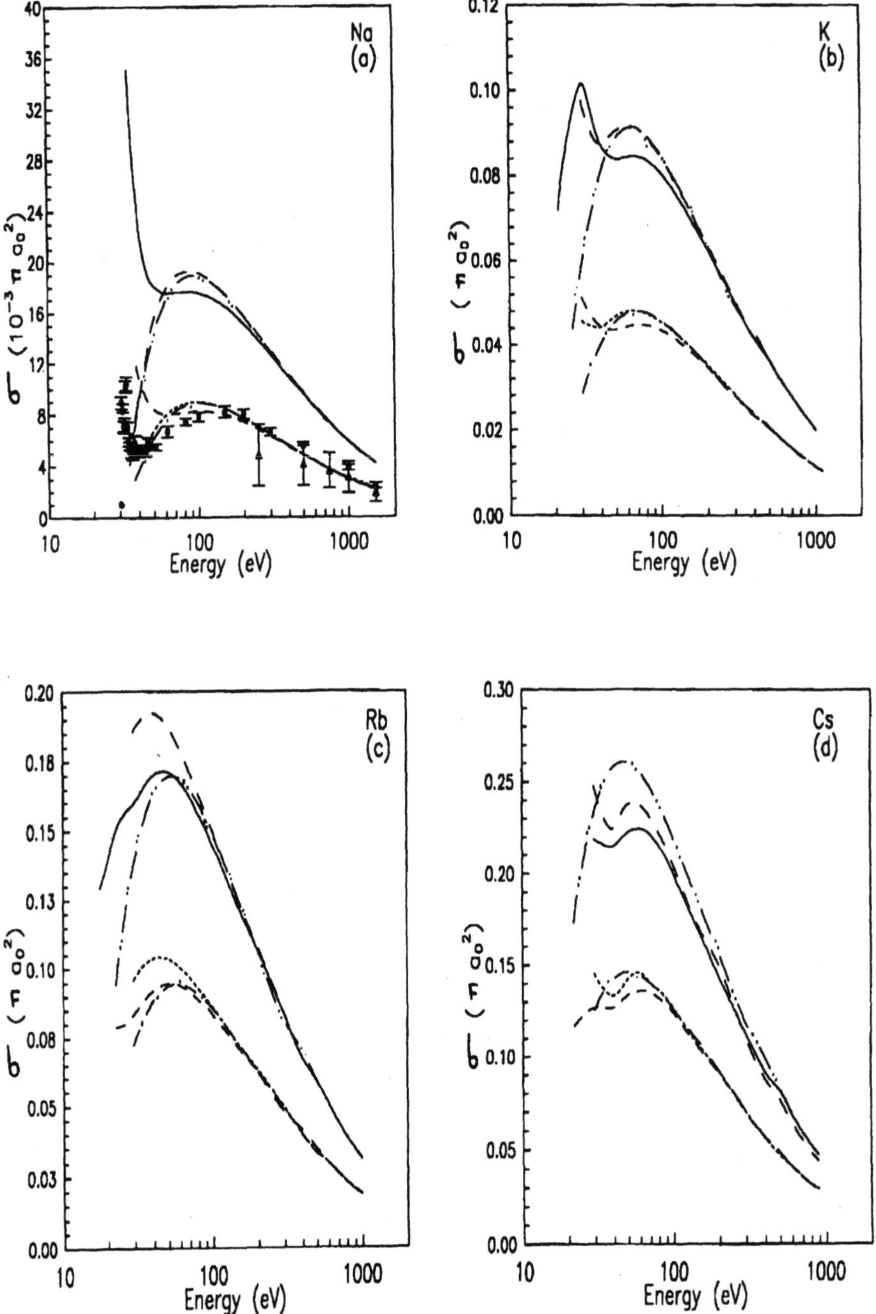

Figure 1. Comparison of total cross section $\sigma(\pi a_0^2)$. Theory: GE type calculation: ————,
DWE; — — —, DWD; – •• –, BA; GG type calculation: – – –, DWE; - - - -,
DWD; – • –, BA. Experiment: •, Feuerstein *et al* [6]; Δ, Kessler [18].

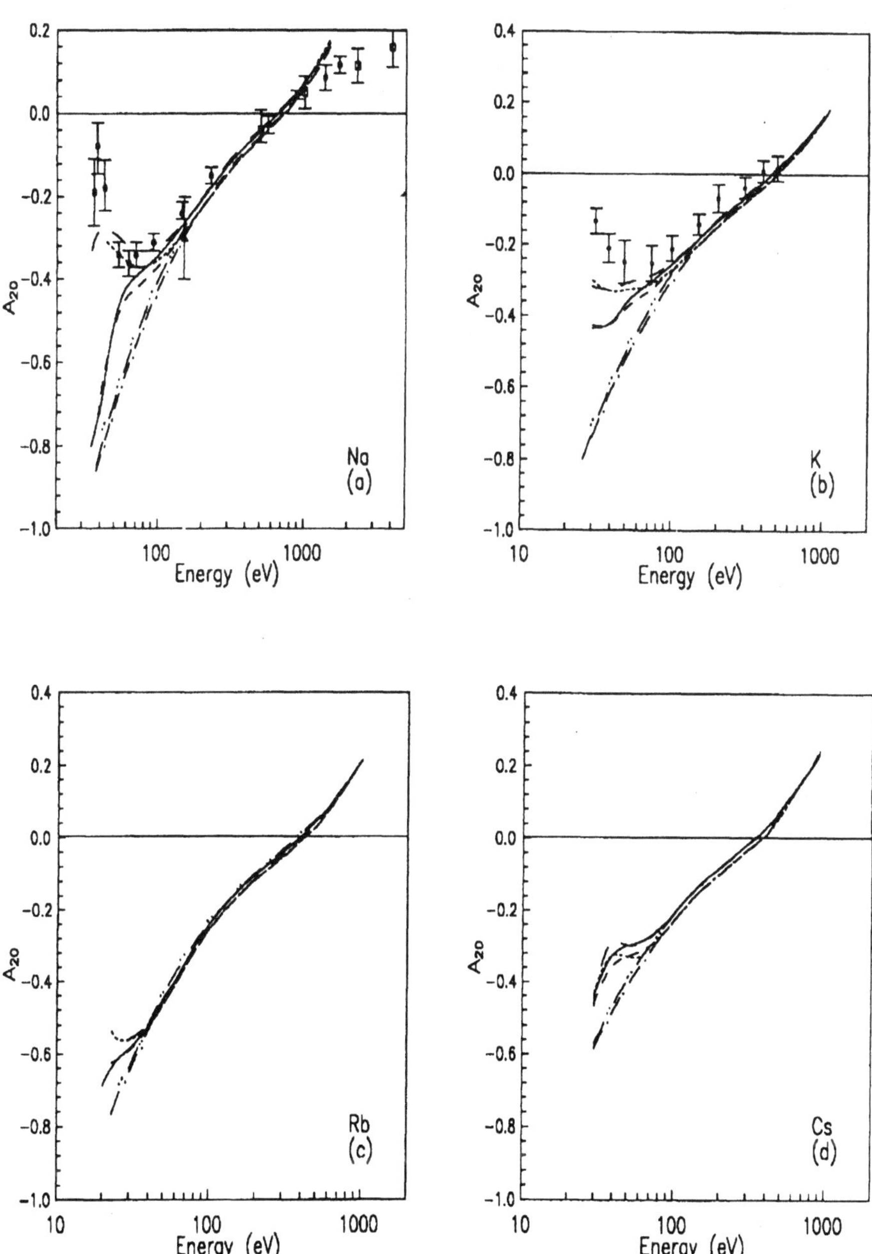

Figure 2. Same as figure 2 but for the alignment parameter $A_{20}(E)$. Experiment : •, DuBois *et al* [15]; □, Breuckmann *et al* [17]; Δ, Ross *et al* [16].

differ significantly at low and intermediate energies showing that an important contribution of exchange effect or distortion potential for each atom results in different manner in both the GG and GE type calculations. On the other hand comparing the same kind of results in both the GG and GE type calculation we find cross-sections differ by nearly a factor of two. We find also that these two types of calculations are not meeting even asymptotically at such a high energy of 1.5 keV. The reason for this is obviously as discussed by Feuerstein *et* al [6] is the use of non-orthogonal wavefunctions for the inner orbitals in the case of GE type calculations. In fact in GG type calculation as described earlier though the inner and outer orbitals are ortho-normal but then they are taken from the description of the ground state wavefunction which may not be justified. In this situation it would be naturally desirable and ideal to carry out calculation where we have inner orbitals orthogonal (as in GG) and also that ground and excited state wavefunctions are separately obtained (as in GE). This we have considered in our RDW calculations. However, we feel that as reported by earlier many workers including PS that the GG type calculations may be considered more reliable than the GE type calculation. This will be further clear when we compare GG type calculation with the RDW results later where the GG type calculations are found closer to the RDW as compared to the GE type calculations.

Let us now focus our attention in Figure 2, for $A_{20}(E)$ parameter which surprisingly shows that hardly there is any difference in the GG and GE type calculations, even for the individual three different kind of calculations i.e. DWE, DWD and BA which agree very well. Thus this can be quite misleading also if we judge the difference between the GG and GE type calculations only by comparing $A_{20}(E)$ parameter. In fact, an important part of the excitation mechanisms cancels in the ratio of the excitation cross sections for different magnetic substates as they enter in the alignment parameter $A_{20}(E)$. From the Figure 2 on comparing our DWE, DWD and BA calculations for Na and K atoms at lower energies we find that DWD results compare with experiment [4, 15-18] better than the DWE while BA calculations totally fail. It is interesting to note that the DWE calculation which should be better than the DWD having exchange in it does not show good agreement with experiment at low energies. May be exchange treatment in DWE should be done in somewhat better manner as the transition is sensitive to exchange. We also observed this point when we compare our RDW calculations in which exchange treatment is though done in the best possible way but still such a feature exists. We also note for $A_{20}(E)$ that at higher energies though all calculations should agree however, for Na we find present DWE results diverge from experiments at such high energies.

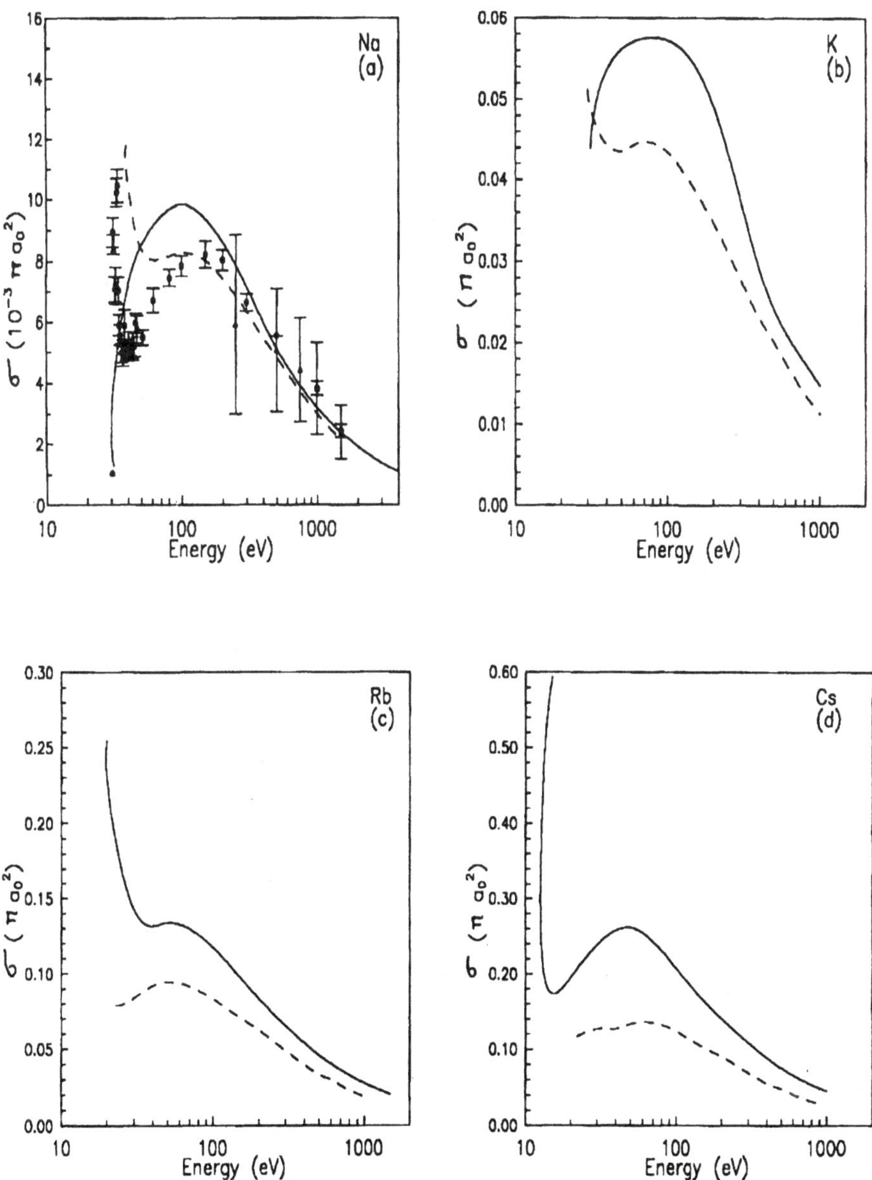

Figure 3 Comparison of RDW calculation with GG type of calculation for the total cross section $\sigma(\pi a_0^2)$. Theory: ——, RDW; – – –, DWE. Experiment: •, Feuerstein *et al* [6]; △, Kessler [18].

156

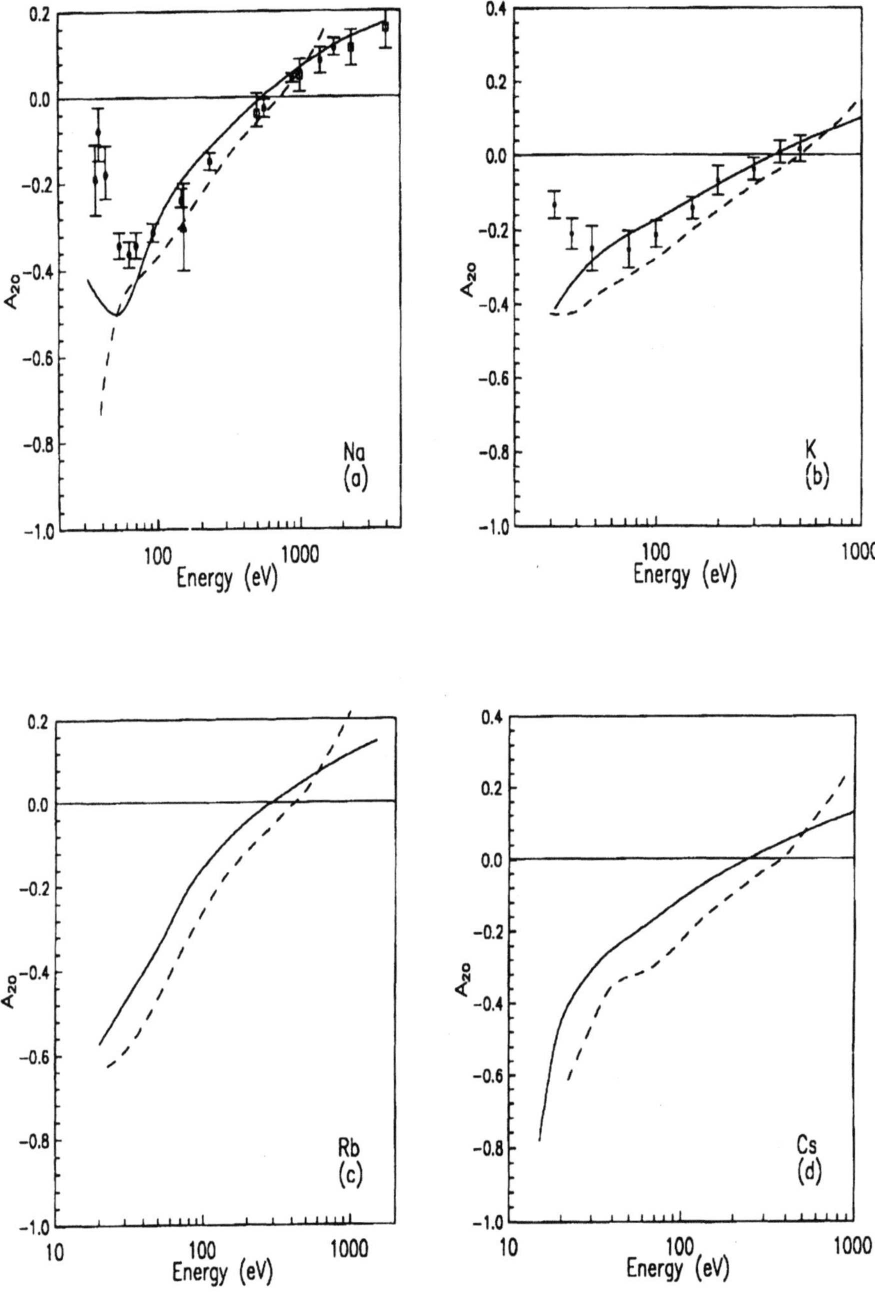

Figure 4. Comparison of RDW calculation with GG type of calculation for the alignment parameter $A_{20}(E)$. Experiment: •, DuBois *et al* [15]; □, Breuckmann *et al* [17]; Δ, Ross *et al* [16].

Finally in Figures 3 and 4 we compare the present non relativistic (DWE) calculation of the GG type with the relativistic RDW calculation [1, 2] for the total cross section of unresolved autoionizing ^{2}P state and $A_{20}(E)$ parameter for all the alkalis i.e. Na, K, Rb and Cs atoms. In these figures the GE type results are discarded being not good even though they give almost same results for $A_{20}(E)$ as obtained in GG type calculation but the total cross section obtained in GE type calculations overestimate the results and has no proper asymptotic behaviour. The experimental results of Na from Feuerstein *et al* [6] are the same as shown in Figures 1 which are normalized to the RDW results. Also same experimental data as compared earlier in figure 2 for $A_{20}(E)$ are shown in figure 4.

We observe from figure 3 that for Na both the RDW and DWE calculations agree above 400 eV and they both give qualitative agreement with experiment. Out of all the alkalis the agreement between both the theories is best for Na as expected it being the lightest atom with less dominant relativistic effects. The agreement of results for K, Rb and Cs seems only qualitative and results tend to approach each other asymptotically. We see for Na and K threshold peaks are dominant in DWE (and similar is expected for Rb and Cs as our computer code showed numerical problems in going near threshold energy for these atoms) while the RDW calculations show prominent peaks at threshold only in Rb and Cs. From Figure 4 where we have shown $A_{20}(E)$ results we observe the agreement of the RDW results with experiment is better than to the DWE results for Na and K in the entire range. However, the disagreement of both the theories at much lower energies with experiment suggest either the need of further improvement of the theories or obvious inadequacy of distorted wave method itself at such low energies. In general we find for $A_{20}(E)$ incorrect asymptotic behaviour with respect to the experiment is predicted from the DWE calculation as compared to RDW. We feel that more experimental data if made available can throw better light on the present study.

ACKNOWLEDGMENTS

The work presented here has been carried out in collaboration with Dr. S. Kaur and Dr. S. Verma, University of Roorkee, Roorkee. I would like to thank Prof. W. Mehlhorn, Freiburg (Germany) and his group members for the fruitful discussion and for sending their unpublished data. Financial assistance from the University Grants Commission, New Delhi and Department of Atomic Energy, Govt. of India is gratefully acknowledged.

158

REFERENCES

1. E. G Berezhko. and N. M. Kabachnik, J. Phys. B: At. Mol. Opt. Phys. **10,** 2467 (1977).

2. B. Cleff and W. Mehlhorn , J. Phys. B: At. Mol. Opt. Phys. **7,** 593 (1974).

3. A. W. Pangantiwar and R. Srivastava, J. Phys. B: At. Mol. Opt. Phys. **20,** 5881 (1987).

4. B. Matterstock, R. Huster, B. Paripas, A. N. Grum-Grzhimailo and W. Mehlhorn, J. Phys. B: At. Mol. Opt. Phys. **28,** 4301 (1995).

5. C. E. Theodosiou, J. Phys. B : At Mol. Opt. Phys. **10,** L253 (1977).

6. B. Feuerstein, A. N. Grum-Grzhimailo and W. Mehlhorn, J. Phys. B: At. Mol. Opt. Phys. **31,** 593 (1998)

7. V. I. Ochkur , Sov. Phys.-JETP **18,** 503 (1964)

8. D. H. Madison and K. Bartschat, *The distorted-wave method for elastic scattering and atomic excitation, Computational Atomic Physics: Electron and Positron Collisions with Atoms and Ions,* ed. Bartschat K., (Springer-Verlag Berlin Heidelberg 1996), p.65-86

9. S. Verma and R. Srivastava, J. Phys. B: At. Mol. Opt. Phys. **28,** 4823 (1995)

10. J. B. Furness and I. E. McCarthy., J. Phys. B: At. Mol. Opt. Phys. **6,** 2280 (1973)

11. C. Froese Fisher, Comput. Phys. Comm. **1** 151 (1969)

12. S. Kaur and R. Srivastava, J. Phys. B: At. Mol. Opt. Phys. (to be published).

13. I. P. Grant, B. J. McKenzie, P. H. Norrington, D. F. Mayers and N. C. Piper, Comput. Phys. Commun. **21,** 207 (1980).

14. S. Kaur, Ph. D Thesis, University of Roorkee, Roorkee (1998).

15. R. D. DuBois, L. Mortensen and M. Rødbro, J. Phys. B: At. Mol. Opt. Phys. **14,** 1613 (1981).

16. K. J. Ross, T. W. Ottley, V. Pejcev and D. Rassi, J. Phys. B: At. Mol. Opt. Phys. **9,** 3237 (1976).

17. E. Breuckmann, B. Breuckmann, W. Mehlhorn and W. Schmitz, J. Phys. B: At. Mol. Opt. Phys. **10,** 3135 (1977).

18. G. Kessler, Diploma Thesis, (Freiburg, 1984).

Depopulation of Low-Rydberg Atoms: A Semi-Classical Study

Anil Kumar[*]

Department of Physics
Florida A & M. University, Tallahassee, FL – 32307, U.S.A.

[*]Department of Physics,
J. P. University,Chapra - 841 301, India

Total depopulation of low-Rydberg atoms colliding with neutral perturbers at thermal energies has been investigated by employing the semi-classical impact-parameter method in the Molecular Orbital (MO) formalism. The interacting system is represented by a quasi-molecule which encompasses the entrance and various exit channels of the reaction. The method of pseudo potential is invoked to account for the effective binding of the active electron in the transient quasi-molecule. The internuclear movement is described classically and the electron's motion is accounted for in terms of its wave function. A large number of coupled equations are solved numerically to obtain the probability of transition from the initial state. The collision dynamics, and hence the individual depopulation cross sections, are found to be significantly different for different perturbers. No simple approach such as the Free Electron Model can be used to explain the details of these state-changing reactions.

1.   INTRODUCTION

Highly excited atoms (A^*), popularly referred to as Rydberg atoms, possess a wealth of exotic properties. Due to their huge size, which scales according to n^2 (n being the principal quantum number), they are sometimes known as *Giant atoms*. The distant electron of the Rydberg atom is so weakly bound that even a small disturbance due to the incoming atomic/ molecular *perturber* (**B**) can open up one of the numerous exit channels,

Trends in Atomic and Molecular Physics,
Edited by Sud and Upadhyaya. Kluwer Academic/Plenum Publishers, New York, 2000.   159

sometimes even in the limit of zero impact energy. With the development of tunable dye laser study on these atoms has multiplied; wide applications in various fields of current interest are also partly responsible for this.

The distant electron of the highly excited atom interacts very weakly with the inner-shells of the target, and, thus, exhibit hydrogenic character. This facilitates a virtual splitting of the Rydberg atom into a *weakly bound electron* and the *core* that encompasses the target nucleus along with inner-shells electrons, if any. Such a simplification has led to the so called Free Electron Model of Fermi [1] which successfully accounts for various reactions involving Rydberg atoms. In this model, the incoming *perturber* is considered to be interacting with the *core* and the *electron* quite independently (Fig. 1). But this simplified picture is found to be inadequate when we go down the Rydberg series, because for lowly excited atoms the two sets of interactions are bound to overlap. In such a situation a complete three-body interaction needs to be taken into account. The semi-classical approximation in the impact parameter formulation [2, 3] is found to provide a convenient way of investigating various interactions involving low-Rydberg atoms.

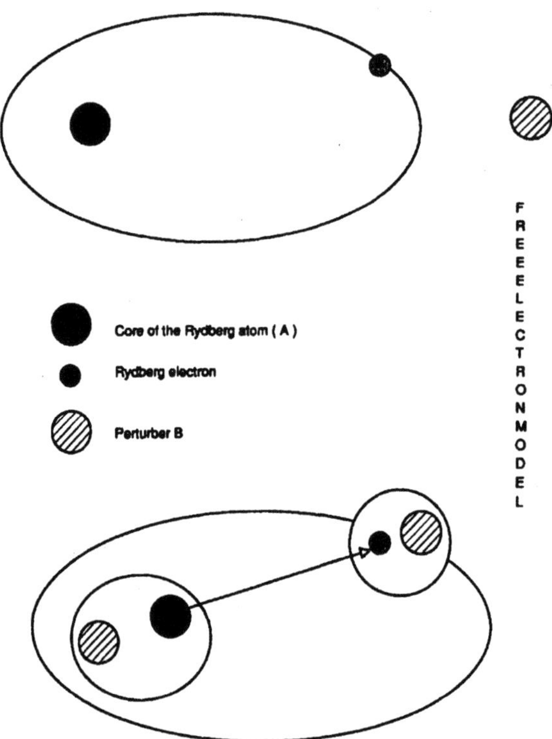

Figure 1. Free electron model of the Rydberg-perturber interaction.

In the present article we will discuss the application of this approximation to investigate the total depopulation of low-Rydberg alkali (A $\equiv$ Na) due to impact of noble gases (B $\equiv$ He, Ne and Ar):

$$A(n\,l)+ B \rightarrow A(n'\,l')+ B \tag{1}$$

The outline of the paper is as follows: we provide a brief description of the theoretical method in the following section. In section 3 we present results for various colliding systems. A comparative study of different colliding pairs is also presented in the same section, enabling us to visualise the role played by the perturbers in such interactions. Throughout the article we use atomic units, unless stated otherwise.

2.   THEORETICAL METHOD

The low-energy collision between an excited atom and an incoming compact perturber can be visualised through a Molecular Orbital (MO) approach. In this model, the colliding pair is supposed to form a transient *diatomic quasi-molecule* whose two centres are the Rydberg core (A^+) and the incident perturber (B). The internuclear separation (R) of this quasi-molecule changes with time as the colliding partners approach and separate. Consequently, the active electron is subjected to a slow varying effective field applied jointly by the two centres. In the asymptotic region $(R \rightarrow \infty)$ this quasi-molecule *disintegrates* to yield the incident and various exit channels of the state-changing reaction.

The relative nuclear motion in this quasi-molecule is described through a well defined classical trajectory. The method of pseudo potential [4] is invoked for simulating the effective binding of the active electron inside the molecule. In this approach the core orbitals of the Rydberg atom need not be included explicitly. The effective potential due to any core is generally represented as a sum of two terms:

$$V_p(\mathbf{r})= V_p^{SR}(\mathbf{r})+V_p^{LR}(\mathbf{r}) \tag{2}$$

The short- and long- range terms, $V_p^{SR}(\mathbf{r})$ and $V_p^{LR}(\mathbf{r})$, in the above expression are given, respectively, by

$$V_p^{SR}(\mathbf{r})= \sum_{l,m} A_l \exp\left(-\xi_l\, r^2\right) |Y_{lm}\rangle \langle Y_{lm}| \tag{3}$$

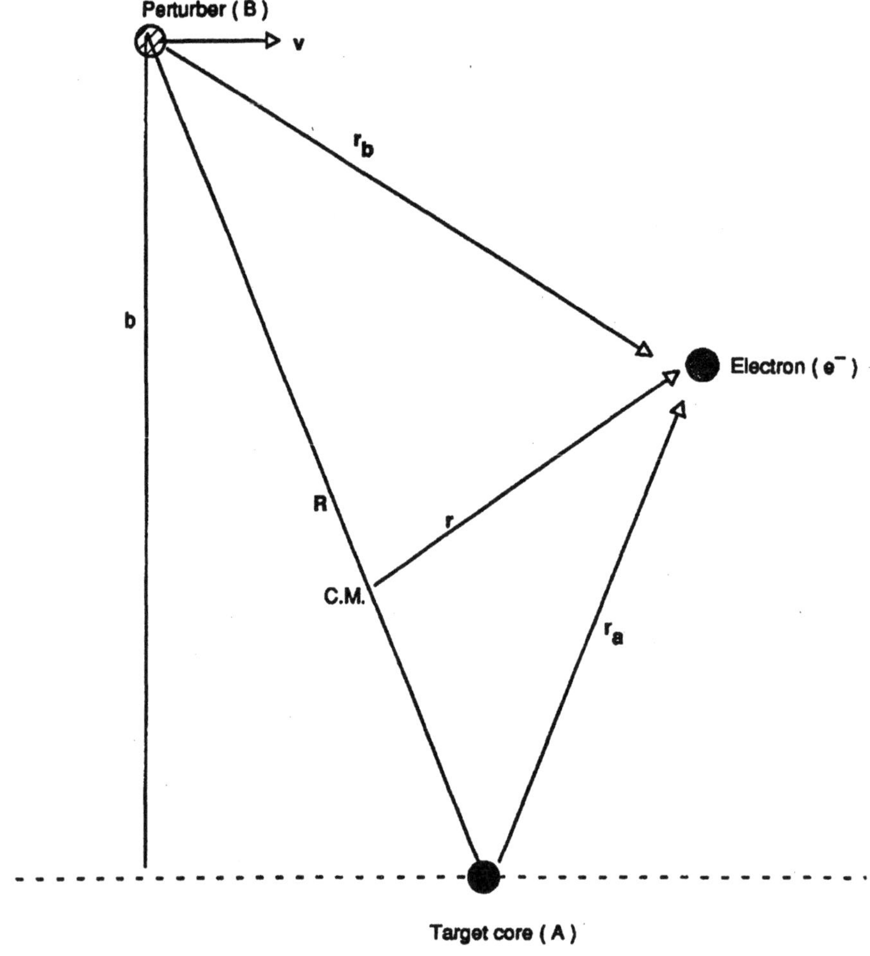

Figure 2. Collision geometry of the Rydberg-perturber interaction.

and

$$V_p^{LR}(\mathbf{r}) = -\frac{0.5\,\alpha_d}{X^2} - \frac{0.5\,\alpha_q}{X^3} - \frac{Z_{eff}}{r} \tag{4}$$

where $X^2 = r^2 + d^2$ and $Z_{eff} = 1.0$. The l-dependent parameters, A_l and ξ_l, are obtained by fitting spectroscopic data, and have been given by

Bardsley [4] and Pascale [5], respectively, for *alkalis* and He; for Ne and Ar we have used these parameters provided by Czuchaj et al [6]. In the long-range interaction term, Z_{eff} is the net effective charge of the core as seen by the electron at infinite distance, and α_d and α_q are the dipole and quadrupole polarizabilities, respectively, and d is a cut-off parameter. In the above expressions $\vec{r}$ stands for both $\vec{r}_a$ and $\vec{r}_b$ (Fig. 2), the position vectors of the electron with respect to the target core (A^+) and the perturber (B). The one-electron Hamiltonian is exprssed as

$$H_{el} = -\frac{1}{2}\nabla_r^2 + V_p(\mathbf{r}_a) + V_p(\mathbf{r}_b) + V_{ct}(\mathbf{r}_b, \mathbf{R}) + V_{cc}(\mathbf{R}) \qquad (5)$$

V_{ct}, generally referred to as the cross term or the three body interaction term accounts for the polarization of the incident perturber (B) by the target core and the Rydberg electron; V_{cc} is the interaction between alkali-core and the noble gas atom [5].

The time-dependent Schrödinger equation for the colliding system, with the electronic Hamiltonian defined above, is solved variationally. For this we construct a Born-Oppenheimer (BO) wave function, $\Phi_n^{BO}(\mathbf{r}, \mathbf{R})$, in which Slater-type Orbitals (STOS) centred on the two cores are employed as basis sets:

$$\Phi_n^{BO}(\mathbf{r}, \mathbf{R}) = \sum_l a_l \chi_l^{STO} + \sum_m b_m \chi_m^{STO} \qquad (6)$$

The total wave function of the colliding system is expanded in terms of products of the adiabatic electronic wave functions $\Phi_n^{BO}(\mathbf{r}, \mathbf{R})$, and phase factors that include the Electron Translation Factors (ETF). The later is essential in order to achieve the Galilean invariance of the resulting coupled equations, and also to ensure the correct asymptotic form of various couplings (for details see refs. [2, 7]). Substituting the above wave function in the time-dependent Schrödinger equation we finally obtain a set of first order coupled equations:

$$i\dot{a}_n = \sum_{k \neq n} \mathbf{v}.(\mathbf{P} + \mathbf{A})_{kn} a_k + \varepsilon_n a_n \qquad (7)$$

where different terms have their usual significance [2, 3]. These coupled equations are solved numerically for each contributing impact parameters (b) to yield the probability of transition to any final k state from an initial n state:

164

$$P_k(E,b) \;=\; \left|a_k(+\infty,b)\right|^2 \qquad (8)$$

where the usual boundary condition holds good:

$$a_k(-\infty) \;=\; \delta_{nk} \qquad (9)$$

Integrating the product of the transition probability and the impact parameter on all contributing values of b, we get the partial cross section for the corresponding transition:

$$Q_k(E) = 2\pi \int_0^\infty db.b.P_k(E, b) \qquad (10)$$

These partial cross sections are added to fetch the cross sections for total depopulation of the parent Rydberg state.

3.   RESULTS AND DISCUSSION

The above method has been used to calculate the total depopulation cross sections of various low-lying Rydberg states of Li, Na and Rb colliding with a number of perturbers such as, He, Ne and Ar [3, 8-13]. In the following we, however, present a few representative calculations for Na(9s) colliding with He, Ne and Ar only. All these calculations have been carried out at thermal energies, because most of the experimental measurements are performed in that energy region. Before we present our results for these systems, it is worth pointing out that we have taken the mean Maxwellian velocity corresponding to a cell temperature 425K as a standard reference for the sake of comparison. This velocity for Na-Ar, Na-Ne and Na-He pairs amounts to 3.6, 4.2 and 7.4 ($\times\,10^{-4}$ a.u.) respectively.

A fairly large basis set of STOs has been employed to carry out the molecular structure calculations. Altogether 35 and 23 configuration interactions (CI) have been used for the Σ and Π states respectively; excitation up to 11s of the target atom (Na) is included in this expansion. It may, however, be pointed out that STOs with $\ell \geq 3$ are not included in these MO calculations. But in the light of the fact that these substates have very small energy defects with respect to the $\ell = 3$ substate, their non-inclusion is not expected to cause any signifiant change in the final outcome of the calculation. In fact through a test calculation [13] it has already been established that in such a scenario the molecular state correlating to n, $\ell = 3$

adjusts itself to represent the molecular channels correlating to the n, $\ell \geq 3$ manifold.

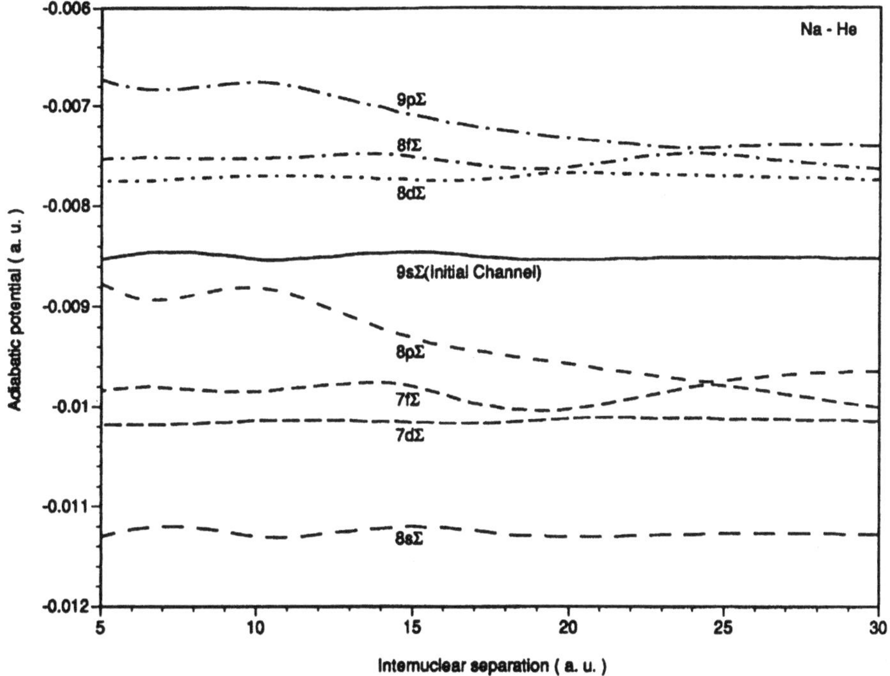

Figure 3. Adiabatic potential surfaces for the colliding pair Na(9s)-He-various Σ channels are shown separately.

The adiabatic potential energy surfaces for the quasi-molecule the initial and various final channels of reactions under consideration are presented in Figs. 3-5. We show all the Σ molecular states (Π states are left out to maintain the clarity of the curves) for the Na-He pair in Fig. 3; the other two pairs differ in details, which we discuss in the following. The most important coupling that characterises these colliding systems is between the initial channel 9sΣ and the immediate endoergic channel 8pΣ. Equally important is the near crossing between the 8pΣ and 7fΣ states at larger R values. The strength of these couplings changes significantly as we change the perturber. The details of these collision dynamics, as we will see later, have important implications on the process of depopulation of the parent Rydberg atom. To visualise these details we present, in Fig. 4, the important avoided crossings among the above discussed Σ states for all the pairs.

166

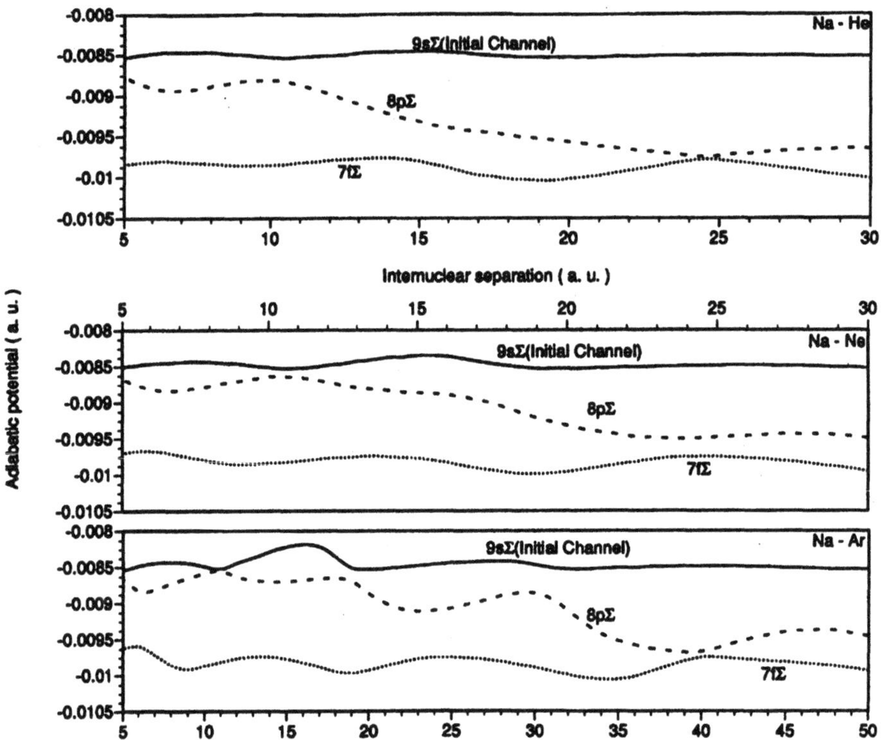

Figure 4. Adiabatic potential energy surfaces for Na(9s)-He/Ne/Ar-various Σ channels responsible for de-excitation of the initial state are shown separately.

In all cases, the $9s\Sigma$ and $8p\Sigma$ channels are found to have strong avoided crossing around $R = 11$ a_0; the strength of this coupling, however, increases as we pass from He towards Ar. The Na-Ar pair is conspicuous by the presence of another near crossing around $R \approx 19$ a_0 between these channels. The presence of an additional coupling turns out to be a key factor in enhancing the depopulation cross section for this pair. The subsequent avoided crossing, between the $8p\Sigma$ and $7f\Sigma$ states, also manifest appreciable change as the perturber is altered. In case of Na-He this coupling is very strong; the energy defect (ΔE) for this transition at $R \approx 26$ a_0 is around 4.4×10^{-5} a.u. In the same R region, the $8p\Sigma$ and $7f\Sigma$ states of Na-Ne pair exhibit much weaker coupling, the closest approach between these two (at $R \approx 24$ a_0) is larger by almost an order of magnitude ($\Delta E \approx 3.4 \times 10^{-4}$ a.u.). In the Na-Ar, these states do not show any significant coupling in the above noted neighbourhood; but at larger R values (≈ 40 a_0) they show moderate coupling ($\Delta E \approx 1.4 \times 10^{-4}$ a.u.). A shift towards larger R values makes this coupling less effective for the Na-Ar pair, at least in the investigated thermal energy region. These minute changes in the collision dynamics governing the endoergic transitions have profound effects on the individual steps responsible for depopulating the initial Rydberg state.

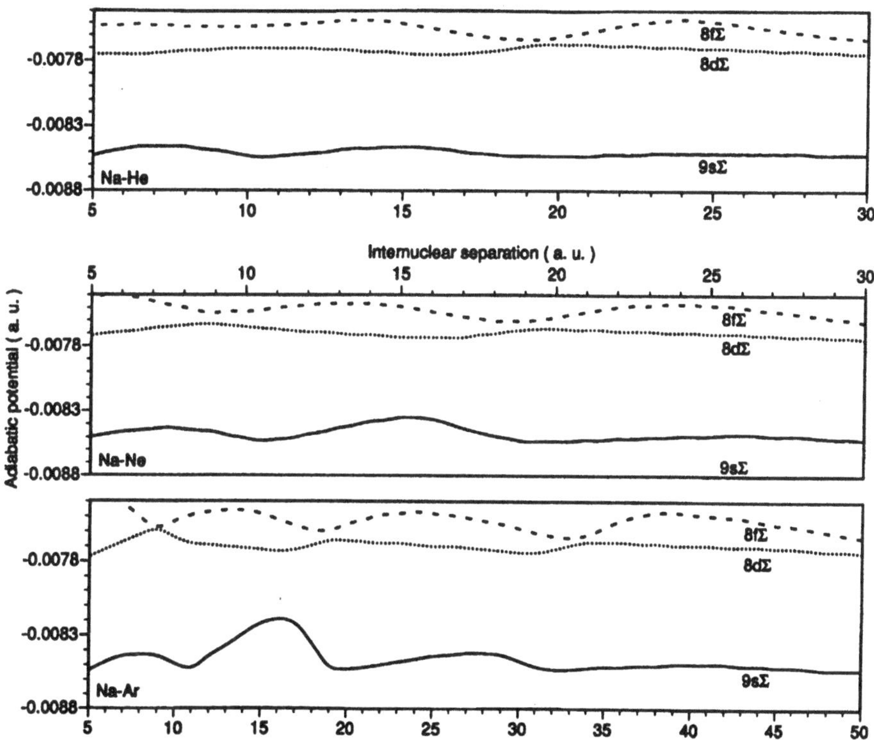

Figure 5. Adiabatic potential energy surfaces for Na(9s)-He/Ne/Ar-various Σ channels responsible for excitation of the initial state are shown separately.

The exoergic transitions are obviously governed by the coupling that exists between the initial channel 9sΣ and the immediate upper channel 8dΣ. The strength of this coupling also depends on the choice of the perturber; to visualise this aspect of the collision we have presented the 9sΣ, 8dΣ and 8fΣ molecular states together for all the pairs in Fig. 5. As is evident from these curves, the exoergic transitions are bound to make successively larger contributions as we move from He to Ar. This results into larger contribution of excitation in the process of total annihilation of the parent state. Since these couplings are comparatively weaker that those responsible for de-excitation processes, they turn increasingly more effective only with the rising velocity. Although we do not show the molecular Π states in these figures, it is worth mentioning at this point that the rotational couplings responsible for excitation are much more effective than those leading to the downward transitions for all the systems we have investigated.

At the same time radial couplings among various Π states are weak in all cases.

Following the above mentioned molecular structure calculations we couple 14 channels (eight Σ and six Π) to estimate the partial and total depopulation cross sections for the colliding pairs under investigation. As noted above, the presence of two avoided crossings between the $9s\Sigma$ and $8p\Sigma$ states in the Na-Ar pair results into appreciably large total depopulation cross sections in the investigated thermal energy region. The cross sections for total quenching of Na(9s) in collisions with the other two perturbers are comparatively much smaller (Fig. 6) because only one set of avoided crossing characterises those pairs. In this context it is worth noting that Valiron et al [14], while looking into thermal depopulation of Na(6p) colliding with He/Ne, have also observed two avoided crossings between the initial and final channels in their model calculations. But the similarity between our Na-Ar and the above mentioned colliding system does not go any further: they found that it was only one of the two near crossings that was responsible for depopulating the Na(6p), whereas for the Na-Ar pair both avoided crossings are found to contribute significantly in the process of total quenching of the parent state. In fact, we carried out a few test calculations by artificially switching off one of these interactions between the $9s\Sigma$ and $8p\Sigma$ states, and found that the large cross sections, also measured by Gallaghar and Cooke [15], can not be accounted for in such a situation. Another important aspect revealed by our comparative study is that the depopulation cross sections for the Na-Ne pair are smaller than those for the Na-He at lower velocities, even though the radial coupling between the $9s\Sigma$ and $8p\Sigma$ states is much stronger for the former. It is only at the higher impact velocities that the strength of this coupling is visualised for the Na-Ne colliding system. The reason behind this apparent discrepancy is the following. The probability of total annihilation of any low-Rydberg atom is determined not only by the most important individual interaction but by the total collision dynamics of the interacting pair. More emphasis is, therefore, needed on the individual partial cross sections to understand the intricacies of such collisions.

When we look into the individual processes, both leading to excitation and de-excitation of the initial state, we find that the three pairs differ significantly. In all cases excitations to higher states are found to make much smaller contributions to the process of total depopulation. But as we go towards the heavier noble gases the exoergic transitions gain in magnitude. In this context it may be pointed out that the rotational couplings are also found to be more effective in such transitions. Another similarity between the three pairs is the existence of a propensity rule $\Delta\ell \geq 3$ in the excitation. As the initial channel $9s\Sigma$ does not possess very strong couplings with the $8f\Sigma/\Pi$ states, these transitions proceed via two-step processes. The flux is initially transferred to the $8d\Sigma/\Pi$ states which subsequently pass on flux to the molecular states responsible for the population of Na(8f). In this multi-step process both radial and rotational couplings are found to make important contributions.

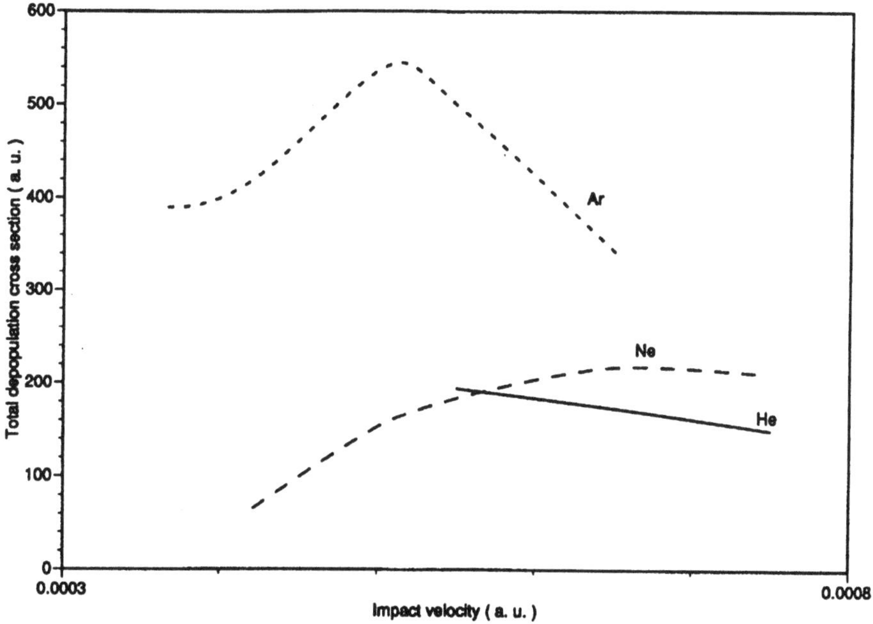

Figure 6. Depopulation cross sections for Na(9s)-He/Ne/Ar-curves for different pairs are shown separately.

But in the process of de-excitation the two-step transition is limited only for the Na-He pair, at least in the investigated energy region. It is the strong radial coupling between the $8p\Sigma$ and $7f\Sigma$ states around R = 25 a_0 that stimulates this transition. In other two pairs this coupling is not that strong. As the process of de-excitation dominates over that of excitation for all pairs, one can realise the importance of this coupling in total depopulation. It is due to the stronger radial coupling between the $8p\Sigma$ and $7f\Sigma$ states that the total depopulation cross sections are found to be larger in Na-He at lower velocities. In other words, this coupling not only changes the individual transitions responsible for depopulating the Rydberg atom, but also significantly affects the overall collision dynamics of the interacting pair. As noted earlier, the rotational couplings are not that important in these downward transition. The Na-Ar pair does exhibit a comparatively stronger radial coupling between the $8p\Sigma$ and $7f\Sigma$ states around R = 40 a_0, but it seems to be less effective in the investigated low energy region. In short, the propensity rule $\Delta\ell \geq 3$ holds only for the Na-He pair in the de-excitation process. But the partial cross sections for the population of Na(7f) are quite significant even for these two pairs. It is expected that if we go further high in the Rydberg series the above propensity rule would be recovered for these systems as well, first for the Na-Ar pair and subsequently for Na-Ne case. This also fits well with the observations made Matsuzawa and co-workers [16, 17] for higher Rydberg atoms.

Since we have calculated our cross sections in a very limited energy region it is perhaps not advisable to estimate the depopulation rates from them. As a result any direct comparison with the reaction rate measurements of Gallaghar and Cooke [15] may not be possible. We can, however, compare our total depopulation cross sections at the mean Maxwellian velocity corresponding to 425K of cell temperature for the three pairs; this will give us some confidence about their relative accuracy. Our cross section at this velocity (3.6×10^{-4} a.u.) for the Na-Ar pair is found to be 2.6 times as large as that for the Na-He pair. This is in good agreement with the observation of Ref. [15] where the former is about 2.58 times larger than the later. Also the relative magnitudes of these cross sections at the mean Maxwellian velocity are found to obey the following inequality:

$$\sigma_{Ne} < \sigma_{He} < \sigma_{Ar} \tag{11}$$

This agrees well with the earlier observations of Hugon et al [18] made for the thermal-energy quenching of Rb(ns) colliding with rare gases.

4.   CONCLUSIONS

We have employed a semi-classical approximation in the MO formulation to look into the total annihilation of low-Rydberg alkalis colliding with noble gases at thermal energies. A complete three body interaction is needed to study such collisions as any simplifying approach in terms of the asymptotic energy defects are bound to meet with little success. With change in the nature of the perturber the entire collision dynamics changes significantly which not only affects the individual transitions but also the probability of total depopulation of the low-Rydberg alkalis. These colliding systems are characterised by a propensity rule $\Delta\ell \geq 3$ in the excitation process; the same propensity rule is expected to hold good in de-excitation, as well, for all pairs if we move slightly up in the Rydberg series. It may also be pointed out that our calculated cross sections, specially the partial one, are sensitive to any change in the shape of the adiabatic potential energy surfaces and the resulting couplings. But we do believe that the present study of ours provides a correct picture of the interactions we have looked into.

ACKNOWLEDGMENTS

The author is grateful to Prof. Neal F. Lane and Dr. B. C. Saha for many helpful discussions and suggestions. This work was partly done at the Physics Department, Florida A. & M. University, Tallahassee, Florida, U.S.A.; the hospitality provided by the department is also thankfully acknowledged.

REFERENCES

1. E. Fermi, Nuovo Cimento **11**, 157 (1934).

2. M. Kimura and N. F. Lane, Case St. At. Mol. Opt. Phys. **26**, 79 (1989).

3. A. Kumar, N. F. Lane and M. Kimura, Phys. Rev. A **39**, 1020 (1989).

4. J. N. Bardsley, Case St. At. Phys. **4**, 299 (1974).

5. J. Pascale, Phys. Rev. A **28**, 632 (1983).

6. E. Czuchaj, J. Sienkiewicz and W. Miklaszewski, Chem. Phys. **116**, 69 (1987).

7. J. B. Delose, Rev. Mod. Phys. **53**, 287 (1981).

8. A. A. Khan, K. K. Prasad, S. K. Verma, V. Kumar and A. Kumar, Pramana – J. Phys. **46**, 373 (1996).

9. S. K. Verma, V. Kumar, A. A. Khan and A. Kumar, J. Phys. B **29**, 1237 (1996).

10. B. C. Saha, Phys. Rev. A **56**, 2909 (1997).

11. B. C. Saha and A. Kumar, J. Phys. B (To be published).

12. B. C. Saha and A. Kumar, J. Phys. B (Communicated).

13. B. C. Saha and A. Kumar, Phys. Rev. A (Communicated).

14. P. Valiron, A. L. Roche, F. Masnou-Seeuws and M. E. Dolan, J. Phys. B **17**, 2803 (1984).

15. T. F. Gallaghar and W. E. Cooke, Phys. Rev. A **19**, 2161 (1979).

16. Y. Sato and M. Matsuzawa, Phys. Rev. A **31**, 1366 (1985).

17. T. Yoshizawa and M. Matsuzawa, J. Phys. Soc. Jpn. **54**, 918 (1985).

18. M. Hugon, B. Sayer, P. R. Fournier and F. Gounand, J. Phys. B. **15**, 2391 (1982).

Chemical Binding and Electron Correlation Effect Studied by Inelastic X-Ray and High Energy Electron Spectroscopy

A. N. Tripathi

Department of Physics
University of Roorkee, Roorkee - 247 667, India.

The recent technical advances in high energy electron and the X-ray scattering have made it possible to measure intensities with high accuracy. In particular, the prospect of obtaining a very reliable separate measurement of inelastic X-ray scattering intensities using the recent developments of synchrotron radiation sources, is very exciting. Here we present an overview of the theoretical developments of electron correlation effect on scattering intensities. Examples of experimental and theoretical intensity for a few molecular systems are given and compared. Based on these intensities, electron pair distribution, the exchange and correlation holes are briefly discussed.

1.   INTRODUCTION

The study of electron-electron correlation has been one of the central issue in theoretical solid state physics and molecular science. Considerable efforts have been made in recent years to deal with this problem. However, experimental observables that reflect the correlation effects to assess such theoretical calculations are scarce. It has long been pointed out that generalised oscillator strength (GOS) is one such quantity which can be used to calculate the x-ray inelastic scattering factor $S(\mathbf{k})$ which is related to two-electron density matrix and is very sensitive to electron correlation effects [1].

The measured quantity GOS is of fundamental importance in understanding many physical properties of substance like optical, dielectric

Trends in Atomic and Molecular Physics,
Edited by Sud and Upadhyaya. Kluwer Academic/Plenum Publishers, New York, 2000.

174

as well as interactions of charged particles with matter [2]. The GOS, $df(\mathbf{k}, E)/dE$ is determined from inelastic scattering experiments where momentum k and energy E are transferred. It is given in terms of dynamic structure factor $S(\mathbf{k}, E)$, as

$$\frac{df(\mathbf{k}, E)}{dE} = \frac{E}{R(k a_0)^2} S(\mathbf{k}, E) \tag{1}$$

$$\text{with } S(\mathbf{k}, E) = \sum_f |\langle f | \sum_{j=1}^{N} \exp(i\mathbf{k} \cdot \mathbf{r}_j) | i \rangle|^2 \delta(E - E_{fi}) \tag{2}$$

Here $| i \rangle$ and $| f \rangle$ are initial and final states, a_0 Bohr radius, r_j the coordinate of the jth electron and R is Rydberg constant. E_{fi} is the energy difference between $| i \rangle$ and $| f \rangle$ states. It is a normalised quantity and can be made absolute by applying the Bethe sum rule:

$$\int_0^\infty \frac{df(\mathbf{k}, E)}{dE} dE = N \tag{3}$$

where N is the total number of electrons in the target.

Electron energy loss spectroscopy (*EELS*) [3-7] has been quite commonly employed to study GOS for gas phase atoms and molecules. Several other properties of matters derived from GOS, such as stopping powers, dielectric response functions, polarizability etc. are, however, more important for aggregates rather than for low pressure gases. Inelastic x-ray scattering spectroscopy (*IXSS*) [8-10] is an alternative to obtain GOS. The emergence of this technique has proved more useful to extract various properties of matters in condensed phases as mentioned above. The double differential cross section of the inelastic X-ray scattering $d^2\sigma/d\Omega\, dE$ is also expressed in terms of $S(\mathbf{k}, E)$.

The static structure factor $S(k)$ is related by

$$S(k) = \int_0^\infty \langle S(\mathbf{k}, E) \rangle_\Omega dE \tag{4}$$

The *IXSS* technique is free from several experimental difficulties with which *EELS* suffers: multiple scatterings, a need of vacuum and charge up phenomena. In past its study has been restricted because of weak scattering intensities. In fact, it has been almost impossible to measure the inelastic X-ray scattering intensities separately from the elastic part because of the small cross sections of the inelastic part and hence the trend has been to study the total intensities [11-13]. Recent developments of insertion devices at Synchrotron radiation (SR) facilities have provided a possibility to extract the inelastic scatterings from the total scattering. Schulke et al [14] measured the inelastic X-ray scattering spectra using single crystal Si with 1.6 eV resolution and obtained $S(\mathbf{k}, E)$ using sum-rules for many simple solids and analysed the results in terms of the homogenous electron gas models. Watanabe et al [8] have analysed the inelastic spectra of liquid water over a wide range of momentum transfer with 2.0 eV resolution and derived Bethe surface, dielectric response function, and static structure factor. For gaseous water, the GOS, has also been obtained from $S(\mathbf{k}, E)$ using *EELS* method in past and it agrees well in absolute scale with that of the measured spectra of liquid water by using *IXSS*. Recently they have extended their measurements of inelastic scattering spectra and $S(k)$'s for several organic compounds [10] and obtained good agreement with their calculations employing configuration interaction wave functions.

These measurements of both elastic and inelastic intensities separately permit a more direct analysis of Coulombic interactions in the molecules to be studied. In this way they become an effective measure of the quality of wave functions to be chosen. Due to the development of high-speed computers, *ab initio* calculations for such cross sections have become quite popular. We refer the reader to various earlier reviews [1, 15, 16] and papers which cover many theoretical investigations of this type. The elastic intensity can be related to the Fourier transform of the diagonal one-electron density matrix, which is rather insensitive to electron correlation but very sensitive to electron bond formation effects in molecules. The inelastic scattering, on the other hand, is related to two-electron density matrix, are sensitive to electron correlation. For the sake of clarity a few words on chemical binding and electron correlation effects are in order. During the formation of a molecule, electron distribution of the constituent atoms will be distorted and this change will be reflected in the scattering intensities. A common practice is to define the changes between the Hartree-Fock (HF) treatment and independent atom model (IAM) as the chemical binding effects. Similarly the electron correlation refers to everything left out of a Hartree-Fock calculations and thus one estimates its effect through the difference between the scattered intensities from correlated wave functions and from the self consistent wave function (SCF) calculations assuming that the correlated wave function yield more accurate results. In recent past Wang et al [17] studied chemical binding and electron correlation effects of several ten-electron molecules on high-energy electron and X-ray scattering intensities based on configurations singles and doubles (CISD). Meyer et al [18, 19], Hoffmeyer et al [20] have also reported similar cross sections using highly correlated MR (multi reference) – CI wave functions for a number of

176

small and medium sized molecules that are expected to be most accurate ones obtained so far.

2.   THEORETICAL METHODOLOGY

The total (I_t^{xr}) and elastic (I_{el}^{xr}) scattered intensities for x-rays can be expressed in terms of density matrices [15]

$$I_t^{xr}(\mathbf{k})/I_{cl} = N + 2\Big\langle \int \exp[i\mathbf{k}\cdot(\mathbf{r}_1 - \mathbf{r}_2)] \times \Gamma^2(\mathbf{r}_1, \mathbf{r}_2 \,|\, \mathbf{r}_1, \mathbf{r}_2)\, d\mathbf{r}_1\, d\mathbf{r}_2 \Big\rangle \quad (5)$$

$$\text{and } I_{el}^{xr}(\mathbf{k})/I_{cl} = \langle\, |\, F(\mathbf{k})\,|^2 \,\rangle \qquad (6)$$

$$\text{with } F(\mathbf{k}) = \int \exp(i\mathbf{k}\cdot\mathbf{r}_1)\gamma\,(\mathbf{r}_1\,|\,\mathbf{r}_1)\, d\mathbf{r}_1 \qquad (7)$$

The operators appearing in Eqs. (5) and (7) are spin-free operators and of strictly multiplicative nature. $\Gamma^2(\mathbf{r}_1, \mathbf{r}_2\,|\,\mathbf{r}_1', \mathbf{r}_2')$ and $\gamma(\mathbf{r}_1\,|\,\mathbf{r}_1')$ are spin-traced two- and one- particle density matrices with Lowdin normalization $N(N-1)/2$ and N, respectively.

The total intensity I_t^{xr} is composed of elastic and inelastic components. The inelastic scattered intensity (I_{in}^{xr}) is determined by

$$I_{in}^{xr}(\mathbf{k})/I_{cl} = S(\mathbf{k}) = I_t^{xr}(\mathbf{k})/I_{cl} - I_{el}^{xr}(\mathbf{k})/I_{cl} \qquad (8)$$

In the above equations, $\mathbf{k}$ is the momentum transfer and $F(\mathbf{k})$ and $S(\mathbf{k})$ are called the coherent and incoherent scattering factors, respectively. I_{cl} is the Thompson X-ray scattering constant.

In an analogous manner, we can also define the total (I_t^{ed}) and elastic (I_{el}^{ed}) electron scattered intensities within the framework of the first Born approximation and the expressions are given elsewhere [17].

Furthermore, $\Gamma^2(\mathbf{r}_1, \mathbf{r}_2\,|\,\mathbf{r}_1', \mathbf{r}_2')$ is a function of six variables and this dimensionality can be reduced without losing its two-electron character by introducing interelectronic (intracule) coordinate $\mathbf{r}_{12} = \mathbf{r}_1 - \mathbf{r}_2$ and centre-of-mass (extracule) coordinate $\mathbf{R} = \frac{1}{2}(\mathbf{r}_1 + \mathbf{r}_2)$. Equation (5) is rewritten in the following way:

$$I_t^{xr}(k)/I_{cl} = N + 2\langle \int h(\mathbf{r}_{12})\exp(i\mathbf{k}\cdot\mathbf{r}_{12})\,d\mathbf{r}_{12}\rangle, \qquad (9)$$

where $h(\mathbf{r}_{12})$ is related to the electron-electron distribution function $p(\mathbf{r}_{12})$ by

$$p(r_{12}) = 4\pi\, r_{12}^2\, h(r_{12}) \qquad (10)$$

Combining Eqs.(9) and (10), one obtains

$$p(r_{12}) = \frac{r_{12}^2}{\pi}\int_0^\infty [I_t^{xr}(\mathbf{k})/I_{cl} - N]\,j_0(\mathbf{k}r_{12})\,k^2\,dk \qquad (11)$$

It is easy to show that the electron pair density has the following properties:

$$\int_0^\infty p(r_{12})\,dr_{12} = \frac{N(N-1)}{2} \qquad (12)$$

and $\displaystyle\int_0^\infty p(r_{12})/r_{12}\,dr_{12} = V_{ee} = \sum_{i\leq j}\langle\frac{1}{\mathbf{r}_{ij}}\rangle$ $\qquad (13)$

where V_{ee} is the electron-electron repulsion energy.

It has been shown by Tavard [21] and Bonham [22] that the total molecular potential energy $\overline{U}$ is connected with the total scattered intensities for electrons

$$\overline{U} = \frac{1}{\pi}\int_0^\infty [\,\mu^4 I_t^{ed}(k)/I_e - \sum_A (Z_A^2 + Z_A)\,]\,dk \qquad (14)$$

This elegant relation enables the potential energy and total energy (by invoking the virial theorem) of a molecular system to be obtained through an electron scattering experiment.

Now consider the true ground state wave function written as a linear combination of Slater determinants in terms of an orthogonal spin orbital set $\{\psi\}$,

$$\Phi = \sum_p C_p \det|\psi_1\dots\psi_N|_p \qquad (15)$$

If the orbital set is complete, the use of all possible determinants in the above wave function, termed complete CI, yields the exact eigenvalues and eigenfunctions. However, in practice a number of schemes are available to truncate the expansion to a finite number of terms. The most common one considers the Hartree-Fock determinant and constructs other configuration state functions (CSFs) by successively substituting one, two, ..., electrons from the occupied Hartree-Fock molecular orbitals (MOs) to the unoccupied MOs . Since only singly and doubly substituted CSFs can interact with the zeroth order wave function, it is natural to truncate the expansion at this level as a first approximation. We thus obtain the single and double substitution wave function from HF (denoted as SDCI).

The spin-traced two electron density matrix occurring in Eq. (5) is given as

$$\Gamma^2(\mathbf{r}_1, \mathbf{r}_2 \mid \mathbf{r}_1', \mathbf{r}_2') = \frac{N(N-1)}{2} \int \Phi^*(\mathbf{r}_1', \mathbf{r}_2', \mathbf{r}_3, \ldots, \mathbf{r}_N) \times \Phi(\mathbf{r}_1, \mathbf{r}_2, \mathbf{r}_3, \ldots, \mathbf{r}_N) d\mathbf{r}_3 \ldots d\mathbf{r}_N \quad (16)$$

Inserting the wave function Φ from Eq.(15) into Eq.(16), we obtain the following:

$$\Gamma^2(\mathbf{r}_1, \mathbf{r}_2 \mid \mathbf{r}_1', \mathbf{r}_2') = \sum_{ijkl}^{N} \lambda_{ijkl}\, \psi_i(\mathbf{r}_1)\psi_j^*(\mathbf{r}_1')\psi_k(\mathbf{r}_2)\psi_l^*(\mathbf{r}_2') \quad (17)$$

and similarly, the spin-traced one electron density matrix becomes

$$\gamma(\mathbf{r}_1 \mid \mathbf{r}_2') = \sum_{ij}^{N} \lambda_{ij}\, \psi_i(\mathbf{r}_1)\psi_j^*(\mathbf{r}_1'), \quad (18)$$

where λ_{ij} and λ_{ijkl} are the elements of the spin traced one- and two-electron density matrices.

In the standard practice, the scattered intensities of a molecule is compared with that of its IAM counterpart. In an IAM model [1], the scattering factors have been given as follows:

$$F(\mathbf{k}) = \sum_{A} f_A(k)\exp(i\mathbf{k}\cdot\mathbf{R}_A) \quad (19)$$

and

$$S(\mathbf{k}) = \sum_{A} s_A(k) \quad (20)$$

where $f_A(\mathbf{k})$ and $s_A(\mathbf{k})$ are atomic form factors and incoherent factors discussed earlier and tabulated at the Hartree-Fock level [23].

The molecular wave functions at a self-consistent-field (SCF) approximation to the Hartree-Fock wave function, based on experimental geometries were constructed using the HONDO [24] program. Subsequently, the same SCF wave functions were further extended to obtain the variational CI wave function with the same package. In the CI calculation, all the singly and doubly substituted configurations generated within the core and valance orbitals from the single determinant reference have been included. In order to assess the basis set dependence, three types of Gaussian type basis sets were used in the present work, namely, the double zeta valence (DZV), double zeta plus the polarization (DZP), and double zeta augmented with both the polarization and the diffuse functions (DZP++) basis sets. The choice of basis set was guided by the fact that it should yield results which are comparable with larger basis sets while remaining computationally manageable. Spherical averages of the scattered quantities were carried out analytically. This not only avoids the "extremely time-consuming" numerical algorithm, but also ensure the accuracy of the calculated quantities. The related formalism may be found elsewhere [25, 10].

3. RESULTS AND DISCUSSION

In this section the theory outlined above is applied to the description of GOS, S(k) for a few selective molecular systems as obtained by us and other workers and compared with recent experimental measurements.

3.1 Observed S(k, E) and S(k)

Watanabe et al [8-10] have reported a detailed measurements of S(k, E) for liquid water and many organic compounds using SR sources. These S(k, E) are then transformed to GOS with the help of Equations. (1-3) and a Bethe surface for different values of momentum and energy transferred were constructed. Subsequently from these, number of properties were extracted.

Figure (1) shows the comparison of liquid phase GOS of water as measured by Watanabe et al [8] with the gas phase GOS of Lassettre and White [3] at several energies. It is evident from the figure that the two measurements agree quite well in absolute scale with each other. It is rather surprising, considering that they are derived from two completely different

180

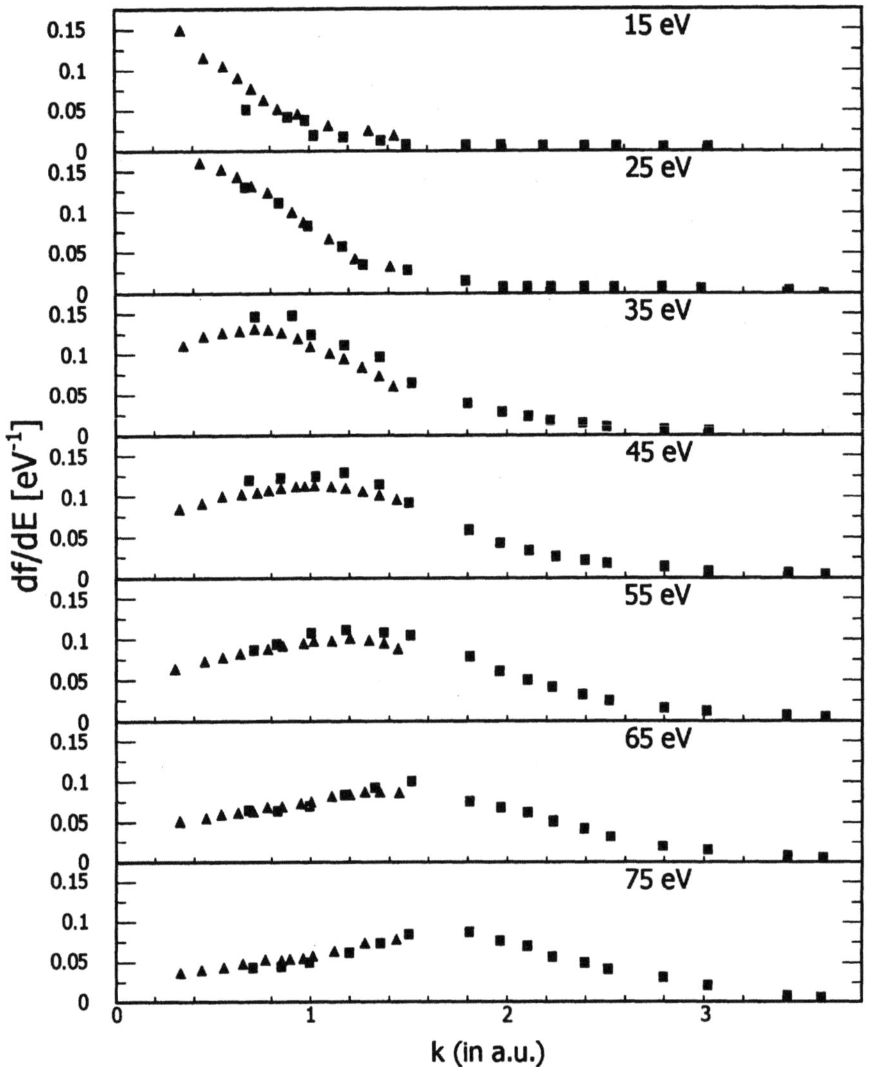

Figure 1. Experimental GOS (df/dE in eV^{-1}) of water as a function of momentum transfer (k in a.u.) for different energy losses: ■ - Watanabe et al [8], ▲ - Lassetre and White [3].

experimental techniques. Slight discrepancies are, however, observed for energy losses of 35 and 45 eV. Watanabe et al's results are systematically higher by 10-20%. Thus it clearly shows that the GOS of liquid water in k-E range studied is determined by individual molecular properties and is little affected by intermolecular interactions. Further, inelastic X-ray scattering spectra for a number of organic compounds are also reported by these authors (not shown here). At small **k**, these spectra are far from symmetric parabolas which are characteristic for a Compton profile. Thus it clearly shows the breakdown of impulse approximation at small **k** which has been always central to the analysis of Compton scattering [26]. With increase in **k**, however, the peak energies drift to the higher energy loss and shape of spectra gets closer to symmetric, that is, more Compton profile like as can be seen from figure 2 for C_6H_6 molecule [10].

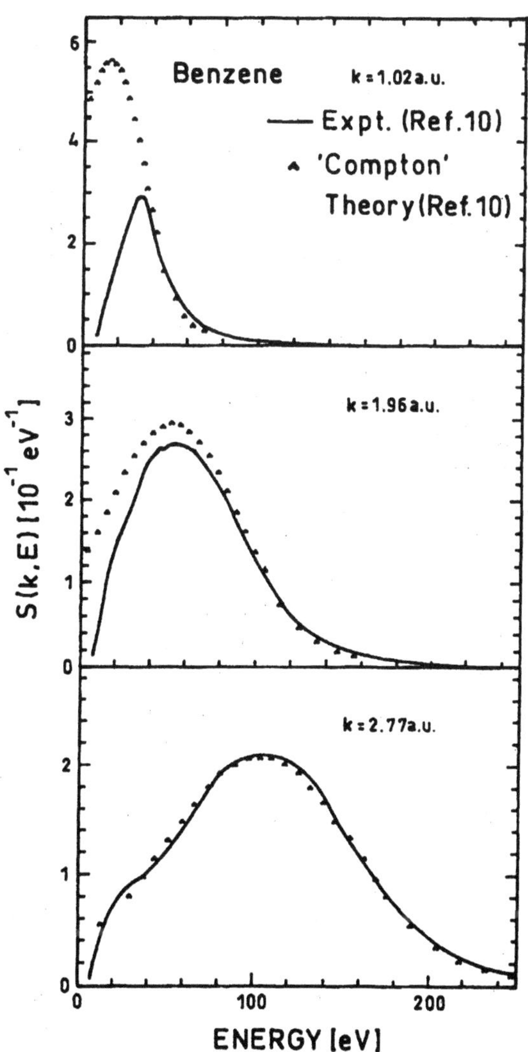

Figure 2. Dynamic structure factor S(**k**, E) of Benzene as a function of energy loss for three different momentum transfers (**k**). —— experimental results [10], ▲ Compton theory [10].

3.2   Chemical binding and electron correlation effects

The effects of electron correlation and chemical binding have been examined in the conventional manner by constructing the difference between the various intensities (from both the SCF and SDCI treatments) with DZP++ basis set and those of IAM. The IAM intensities were calculated using the coherent and incoherent scattering factors and Eqs. (19) and (20). These calculated difference functions for the inelastic X-ray scattered intensities have been displayed in Figs. (3) and (4) for water, methanol and benzene molecules respectively. The results are compared with the measured values of S(k) of Watanabe et al and Shibata et al [27]. It is seen that, while elastic scattering intensities do not differ very much between HF and CI levels (see ref. 17), inelastic scattering intensities are very sensitive to the inclusion of CI. Figure 3 shows, our calculations for water. It clearly shows very good agreement with the experimentally observed S(k). The overall shape and the position of extrema of the calculated curve at CI level almost coincide with the experimental curve. Although Watanabe et al's experimental values are a little deeper in figure but it may be considered to be within experimental errors. Shibata's et al measured the momentum transfer dependence of the elastic and total differential cross-section (TDC) for high-energy electron scattering over a wide range of momentum transfer. They found that the elastic scattering cross-sections by SCF-MO calculations were in good agreement with their measurements. On the other hand, the calculated values for the total cross-section by IAM deviate significantly from the experimental ones. Later on Takeuchi et al [13] calculated total (elastic + inelastic) differential cross-sections at SCF-CI levels and noted that the calculated values are in good agreement with the experimental once. The experimental S(k) as shown here is extracted from Shibata's TDC measurements by subtracting the elastic components. It is seen that the two experimental values obtained using two different techniques compare well among themselves, but Watanabe et al's separate measurements for S(k) are closer to the calculated values.

Fig. (4) shows similar calculations of S(k)'s at CISD level using various basis sets for methanol (CH_3OH) and benzene (C_6H_6) molecules as obtained by Watanabe et al. For methanol, the calculations were done both at DZP and TZP+++ but due to limitation in computing time, the calculations for benzene were obtained only with DZP basis set. It is again observed that the S(k) from TZP+++ for methanol is closer to the experimental results but the differences are quite large for benzene.

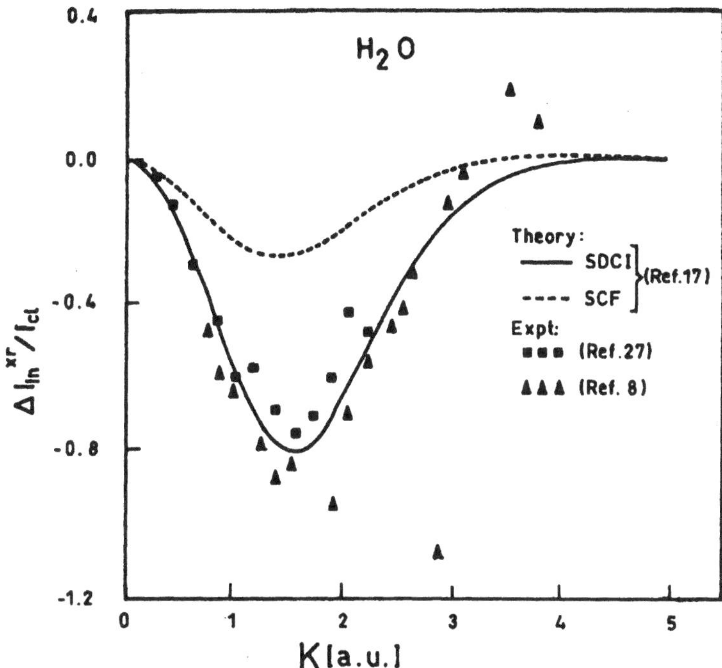

Figure 3. The difference function $\Delta I_{in}^{xr}/I_{el}$ as a function of momentum transfer (in a.u.) of water. – – – SCF treatment [17], ——— SDCI [17], ■ Shibata et al [27], ▲ Watanabe et al [8].

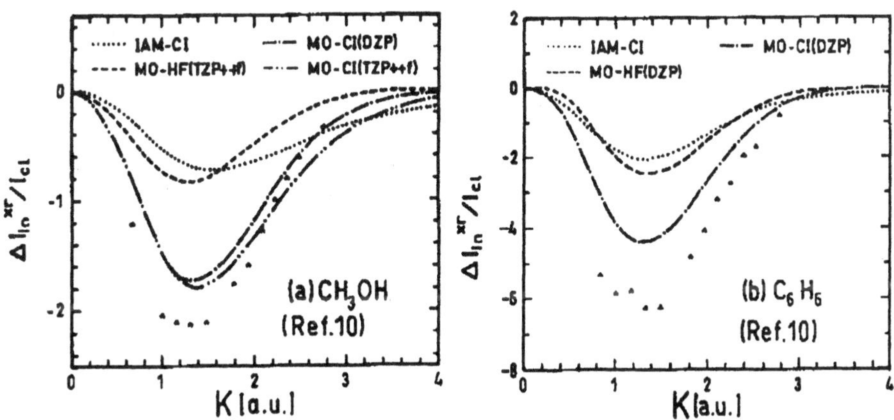

Figure 4. The difference function $\Delta I_{in}^{xr}/I_{el}$ as a function of momentum transfer (in a.u.) for (a) CH$_3$OH and (b) C$_6$H$_6$ (Watanabe et al [10]|). ···· IAM-CI, -•- MO-CI (DZP), – – – MO-HF (TZP++f) for (a) and – – – MO-HF (DZP) for (b).

184

In recent years systematic CI calculations of high-energy electron and X-ray scattering intensities by variety of molecules have been reported. The results suggest that almost 90% of the correlations effects are recovered by SDCI calculations. The effects of higher substitution can be recovered satisfactorily by MR-SDCI computations, possibly with perturbative corrections. Agreement between the theoretical S(k)'s and experimental ones from X-ray inelastic scattering observed demonstrate that the calculations at the CI level and its variants can predict S(k) fairly well even for larger molecules. However, there still remain unresolved discrepancies between measured and calculated values of intensities for high-energy electron scattering by Meyer et al [19] on N_2. They have analysed the possible sources of the discrepancies between calculated and measured intensities for electron scattering. They concluded that the discrepancies are due to either experimental errors or to the neglect of higher order Born corrections.

For molecular systems, the relationship given by Eq. (11) offers an alternative way to analyze chemical binding and electron correlation, i.e. from electron distributions rather than scattered intensities. If the electron-electron pair distribution function $p(r_{12})$ [or $h(r_{12})$] is calculated with HF and CI wave functions, the changes will provide an effective measure of the Coulomb correlation.

$$\Delta P(r_{12}) = p_{CI}(r_{12}) - p_{HF}(r_{12}) \tag{21}$$

Similar to this, the effect of chemical bonding can also be investigated by

$$\Delta P(r_{12}) = p_{HF}(r_{12}) - p_{IAM}(r_{12}) \tag{22}$$

where $p_{IAM}(r_{12})$ is the probability density for r_{12} computed using the quasi-exact atomic HF wave function (for the IAM molecule).

Figure 5 displays the electron correlation effect on the radial intracule distributions of the ten-electron non-linear molecule (CH_4, NH_3 and H_2O) and linear molecule HF. Each curve shows a pronounced minimum and maximum. The locations of the extrema shift to the left as the nuclear charge of the heavier atom increases. The curves come across the zero $\Delta P(r_{12})$ line and the corresponding r_{12} values can be defined as the Coulomb correlation radii. The correlation radius decreases from CH_4 to HF (Fig. 4). This observation is consistent with the earlier studies carried out in atomic systems by Boyd [28]. Boyd has shown that as the nuclear charge of the atom increases, the charge density tends to be more compact and the correlation radius will shift to smaller r_{12} regions. Also, it is clear from Fig.

4 that electron correlation has a far larger effect in HF than in CH_4 in spite of the small correlation radius in the former system.

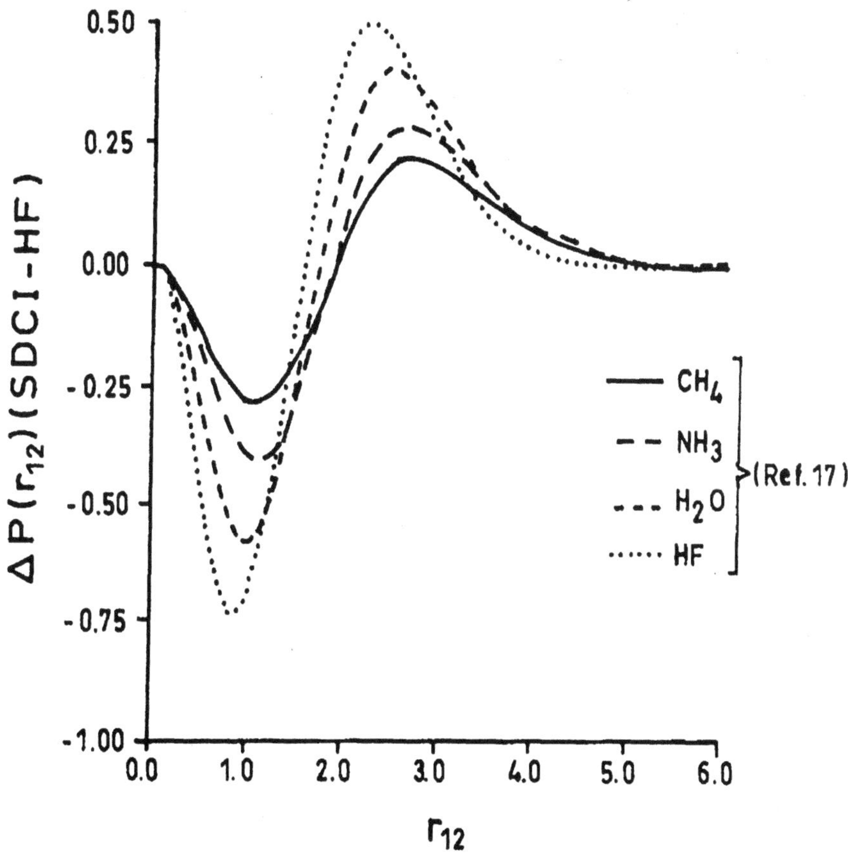

Figure 5. Electron corrrelation effect ($\Delta P(r_{12})$) on radial intracule distribution of the ten-electron molecules (CH_4, NH_3, H_2O and HF).

In Eq. (11) above, we have outlined the relationship between the total X-ray scattered intensity and the radial intracule distributions. By analogy, the classical radial electron pair distribution [29, 30] is defined as the Fourier transformation of the elastic scattered intensities

$$p_c(r_{12}) = \frac{r_{12}}{\pi} \int_0^\infty I_{el}^{xr}/I_{cl} \sin(r_{12}\, k)\, k\, dk \qquad (23)$$

Note that the thus defined electron pair distribution function is normalized to $N^2/2$, i.e.,

$$\int_0^\infty p_c(r_{12})\, dr_{12} = N^2/2 \tag{24}$$

After introduction of the classical electron pair density, one can conveniently study statistical electron correlation. Thakkar et al [30] defined a function $c(r_{12})$, which is given

$$c(r_{12}) = \frac{2}{N(N-1)} p(r_{12}) - \frac{2}{N^2} p_c(r_{12}). \tag{25}$$

Breteinstein et al [29] also introduced a one-dimensional pair correlation function

$$g(r_{12}) = p(r_{12})/p_c(r_{12}) \tag{26}$$

as a measure of electron correlation.

Since this $g(r_{12})$ displays the exchange and correlation holes for a given system, studying this function may provide valuable information on the design of exchange and correlation functionals in molecular systems. As the functional form of $g(r_{12})$ is in general experimentally accessible through scatterings (for details see Ref. 17), more study on this function is highly desirable.

From the results presented here it is clear that the incoherent intensities are very sensitive to electron correlation effects and can be used to assess the quality of wave functions. The results of S(k) and GOS derived using two different techniques EELS and IXSS compare well among themselves. Therefore, indicates that the IXSS can serve as an effective tool for understanding the electron correlation effect in matter. Further, the knowledge of various exchange and correlation function can also find possible applications to the density functional theory.

ACKNOWLEDGMENTS

The work has been supported by the research in Nuclear Sciences, Department of Atomic Energy, Govt. of India. I also thank Prof. V. H. Smith Jr., Prof. A. J. Thakkar and Dr. J. Wang for their collaboration and help. Author also thanks Dr. Watanabe for giving permission to use his figures.

REFERENCES

1. R. A. Bonham and M. Fink, *High energy electron scattering* (Van Nostrand-Reinhold) New York, (1974).

2. M. Inokuti, Y. Itikawa and J. E. Turner, Rev. Mod. Phys. **50**, 23 (1978); *ibid*, **43**, 297 (1971).

3. E. N. Lassettre and E. R. White, J. Chem. Phys. **60**, 2464 (1974).

4. T. C. Wong, J. S. Lee, H. F. Wellenstein and R. A. Bonham, Phys. Rev. A **12**, 1846 (1975); see also J. Chem. Phys. **60**, 103 (1974).

5. A. Lahman-Bennani, A. Duget and H. F. Wellenstein and M. Roualt, J. Chem. Phys.**72**, 6398 (1980); see also Chem. Phys. Lett. **60**, 407 (1979).

6. R. S. Barbieri and R. A. Bonham, Phys. Rev. A **45**, 7929 (1992).

7. J. F. Ying and K. T. Leung, Phys. Rev.A **53**, 1476 (1996); see also J. Chem. Phys.**100**, 7120 (1994); **101**, 8333 (1994).

8. N. Watanabe, N. Hayashi and Y. Udagawa, Bull Chem. Soc. Jpn. **70**, 719 (1997).

9. H. Hayashi, N. Watanabe, Y. Udagawa and C. C. Kao, J. Chem. Phys. **108**, 823 (1998).

10. N. Watanabe, H. Hayashi, Y. Udagawa, S. Ten-no and S. Iwata, J. Chem. Phys. **108**, 4545 (1998).

11. G. E. Ice, M. H. Chen and B. Crasemann, Phys. Rev. A **17**, 650 (1978).

12. N. Nishikawa and T. Iijima, J. Chem. Phys. **87**, 3753 (1987); see also Chem. Phys. Lett. **109**, 195 (1984).

13. H. Takeuchi, M. Nakagawa, T. Saito, T. Egawa, K. Tanaka, S. Konaka and T. Mitsuhashi, Int. J. Quantum Chem. **52** ,1339 (1994).

14. W. Schulke, H. Schulte-Schrepping and J. R.Schmitz, Phys. Rev. B **47**, 12462 (1993).

15. R. Benesch and V. H. Smith Jr., Acta Crystallogr. Sect. A **26**, 579 (1970).

16. A. N. Tripathi and V. H. Smith Jr., in comparison of Ab initio Quantum Chemistry with experiment: State-of-the-art, R.Bartlett, Ed. (Reidel Dordrecht) pp.439-462 (1985).

17. J. Wang, A. N. Tripathi and V. H. Smith Jr., J. Chem. Phys. **101**, 4842 (1994).

18. H. Meyer, T. Muller and A. Schweig, Chem. Phys. Lett. **236**, 497 (1995).

19. H. Meyer, T. Muller and A. Schweig, Chem. Phys.**191**, 213 (1995); see also Mol. Phys. **88**, 1563 (1996).

20. R. E. Hoffmeyer, P. Bundgen and A. J. Thakkar, J. Phys. B: At. Mol. Opt. Phys (in press) (1998).

21. C-Taward, Cah. Phys. **20**, 397 (1965).

22. R. A. Bonham, J. Phys. Chem. **71**, 856 (1967).

23. J. Wang, R. P. Sagar, H. Schmider and V. H. Smith Jr., At. Data Nucl. Data Tables **53**, 233 (1993).

24. M. Dupuis, A. Farazdel, S. P. Karna and S. A. Maluendes in *MOTECC: Modern Techniques in Computational Chemistry* edited by E. Clementi (ESCOM, Leiden), p. 277 (1990).

25. J. Wang and V. H. Smith Jr., Int. J. Quantum Chem. **52**, 1145 (1994).

26. A. N. Tripathi, V. H. Smith Jr., P. Kaijser and G. H. F. Diercksen, Phys. Rev. A **41**, 2468 (1990); see also Phys. Rev. A **55**,4074 (1987).

27. S. Shibata, F. Hirota, W. Kakuta and T. Muramatsu, Int. J. Quantum Chem. **18**, 281 (1980).

28. R. J. Boyd, Can. J. Phys. **53**, 592 (1975).

29. M. Breitenstein, H. Meyer and A. Schweig, Chem. Phys.**124**, 47 (1988).

30. A. J. Thakkar, A. N. Tripathi and V. H. Smith Jr., Phys. Rev.A **29**, 1108 (1984).

Quantum Optical Resonance

S. Mathur and U. N. Upadhyaya

Department of Physics
College of Science, M. L. Sukhadia University, Udaipur - 313 001, India

Jaynes-Cummings model of quantum optical resonance is investigated using Tomonaga-Schwinger formalism and without using the rotating wave approximation. Detailed study of the fluctuations in the initial number state and coherent state is made as a function of the interaction strength, frequency detuning and the initial conditions. It is found that the counter rotating terms play an important role in the evolution of the quantum state and the entanglement between the atom and the field.

1. INTRODUCTION

The field of Quantum optical resonance has been a growing and intensely active area of research. The Jaynes-Cummings model is the simplest but important theoretical model of quantum optical resonance and it is the basic building block for much of quantum optics. It consists of a single mode of quantized electromagnetic field interacting with a single two level atom. The interaction Hamiltonian between an atom and a classical field is given in the dipole approximation by [1],

$$H_{\text{int}} = -e\left(\underline{r} \cdot \underline{E}\right) \tag{1}$$

where $\underline{E}$ is the electric field and $-e\,\underline{r}$ is the atomic dipole moment operator. For a single mode field the interaction Hamiltonian becomes [1],

$$H_{\text{int}} = \eta g\left(b + b^{+}\right)\left(S_{+} + S_{-}\right) \tag{2}$$

Trends in Atomic and Molecular Physics,
Edited by Sud and Upadhyaya. Kluwer Academic/Plenum Publishers, New York, 2000.

190

where S_+ and S_- are the "spin-flip" matrices,

$$S_+ = |e\rangle\langle g|, \quad S_- = |g\rangle\langle e|, \tag{3}$$

and g is the electric dipole matrix element. g can be assumed to be real. b, b^+ are the annihilation and creation operators, respectively of the quanta of the electromagnetic field of frequency Ω. $|e\rangle$ and $|g\rangle$ are the excited and ground state of the two level atom.

The total atom-field Hamiltonian is [1],

$$H = \eta\omega S_z + \eta\Omega b^+ b + \eta g\left(b + b^+\right)\left(S_+ + S_-\right) \tag{4}$$

where ω is the atomic transition frequency and S_z is the "spin angular momentum" operator related to Pauli matrix as $S_z = (1/2)\,\sigma_z$.
The S_α operators can be shown to satisfy the commutation rules,

$$\left[S_\pm, S_z\right] = \mu S_\pm, \quad \left[S_+, S_-\right] = 2S_z, \quad S_+ S_- + S_- S_+ = 1,$$

where,

$$S_\pm = S_x \pm i S_y \tag{5}$$

The Hamiltonian (4) has been extensively studied in a wide variety of physics problems. Jaynes and Cummings [2] used the JCM to study the amplitude and frequency stability in a molecular beam maser. Glauber [3] has considered an equivalent Hamiltonian to (4) for the study of coherence and quantum detection. Stenholm [4] has discussed laser systems with the help of (4). Narozhny, Mondragon and Eberly [5], Puri and Agarwal [6] and J. Gea-Banacloche [7] have studied the collapse and revival of the state vector, Barnett and Knight [8] have studied quantum dissipation in the above model.

The JC Model has been extensively studied in the Rotating Wave Approximation (RWA). In RWA, the terms like bS_- and b^+S_+ which have time dependence of $\exp[\mu\, i\,(\omega + \Omega)\,t]$ are dropped because they tend to zero in the course of a few optical periods but bS_+ and b^+S_- vary slowly near resonance and hence are retained in RWA [1]. RWA, valid only near resonance condition $\Omega \approx \omega$ is equivalent to decomposing the mode field into two opposite circularly polarized waves and keeping only the one rotating in the sense of spin precession.

One great advantage of the above model is that within the rotating wave approximation (RWA), it is exactly solvable, whether the field is treated classically [7] or quantum mechanically [9].

The Hamiltonian (4) can be rewritten in a more convenient form for the study of resonance and frequency detuning behaviour by regrouping the terms as [4],

$$
\begin{aligned}
H &= \eta\Omega\left(S_z + b^\dagger b\right) + \eta\left(\Delta\Omega\right)S_z + \eta g\left(b + b^\dagger\right)\left(S_+ + S_-\right) \\
&\equiv H_0 + V,
\end{aligned}
\tag{6}
$$

with $H_0 = \eta\Omega\left(S_z + b^\dagger b\right)$, $V = \eta\left(\Delta\Omega\right)S_z + \eta g\left(b + b^\dagger\right)\left(S_+ + S_-\right)$, and $\Delta\Omega = \omega - \Omega$.

2. TIME DEVELOPMENT OPERATOR AND THE DENSITY MATRIX

The density matrix operator for the combined atom - field system follows a unitary time evolution generated by the time development operator $U(t, 0)$ [9],

$$
\rho(t) = U(t, 0)\,\rho(0)\,U^+(t, 0)
\tag{7}
$$

where $\rho(0)$ is the initial density matrix operator. In RWA, $U(t, 0)$ has a simple form [11] as H_0 and V in this approximation commute with each other. In the present work H_0 and V do not commute. In the interaction picture, for time independent Hamiltonian, $U(t, 0)$ can be written as [10],

$$
U(t, 0) = \exp\left(iH_0 t / \eta\right)\exp\left\{-iHt / \eta\right\}
\tag{8}
$$

Differentiating (8) with respect to t one obtains

$$
\frac{dU(t, 0)}{dt} = \frac{iH_0}{\eta}U(t, 0) + \exp\left(\frac{iH_0 t}{\eta}\right)\left(-\frac{iH}{\eta}\right)\exp\left\{-\frac{iH_0 t}{\eta}\right\}U(t, 0)
\tag{9}
$$

In the following analysis we use the Baker - Hausdorff lemma [11]:

$$
\exp(B)\,X\,\exp(-B) = \left\{X + \{B, X\} + \frac{1}{2!}\{B^2, X\} + \Lambda + \frac{1}{n!}\{B^n, X\} + \Lambda\right\}
\tag{10}
$$

192

where $\{B, X\} = [B, X]$ and $\{B^n, X\} = [B, \{B^{n-1}, X\}]$

Then,

$$\frac{dU(t,0)}{dt} = \left[\frac{iH_0}{\eta} - \frac{iH}{\eta} + \left\{ \frac{iH_0 t}{\eta}, \left(\frac{-iH}{\eta} \right) \right\} + \frac{1}{2!} \left\{ \left(\frac{iH_0 t}{\eta} \right)^2, \left(\frac{-iH}{\eta} \right) \right\} + \Lambda \right.$$

$$\left. + \frac{1}{n!} \left\{ \left(\frac{iH_0 t}{\eta} \right)^n, \left(\frac{-iH}{\eta} \right) \right\} + \Lambda \right] U(t,0)$$

$$\equiv G(t) U(t,0)$$

$$(11)$$

This is a general matrix differential equation with the initial condition $U(0, 0) = 1$. The general solution of this type of equation is discussed by Weiss and Maradudin [12]. One can write the solution of equation (11) as

$$U(t, 0) = \exp[\Omega(t)] \tag{12}$$

$\Omega(t)$ is obtained by iterating as a series,

$$\Omega(t) = \int_0^t G(\tau) d\tau + \frac{1}{2} \int_0^t \left[G(\tau), \int_0^\tau G(\sigma) d\sigma \right] d\tau + \Lambda \tag{13}$$

In the present work we investigate the effect of only the first term in the above expansion. Thus the time development operator $U(t, 0)$ can be written as

$$U(t,0) = \exp\left[\frac{iH_0 t}{\eta} - \frac{iHt}{\eta} + \frac{(it)(-it)}{2! \eta^2} \{H_0, H\} + \frac{(it)^2(-it)}{3! \eta^3} (H_0^2, H) + \Lambda \right.$$

$$\left. + \frac{(it)^{n-1}(-it)}{n! \eta^n} \{H_0^{n-1}, H\} + \Lambda \right]$$

$$(14)$$

The commutator $[H_0, H]$ is obtained by using the commutation relations (5) and that for b and b^+. Inspection of commutators $\{H_0^n, H\}$ for various n values reveal that alternate commutators have the same operator structure in view of (5) and forms of H_0 and V. Thus one obtains the following general form of even and odd commutators:

$$\{H_0^{2n}, V\} = 2^{2n-1}\,\eta^{2n+1}\,\Omega^{2n}\,g\,[(b+b^+)(S_+ + S_-) + (b^+ - b)(S_+ - S_-)]$$

$$\{H_0^{2n+1}, V\} = 2^{2n}\,\eta^{2n+2}\,\Omega^{2n+1}\,g\,[(b^+ - b)(S_+ + S_-) + (b+b^+)(S_+ - S_-)] \tag{15}$$

Therefore it is appropriate to group even and odd terms separately in (14) as,

$$
U(t,0) = \exp\left[\frac{-iVt}{\eta} - \left(\sum_{n=0}^{\infty}\frac{1}{(2n+2)!}\left(\frac{it}{\eta}\right)^{2n+2}\right)\{H_0^{2n+1}, V\}\right.
$$
$$
\left. - \left(\sum_{n=0}^{\infty}\frac{1}{(2n+1)!}\left(\frac{it}{\eta}\right)^{2n+1}\right)\{H_0^{2n}, V\}\right]. \tag{16}
$$

Using (6) and (15) in (16) one obtains for U (t, 0) as,

$$
U(t,0) = \exp\left\{-it\,[(\Delta\Omega)S_z + g(b+b^+)(S_+ + S_-)]\right.
$$
$$
+ \frac{g}{4\Omega}\{1-\cos(2\Omega t)\}\,[(b^+ - b)(S_+ + S_-) + (b^+ + b)(S_+ - S_-)]
$$
$$
\left. + ig\left\{\frac{t}{2} - \frac{1}{4\Omega}\sin(2\Omega t)\right\}[(b^+ + b)(S_+ + S_-) + (b^+ - b)(S_+ - S_-)]\right\} \tag{17}
$$

Regrouping the terms in the exponent in (17) as coefficients of S_+, S_- and S_z operators one obtains for $U(t, 0)$ as,

$$
U(t,0) = \exp\left\{-it(\Delta\Omega)\,S_z - igt\left(b + b^+ \exp(i\Omega t)\frac{\sin\Omega t}{\Omega t}\right)S_+\right.
$$
$$
\left. - igt\left(b^+ + b\exp(-i\Omega t)\frac{\sin\Omega t}{\Omega t}\right)S_-\right\} \tag{18}
$$

The matrix representation of S_+, S_- and S_z in (18) gives the following expression for the time development operator $U(t, 0)$

$$\hat{U}(t,0) = \exp\left[-i\begin{pmatrix} \dfrac{(\Delta\Omega)t}{2} & gt(\hat{b}+e^{i\Omega t}\gamma\hat{b}^+) \\ gt(\hat{b}^+ +e^{-i\Omega t}\gamma\hat{b}) & -\dfrac{(\Delta\Omega)t}{2} \end{pmatrix}\right] \tag{19}$$

with $\gamma(t) = \dfrac{\sin\Omega t}{\Omega t}$

In view of the above form of $U(t, 0)$ it is more convenient to define new creation and annihilation operators a and a^+ of the quanta of the electromagnetic field as,

$$a(t)=G\left(b+e^{i\Omega t}\gamma\, b^+\right),\ a^+(t)=G\left(b^+ +e^{-i\Omega t}\gamma\, b\right) \tag{20}$$

with $G = 1/(1-\gamma^2)$
Conversely the operators b and b^+ can be expressed in terms of new operators a and a^+ as,

$$b(t) = G(a - e^{i\Omega t}\gamma\, a^+),\ b^+(t) = G(a^+ - e^{-i\Omega t}\gamma\, a) \tag{21}$$

The new operators a and a^+ obey the commutation relations,

$$[a(t), a^+(t)] = 1 \tag{22}$$

In view of the above commutation relation: a, a^+ are alternative and equivalent sets of annihilation and creation operators to describe the excitation of the electromagnetic field. Later on we study the relation between the eigenkets of $a^+ a$ and $b^+ b$.
Thus one can write for $U(t, 0)$ as,

$$U(t, 0) = \exp\left[-i\begin{pmatrix} z & c\,a(t) \\ c\,a^+(t) & -z \end{pmatrix}\right] \tag{23}$$

with $c = g\,t\,(1-\gamma^2)$ and $z = (\Delta\Omega)\,t/2$.
It is convenient to consider the power series expansion of the above expression. The even and odd powers of the expansion of the exponential operator have a very simple structure. Thus one obtains for $U(t, 0)$ as,

$$U(t, 0) = \begin{pmatrix} \cos\hat{\theta}_1(t) - i\,\dfrac{\sin\hat{\theta}_1(t)}{\hat{\theta}_1(t)}\,z & -i\,\dfrac{\sin\hat{\theta}_1(t)}{\hat{\theta}_1(t)}\,ca \\[2ex] -i\,\dfrac{\sin\hat{\theta}_2(t)}{\hat{\theta}_2(t)}\,ca^+ & \cos\hat{\theta}_2(t) - i\,\dfrac{\sin\hat{\theta}_2(t)}{\hat{\theta}_2(t)}\,z \end{pmatrix} \qquad (24)$$

where we have defined new variables $\theta_1(t)$ and $\theta_2(t)$ as,

$$\hat{\theta}_1(t) = (z^2 + c^2 a a^+)^{1/2}, \ \hat{\theta}_2(t) = (z^2 + c^2 a^+ a)^{1/2} \qquad (25)$$

The time development of the density matrix is given by

$$\rho(t) = U(t, 0)\, \rho(0)\, U^+(t, 0) \qquad (26)$$

where $\rho(0)$ is the initial density matrix operator. If we consider that at $t = 0$, the atom is in the excited state and the atom and the field are uncorrelated, one can write,

$$\rho(0) = \begin{pmatrix} \rho_F(0) & 0 \\ 0 & 0 \end{pmatrix} \qquad (27)$$

In the above expression the reduced density operator for the field is given by [9],

$$\rho_F(t) = Tr_A \{ U(t, 0)\, \rho(0)\, U^+(t, 0) \} \qquad (28)$$

where the trace is taken over the atomic variables.
Using the expressions (24) and (27) in (28) one obtains for $\rho_F(t)$ as,

$$\rho_F(t) = [C\, \rho_F(0)\, C^+ + S\, \rho_F(0)\, S^+],$$

with,

$$C = \left(\cos\hat{\theta}_1(t) - i\,\frac{\sin\hat{\theta}_1(t)}{\hat{\theta}_1(t)}\,z \right) \text{ and } S = \frac{\sin\hat{\theta}_2(t)}{\hat{\theta}_2(t)}\,(c\,a^+).. \qquad (29)$$

From the above expression for the time dependence of density matrix we can obtain different statistical properties of the atom-field system. The average value of any observable $\hat{O}$ can be obtained with the help of density matrix ρ by the relation,

$$\langle \hat{O} \rangle = Tr \, (\rho \hat{O}) \tag{30}$$

3.   TRANSFORMATION MATRIX

It is seen from the expression (25) that the operators $\theta_1(t)$ and $\theta_2(t)$ are diagonal in the basis states $|n_a\rangle$. The states $|n_a\rangle$ are the eigenstates of the number operator $a^+ a$.

$$a^+ a \left| n_a \right\rangle = n_a \left| n_a \right\rangle$$

It is therefore useful to expand the eigenstates $|n_b\rangle$ of operator $b^+ b$ in terms of eigenstates $|n_a\rangle$ of operator $a^+ a$.

$$| n_b \rangle = \sum_{n_a = 0}^{\infty} \langle n_a \mid n_b \rangle \mid n_a \rangle \tag{31}$$

We consider below the evaluation of the expansion coefficients $\langle n_a \mid n_b \rangle$.

We take the matrix elements of relation (21) for $b(t)$ in terms of a and a^+ between the ket and bra states $|0_a\rangle$ and $\langle 0_b|$. Operating a, a^+, b and b^+ on the respective eigenkets of the number operators, one obtains,

$$\langle 1_a \mid 0_b \rangle = 0 \tag{32}$$

Next we take the matrix elements between $|0_b\rangle$ and $\langle 0_a|$ in relation (20) for $a(t)$ and obtain,

$$\langle 1_b \mid 0_a \rangle = 0 . \tag{33}$$

Similarly, taking the general matrix elements between the states $|n_a\rangle$ and $\langle n_b|$ of relation (20) one obtains,

$$\sqrt{n_a} \, \langle n_b \mid n_a - 1 \rangle = G \left[\; \sqrt{n_b + 1} \, \langle n_b + 1 \mid n_a \rangle + e^{i\Omega t} \gamma \sqrt{n_b} \, \langle n_b - 1 \mid n_a \rangle \; \right]$$

$$\tag{34}$$

For $n_a = 0$ we reiterate (34) to obtain a general expression for $\langle n_b|0_a\rangle$ in terms of $\langle 0_b|0_a\rangle$ for even values of n_b and $\langle 1_b|0_a\rangle$ for odd values of n_b.

$$\langle 2n_b \mid 0_a\rangle = \left(- e^{i\Omega t}\,\gamma\right)^{n_b} \sqrt{\frac{(2n_b -1)!!}{(2n_b)!!}}\; \langle 0_b \mid 0_a\rangle \tag{35}$$

and $\langle 2n_b + 1 \mid 0_a\rangle = 0$ (36)

where $x!! = x(x-2)(x-4)\mathrm{K}\,2$, if x is even and $x!! = x(x-2)(x-4)\mathrm{K}\,1$ if x is odd.

Equations (35) and (36) show that only the matrix elements with even number of b type photons and $|0_a\rangle$ are non-zero and the matrix elements with odd number of b type photons and $|0_a\rangle$ are zero in view of (33).

Taking the matrix elements between the general states $\langle n_a|$ and $|n_b\rangle$ of relation (21) for $b(t)$ one obtains,

$$\sqrt{n_b}\,\langle n_a \mid n_b - 1\rangle = G\left[\;\sqrt{n_a + 1}\,\langle n_a + 1 \mid n_b\rangle - e^{i\Omega t}\gamma\,\sqrt{n_a}\,\langle n_a - 1 \mid n_b\rangle\;\right] \tag{37}$$

We put $n_b = 0$ in the above expression and following similar steps as above one obtains,

$$\langle 2n_a \mid 0_b\rangle = (e^{i\Omega t}\,\gamma)^{n_a} \sqrt{\frac{(2n_a -1)!!}{(2n_a)!!}}\; \langle 0_a \mid 0_b\rangle \tag{38}$$

and $\langle 2n_a + 1 \mid 0_b\rangle = 0$ (39)

Complex conjugate relation for equations (32), (33), (35), (36), (38) and (39) also hold good.

To evaluate any general expansion coefficient $\langle n_a \mid n_b\rangle$, we rewrite the expression (37) as

$$\langle n_a + 1 \mid n_b\rangle = \left[\;\alpha(n_a + 1)\langle n_a - 1 \mid n_b\rangle + \beta\,(n_a + 1,\, n_b)\langle n_a \mid n_b - 1\rangle\;\right] \tag{40}$$

where $\alpha\,(n_a + 1)$ and $\beta\,(n_a + 1,\, n_b)$ are new parameters defined as,

198

$$\alpha\,(n_a+1) \equiv e^{i\Omega t}\gamma\sqrt{\frac{n_a}{n_a+1}}\,,$$

$$\beta\,(n_a+1,\,n_b) \equiv \frac{1}{G}\sqrt{\frac{n_b}{n_a+1}}$$

$\alpha\,(n_a+1)$ is associated with the expansion coefficient with n_a reduced by two and n_b remaining the same and $\beta\,(n_a+1,\,n_b)$ is associated with the expansion coefficient with n_a and n_b both reduced by one.

For $(n_a+1) < n_b$, and (n_a+1) an even integer, $\langle n_a+1|n_b\rangle$ is reduced to $\langle 0_a|n_b'\rangle$ (where $|n_b'\rangle$ is any eigenket of b^+b operator, lower than $|n_b\rangle$) by successive iterations of (40). It is seen that $\langle n_a+1|n_b\rangle$ can be written in the reduced form as

$$\langle n_a+1|n_b\rangle = \sum_{\substack{x=0 \\ x\,even}}^{\infty} P\big((n_a+1-x)/2\ \alpha\ and\ x\ \beta\big)\langle 0_a\,|\,n_b-x\rangle \quad (41)$$

where $P((n_a+1-x)/2\ \alpha\ and\ x\ \beta)$ has been obtained after some lengthy algebra. Expression (41) shows that $\langle n_a+1|n_b\rangle$ is non-zero only for even values of n_b. Thus $\langle 2n_a|2n_b\rangle$ is obtained by substituting for $P((n_a+1-x)/2\ \alpha\ and\ x\ \beta)$ and using (35) in the above expression as,

$$\langle 2n_a\,|\,2n_b\rangle_{n_a<n_b} = \exp[i\Omega t\,(n_a-n_b)]\,\langle 0_a\,|\,0_b\rangle$$

$$\cdot\sum_{x=0}^{n_a}\left[\frac{(-1)^{(n_b-x)}\left(\dfrac{\sin\Omega t}{\Omega t}\right)^{(n_a+n_b-2x)}\sqrt{2n_a!\,2n_b!}}{G^{2x}\,(2n_a-2x)\,!!\,(2n_b-2x)\,!!\,(2x)\,!}\right] \quad (42)$$

For $n_b < n_a$, we write an expression for $\langle n_b+1|n_a\rangle$ using equation (34) as,

$$\langle n_b+1|n_a\rangle = \left[-e^{i\Omega t}\gamma\sqrt{\frac{n_b}{n_b+1}}\,\langle n_b-1|n_a\rangle + \frac{1}{G}\sqrt{\frac{n_a}{n_b+1}}\,\langle n_b\,|\,n_a-1\rangle\right] (43)$$

Comparing this expression with (40), one observes that (43) yields (40) by the interchange of n_a and n_b in (43) and change of sign of γ.

Thus one obtains an expression for $\langle 2n_a|2n_b\rangle$ with $n_b < n_a$, by interchanging n_a and n_b and changing the sign of γ in (42), then taking its complex conjugate.

$$\langle 2n_a \mid 2n_b \rangle_{n_a > n_b} = \exp[i\Omega t\,(n_a - n_b)]\langle 0_a \mid 0_b \rangle$$
$$\cdot \sum_{x=0}^{n_b} \left[\frac{(-1)^{(n_b-x)}\left(\dfrac{\sin\Omega t}{\Omega t}\right)^{(n_a+n_b-2x)}\sqrt{2n_a!\,2n_b!}}{G^{2x}\,(2n_a-2x)!!\,(2n_b-2x)!!\,(2x)!} \right] \tag{44}$$

For the case of odd-odd matrix elements, $\langle n_a + 1|n_b\rangle$ is expanded as (41) with x being odd and following similar steps as above one obtains,

$$\langle 2n_a +1 \mid 2n_b +1 \rangle_{n_a < n_b} = \exp[i\Omega t\,(n_a - n_b)]\langle 0_a \mid 0_b \rangle$$
$$\cdot \sum_{x=0}^{n_a} \left[\frac{(-1)^{(n_b-x)}\left(\dfrac{\sin\Omega t}{\Omega t}\right)^{(n_a+n_b-2x)}\sqrt{(2n_a +1)!\,[2n_b +1]!}}{G^{2x+1}\,(2n_a-2x)!!\,(2n_b-2x)!!\,(2x)!} \right] \tag{45}$$

$$\langle 2n_a +1 \mid 2n_b +1 \rangle_{n_a > n_b} = \exp[i\Omega t\,(n_a - n_b)]\langle 0_a \mid 0_b \rangle$$
$$\cdot \sum_{x=0}^{n_b} \left[\frac{(-1)^{(n_b-x)}\left(\dfrac{\sin\Omega t}{\Omega t}\right)^{(n_a+n_b-2x)}\sqrt{(2n_a +1)!\,[2n_b +1]!}}{G^{2x+1}\,(2n_a-2x)!!\,(2n_b-2x)!!\,(2x)!} \right] \tag{46}$$

An inspection of equations (42), (44), (45) and (46) shows that a general matrix element $\langle n_a|n_b\rangle$ can be written as,

$$\langle n_a \mid n_b \rangle = \exp\left(\frac{i\Omega t(n_a - n_b)}{2} \right)\langle n_a \mid n_b \rangle'. \tag{47}$$

200

These matrix elements $\langle n_a|n_b\rangle'$ are determined for different n_a and n_b in terms of $\langle 0_a|0_b\rangle$. To determine $\langle 0_a|0_b\rangle$ we use the normalization condition $\langle n_b|n_b\rangle = 1$. Thus, the condition of normalization of set n_a with equation (31) yields,

$$1 = \langle n_b \mid n_b \rangle = \sum_{n_a} |\langle n_a \mid n_b \rangle|^2$$

$$\equiv \sum_{n_a} |\langle n_a \mid n_b \rangle_n|^2 \, |\langle 0_a \mid 0_b \rangle|^2 \tag{48}$$

In the above equation, transformation matrix $\langle n_a|n_b\rangle_n$ is defined as,

$$\langle n_a \mid n_b \rangle = \langle 0_a \mid 0_b \rangle \langle n_a \mid n_b \rangle_n \tag{49}$$

Thus one obtains from (3.44),

$$|\langle 0_a \mid 0_b \rangle|^2 = \left[\sum_{n_a} |\langle n_a \mid n_b \rangle_n|^2 \right]^{-1} \tag{50}$$

Expression (50) determines $\langle 0_a|0_b\rangle$, except for a phase factor. In what follows, we take, for convenience, the undetermined phase factor to be unity.

4. STATISTICAL PROPERTIES

We now derive the expressions for average photon number and variance for different initial conditions.
The average photon number in the cavity is given by [9],

$$\langle b^+ b \rangle = Tr_F \{\rho_F(t) b^+ b\} \tag{51}$$

For the initial density matrix operator $\rho_F(0)$ in the number state basis,

$$\rho_F(0) = | n_b \rangle \langle n_b | \tag{52}$$

The time dependent density matrix operator in the number state basis is therefore written from (29) as

$$\rho_F(t) = C\,|\,n_b\rangle\langle n_b\,|\,C^+ + S\,|\,n_b\rangle\langle n_b\,|\,S^+ \tag{53}$$

It is convenient to simplify (53) by using the expansion of $|n_b\rangle$ given by (31). Thus $\rho_F(t)$ is written as,

$$\rho_F(t) = \sum_{n_a,\,n'_a=0}^{\infty}\langle n_a\,|\,n_b\rangle\langle n_b\,|\,n'_a\rangle \left(C\,|\,n_a\rangle\langle n'_a\,|\,C^+ + S\,|\,n_a\rangle\langle n'_a\,|\,S^+\right) \tag{54}$$

The eigenvalues of operators $\hat{\theta}_1$ and $\hat{\theta}_2$ involved in the above expression are $\alpha_1(n_a)$ and $\alpha_2(n_a)$ respectively.

$$\hat{\theta}_1(t)|\,n_a\rangle = \alpha_1(n_a)|\,n_a\rangle \tag{55}$$

$$\hat{\theta}_2(t)|\,n_a\rangle = \alpha_2(n_a)|\,n_a\rangle \tag{56}$$

where,

$$\alpha_1(n_a) = \left(z^2 + c^2(n_a+1)\right)^{1/2} \text{ and } \alpha_2(n_a) = \left(z^2 + c^2 n_a\right)^{1/2}. \tag{57}$$

Thus,

$$\alpha_2(n_a+1) = \alpha_1(n_a). \tag{58}$$

Using equations (55–58) in (54) one can write for $\rho_F(t)$ for the initial number state basis as,

$$\rho_F(t) = \sum_{n_a, n'_a = 0}^{\infty} \Bigg[\langle n_a | n_b \rangle \langle n_b | n'_a \rangle$$

$$\left\{ \left(\cos\alpha_1(n_a) - i \, \frac{\sin\alpha_1(n_a)}{\alpha_1(n_a)} \, z \right) \left(\cos\alpha_1(n'_a) + i \, \frac{\sin\alpha_1(n'_a)}{\alpha_1(n'_a)} \, z \right) \cdot | n_a \rangle \langle n'_a |\right.$$

$$\left. + c^2 \sqrt{(n_a+1)(n'_a+1)} \, \frac{\sin\alpha_1(n_a)}{\alpha_1(n_a)} \, \frac{\sin\alpha_1(n'_a)}{\alpha_1(n'_a)} \, | n_a \rangle \langle n'_a | \right\} \Bigg].$$

$$(59)$$

The average photon number can now be written from (53) using (59) and (47). It is seen that in the resultant expression, the real part is symmetric while the imaginary part is antisymmetric in the exchange of dummy indices n_a and n_a'. The sum over dummy indices n_a and n_a' thus reduces the imaginary term to zero.

Thus, $\langle b^+ b \rangle_{n_b}$ is given by,

$$\langle b^+ b \rangle_{n_b} = \sum_{n'_b = 0}^{\infty} \sum_{n_a, n'_a = 0}^{\infty} \Bigg[n'_b \langle n'_a | n_b \rangle' \langle n_b | n_a \rangle' \cdot \Bigg\{ \Bigg(\langle n'_b | n'_a \rangle' \langle n_a | n'_b \rangle'$$

$$\cdot \left(\cos\alpha_1(n'_a) \cos\alpha_1(n_a) + \frac{\sin\alpha_1(n'_a)}{\alpha_1(n'_a)} \, \frac{\sin\alpha_1(n_a)}{\alpha_1(n_a)} \, z^2 \right) \Bigg)$$

$$+ \Bigg(-\langle n'_b | n'_a + 1 \rangle' \langle n_a + 1 | n'_b \rangle'$$

$$\cdot \, c^2 \sqrt{(n'_a + 1)(n_a + 1)} \, \frac{\sin\alpha_1(n'_a)}{\alpha_1(n'_a)} \, \frac{\sin\alpha_1(n_a)}{\alpha_1(n_a)} - \Bigg) \Bigg\} \Bigg].$$

$$(60)$$

Next we obtain the average photon number in the initial coherent state basis. Coherent states [3] of the electromagnetic field are characterised by the criteria that all the correlation functions factorize. These conditions also imply those states of the harmonic oscillator with mean energy equal to the classical energy [1]. The Coherent state in the number state basis can be written as [1],

$$|\alpha\rangle = e^{-(1/2)|\alpha|^2} \sum_{n=0}^{\infty} \frac{\alpha^n}{\sqrt{n!}} |n\rangle . \tag{61}$$

where α is a complex number such that $|\alpha|^2 = \langle b^+ b\rangle_{t=0}$, the initial average photon number. Thus the initial density matrix for the coherent state can be written as,

$$\rho_F(0)_{coh} = |\alpha\rangle\langle\alpha| . \tag{62}$$

This can be written from (61) as,

$$\rho_F(0)_{coh} = e^{-|\alpha|^2} \sum_{n_b, \, n_b'=0}^{\infty} \frac{\alpha^{n_b} \alpha^{*n_b'}}{\sqrt{n_b ! \, n_b'!}} |n_b\rangle\langle n_b'| . \tag{63}$$

The time dependent density matrix is now obtained from (29) using (63) as,

$$\rho_F(t)_{coh} = e^{-|\alpha|^2} \sum_{n_b, n_b'=0}^{\infty} \frac{\alpha^{n_b} \alpha^{*n_b'}}{\sqrt{n_b! \, n_b'!}} \left[C | n_b\rangle\langle n_b' | C^+ + S | n_b\rangle\langle n_b' | S^+ \right] \tag{64}$$

The average photon number in the initial coherent state basis is obtained using (64) in expression (30).

$$\langle b^+ b\rangle_{coh} = \sum_{n_b''=0}^{\infty} \langle n_b'' | \rho_F(t)_{coh} \, b^+ b | n_b''\rangle , \tag{65}$$

The above expression is simplified by using the completeness relation for the states $|n_b''\rangle$. The operators C and S involve $\hat\theta_1$ and $\hat\theta_2$. Therefore it is convenient to expand the states $|n_b'\rangle$ in terms of set of states $|n_a\rangle$ and also the operator $b^+ b$ in terms of operators a and a^+ using relations (21). We allow the functions of a and a^+ to operate on the ket $|n_a\rangle$. Then using the relation (47) and expressing the phase factor exponential of the matrix elements as sum of cosine and sine terms, one arrives at a relation for $\langle b^+ b\rangle_{coh}$ as,

204

$$
\langle b^{+}b \rangle_{coh} = e^{-\alpha^2} \sum_{n_a=0}^{\infty} \sum_{n_b,n_b'=0}^{\infty} \frac{\alpha^{n_b+n_b'}}{\sqrt{n_b!\, n_b'!}} \langle n_a \mid n_b \rangle' \frac{1}{G^2}
$$

$$
\cdot \left(\cos\left\{ \frac{\Omega t}{2}(n_b' - n_b) \right\} + i \sin\left\{ \frac{\Omega t}{2}(n_b' - n_b) \right\} \right)
$$

$$
\cdot \Bigg[\langle n_b' \mid n_a \rangle' \Bigg\{ \left(n_a + \gamma^2(n_a+1) \right) \left(\cos^2 \alpha_1(n_a) + \frac{\sin^2 \alpha_1(n_a)}{\{\alpha_1(n_a)\}^2} z^2 \right)
$$

$$
+ c^2 \left((n_a+1) + \gamma^2(n_a+2) \right) \frac{\sin^2 \alpha_1(n_a)}{\{\alpha_1(n_a)\}^2} (n_a+1) \Bigg\}
$$

$$
- \gamma \sqrt{(n_a+1)(n_a+2)} \, \langle n_b' \mid n_a+2 \rangle'
$$

$$
\cdot \Bigg\{ \left(\cos\alpha_1(n_a)\cos\alpha_1(n_a+2) + \frac{\sin\alpha_1(n_a)}{\alpha_1(n_a)} \frac{\sin\alpha_1(n_a+2)}{\alpha_1(n_a+2)} z^2 \right.
$$

$$
\left. + c^2 \frac{\sin\alpha_1(n_a)}{\alpha_1(n_a)} \frac{\sin\alpha_1(n_a+2)}{\alpha_1(n_a+2)}(n_a+3) \right)
$$

$$
+ i\, z \left(\cos\alpha_1(n_a)\frac{\sin\alpha_1(n_a+2)}{\alpha_1(n_a+2)} - \frac{\sin\alpha_1(n_a)}{\alpha_1(n_a)}\cos\alpha_1(n_a+2) \right) \Bigg\}
$$

$$
- \gamma \sqrt{n_a(n_a-1)} \, \langle n_b' \mid n_a-2 \rangle'
$$

$$
\cdot \Bigg\{ \left(\cos\alpha_1(n_a)\cos\alpha_1(n_a-2) + \frac{\sin\alpha_1(n_a)}{\alpha_1(n_a)} \frac{\sin\alpha_1(n_a-2)}{\alpha_1(n_a-2)} z^2 \right.
$$

$$
\left. + c^2 \frac{\sin\alpha_1(n_a)}{\alpha_1(n_a)} \frac{\sin\alpha_1(n_a-2)}{\alpha_1(n_a-2)}(n_a+1) \right)
$$

$$
+ i\, z \left(\cos\alpha_1(n_a)\frac{\sin\alpha_1(n_a-2)}{\alpha_1(n_a-2)} - \frac{\sin\alpha_1(n_a)}{\alpha_1(n_a)}\cos\alpha_1(n_a-2) \right) \Bigg\} \Bigg]
$$

$$
\tag{66}
$$

In writing (66) α has been taken to be real for convenience.

It is seen from the above expression that by relabelling the dummy indices n_a as (n_a+2) and interchanging n_b and n_b' in the terms involving matrix elements $\langle n_b'|n_a-2\rangle'$, the imaginary parts cancel with those of the terms involving matrix elements $\langle n_b'|n_a+2\rangle'$ whereas, their real parts add up. Also, it is seen that the imaginary part of the term involving matrix elements

$\langle n_b'|n_a\rangle$ is identically zero because of the antisymmetric nature of the sine function in the interchange of dummy indices n_b and n_b'. We thus obtain for $\langle b^+b\rangle_{coh}$ as,

$$\langle b^+b\rangle_{coh}=e^{-\alpha^2}\sum_{n_a=0}^{\infty}\sum_{n_b,\,n_b'=0}^{\infty}\frac{\alpha^{n_b+n_b'}}{\sqrt{n_b!\,n_b'!}}\langle n_a|n_b\rangle'\frac{1}{G^2}$$

$$\cdot\left[\cos\left\{\frac{\Omega t}{2}(n_b'-n_b)\right\}\langle n_b'|n_a\rangle'\right.$$

$$\cdot\left\{(n_a+\gamma^2(n_a+1))\left(\cos^2\alpha_1(n_a)+\frac{\sin^2\alpha_1(n_a)}{\{\alpha_1(n_a)\}^2}z^2\right)\right.$$

$$\left.+c^2((n_a+1)+\gamma^2(n_a+2))\frac{\sin^2\alpha_1(n_a)}{\{\alpha_1(n_a)\}^2}(n_a+1)\right\}$$

$$-2\cos\left\{\frac{\Omega t}{2}(n_b'-n_b)\right\}\langle n_b'|n_a+2\rangle'\gamma\sqrt{(n_a+1)(n_a+2)}$$

$$\cdot\left(\cos\alpha_1(n_a)\cos\alpha_1(n_a+2)+\frac{\sin\alpha_1(n_a)}{\alpha_1(n_a)}\frac{\sin\alpha_1(n_a+2)}{\alpha_1(n_a+2)}(z^2+c^2(n_a+3))\right)$$

$$+2\sin\left\{\frac{\Omega t}{2}(n_b'-n_b)\right\}\langle n_b'|n_a+2\rangle'\frac{\sin(\Omega t)}{(\Omega t)}\sqrt{(n_a+1)(n_a+2)}$$

$$\left.\cdot z\left(\cos\alpha_1(n_a)\frac{\sin\alpha_1(n_a+2)}{\alpha_1(n_a+2)}-\frac{\sin\alpha_1(n_a)}{\alpha_1(n_a)}\cos\alpha_1(n_a+2)\right)\right]$$

$$(67)$$

The variance in the photon number is defined as [9],

$$(\Delta n)^2=\langle(b^+b)^2\rangle-\langle b^+b\rangle^2.\tag{68}$$

In the number state basis $\langle b^+b\rangle_{n_b}^2$ is obtained using (30) and (29) as,

$$\langle(b^+b)^2\rangle_{n_b}=\sum\langle n_b'|\rho_F(t)_{n_b}(b^+b)^2|n_b'\rangle,\tag{69}$$

We substitute for $\rho_F(t)$ for the number state basis from (59) and collect the real and imaginary parts in the expression for $\langle b^+b\rangle_{n_b}^2$. It is seen that the imaginary term identically cancels due to its antisymmetric nature in the interchange of dummy indices n_a and n_a'. One can thus write for $\langle (b^+ b)^2 \rangle_{n_b}$ as,

$$
\begin{aligned}
\langle (b^+ b)^2 \rangle_{n_b} = \sum_{n_b'=0}^{\infty} \sum_{n_a, \, n_a'=0}^{\infty} \Bigg[& (n_b')^2 \langle n_a \mid n_b \rangle' \langle n_b \mid n_a' \rangle' \\
& \left\{ \left(\cos\alpha_1(n_a) \cos\alpha_1(n_a') + \frac{\sin\alpha_1(n_a)}{\alpha_1(n_a)} \frac{\sin\alpha_1(n_a')}{\alpha_1(n_a')} z^2 \right) \right. \\
& \langle n_b' \mid n_a \rangle' \langle n_a' \mid n_b' \rangle \\
& + \left(c^2 \sqrt{(n_a+1)(n_a'+1)} \, \frac{\sin\alpha_1(n_a)}{\alpha_1(n_a)} \frac{\sin\alpha_1(n_a')}{\alpha_1(n_a')} \right) \\
& \left. \langle n_b' \mid n_a+1 \rangle' \langle n_a'+1 \mid n_b' \rangle' \right\} \Bigg]
\end{aligned}
$$

$$\tag{70}$$

For the initial coherent state field $\langle (b^+ b)^2 \rangle_{coh}$ is written as,

$$
\langle (b^+ b)^2 \rangle_{coh} = \sum_{n_b'}^{\infty} \langle n_b' \mid \rho_F(t)_{coh} \left(b^+ b \right)^2 \mid n_b' \rangle \tag{71}
$$

Performing a lengthy algebra similar to that done earlier for $\langle b^+ b \rangle_{coh}$ one obtains for $\langle (b^+ b)^2 \rangle_{coh}$ as,

$$
\langle (b^+ b)^2 \rangle_{coh} = \exp(-\alpha^2) \sum_{n_b,\, n_b'=0}^{\infty} \sum_{n_a=0}^{\infty} \left[\frac{\alpha^{\,n_b+n_b'}}{\sqrt{n_b! n_b'!}} \frac{1}{G^4} \langle n_a \mid n_b \rangle'\right.
$$

$$
\left[\left\{ \cos \frac{\Omega t}{2}(n_b' - n_b) \langle n_b' \mid n_a \rangle' \right.\right.
$$

$$
\left\{ (n_a^2 + \gamma^{\,4}(4n_a^2 + 4n_a + 2) + \gamma^{\,4}(n_a+1)^2) \left(\cos^2 \alpha_1(n_a) + \frac{\sin^2 \alpha_1(n_a)}{\{\alpha_1(n_a)\}^2} z^2 \right) \right.
$$

$$
+ c^2(n_a+1) \frac{\sin^2 \alpha_1(n_a)}{\{\alpha_1(n_a)\}^2}
$$

$$
\left. ((n_a+1)^2 + \gamma^{\,2}\{4(n_a+1)^2 + 4(n_a+1)+2\} + \gamma^{\,4}(n_a+2)^2) \right\} \Big\}
$$

$$
-2 \langle n_b' \mid n_a+2 \rangle' \, \gamma \sqrt{(n_a+1)(n_a+2)}
$$

$$
\left\{ \cos \frac{\Omega t}{2}(n_b' - n_b) \right.
$$

$$
\left\{ \left((2n_a+2) + \gamma^{\,2}(2n_a+4) \right) \right.
$$

$$
\cdot \left(\cos \alpha_1(n_a) \cos \alpha_1(n_a+2) + \frac{\sin \alpha_1(n_a)}{\alpha_1(n_a)} \frac{\sin \alpha_1(n_a+2)}{\alpha_1(n_a+2)} z^2 \right)
$$

$$
\left. \left((2n_a+2) + \gamma^{\,2}(2n_a+4) \right) c^2 (n_a+3) \frac{\sin \alpha_1(n_a)}{\alpha_1(n_a)} \frac{\sin \alpha_1(n_a+2)}{\alpha_1(n_a+2)} \right\}
$$

$$
- \sin \frac{\Omega t}{2}(n_b' - n_b) \cdot
$$

$$
\left\{ z \left((2n_a+2) + \gamma^{\,2}(2n_a+4) \right) \right.
$$

$$
\left. \cdot \left(\cos \alpha_1(n_a) \frac{\sin \alpha_1(n_a+2)}{\alpha_1(n_a+2)} - \cos \alpha_1(n_a+2) \frac{\sin \alpha_1(n_a)}{\alpha_1(n_a)} \right) \right\}
$$

$$
+ 2 \langle n_b' \mid n_a+4 \rangle' \, \gamma^2 \sqrt{(n_a+1)(n_a+2)(n_a+3)(n_a+4)}
$$

$$
\cdot \left\{ \cos \frac{\Omega t}{2}(n_b' - n_b) \left\{ -\cos \alpha_1(n_a) \cos \alpha_1(n_a+4) \right.\right.
$$

$$
\left. + \frac{\sin \alpha_1(n_a)}{\alpha_1(n_a)} \frac{\sin \alpha_1(n_a+4)}{\alpha_1(n_a+4)} z^2 + c^2 (n_a+5) \frac{\sin \alpha_1(n_a)}{\alpha_1(n_a)} \frac{\sin \alpha_1(n_a+4)}{\alpha_1(n_a+4)} \right\}
$$

$$
- \sin \frac{\Omega t}{2}(n_b' - n_b) \left(\cos \alpha_1(n_a) \frac{\sin \alpha_1(n_a+4)}{\alpha_1(n_a+4)} z \right.
$$

$$
\left.\left.\left.\left. - \cos \alpha_1(n_a+4) \frac{\sin \alpha_1(n_a)}{\alpha_1(n_a)} \right) \right\} \right]
$$

$$
(72)
$$

5. NUMERICAL RESULTS AND DISCUSSION

The expressions for different statistical properties derived earlier and the expansion coefficients $\langle n_a|n_b\rangle'$ involve the knowledge of $\gamma(t)$ and $G(t)$. It is therefore instructive to analyse the time dependence of these parameters. In figure (1) $\gamma(t)$ and $G(t)$ are plotted as a function of t.

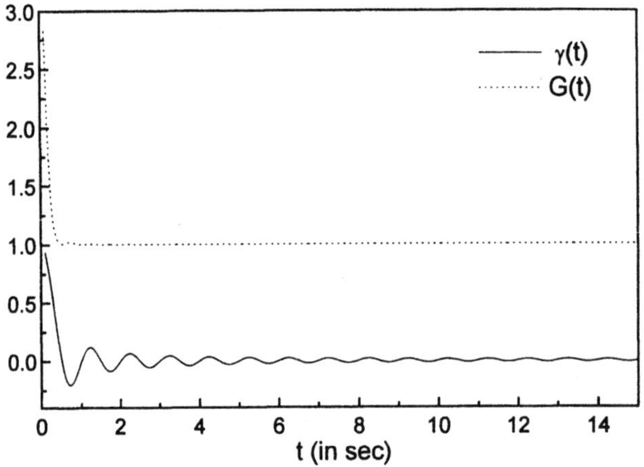

Figure 1. Time depencence of $\gamma(t)$ and $G(t)$ with $\Omega = 2\pi$.

It illustrates the oscillation and rapid convergence of $\gamma(t)$ with increase of t. $\gamma(t)$ becomes less than 0.2 for $t > 0.7$ second and it rapidly becomes small with further increase in t.

The graph also shows that for $t < 0.4$ sec., G is a rapidly decreasing function of t and for $t \geq 0.5$ sec., it becomes practically unity.

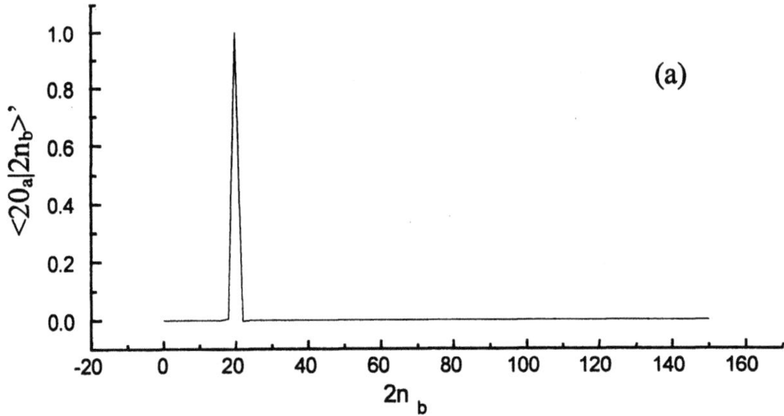

Figure 2. Variation of expansion coefficients as a function of n_a and n_b for (a) large t ($t = 1$ sec) and (b) small t ($t = 0.1$ sec).

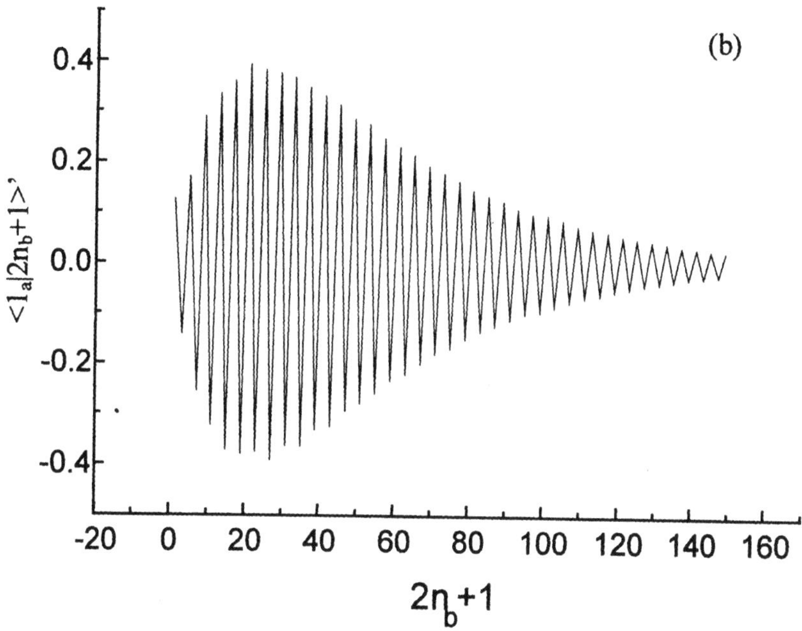

Figure 2. Continued.

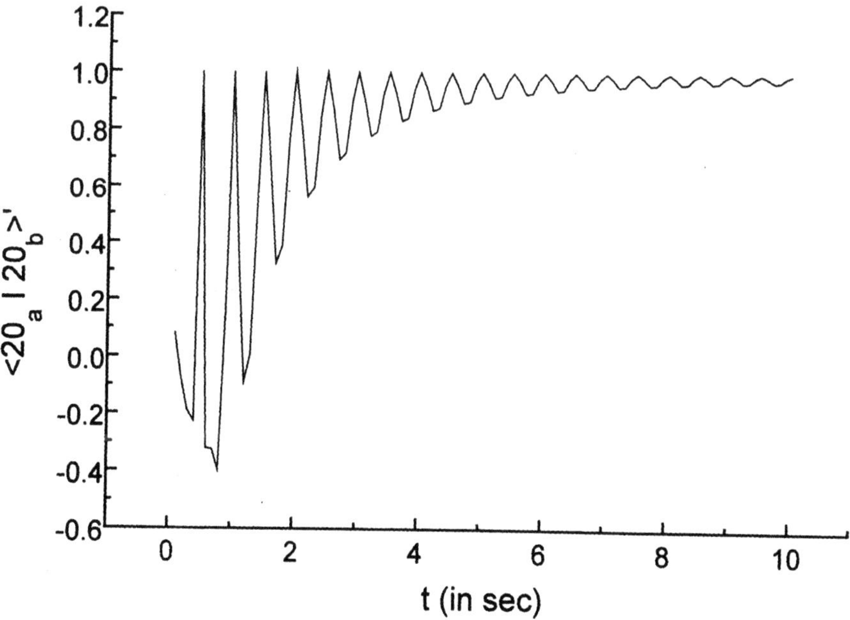

Figure 3. Variation of diagonal matrix elements $\langle 20_a | 20_b \rangle'$ with time.

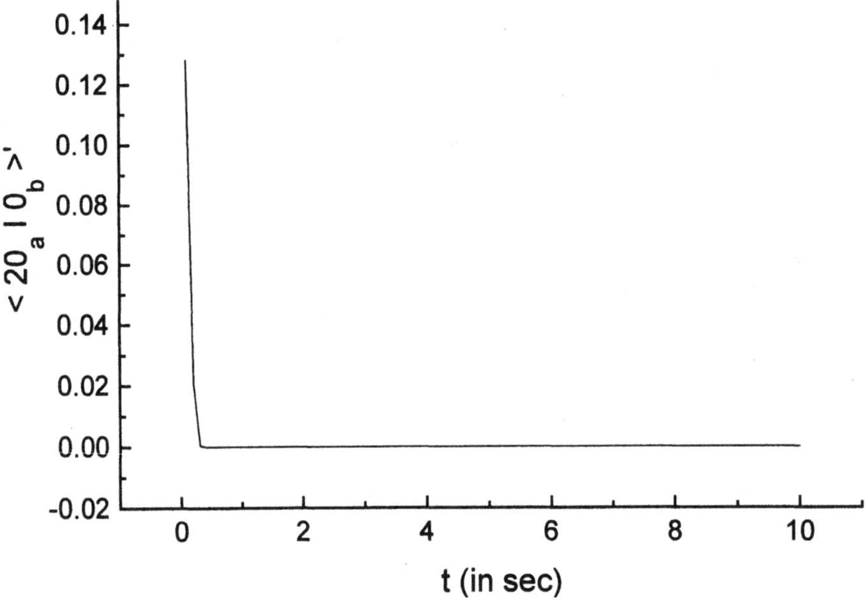

Figure 4. Variation of off-diagonal matrix elements $\langle 20_a|0_b\rangle'$ with time.

Next we consider the variation of $\langle n_a|n_b\rangle'$ as a function of n_a, n_b and t. Figures (2-4) show the numerical evaluation of $\langle n_a|n_b\rangle'$ based on the formulae (42), (44), (45) and (46). One can draw the following conclusions from the perusal of the Figures (2-4):

i) Figure (2a) illustrates that $\langle n_a|n_b\rangle'$ is maximum when $n_a \sim n_b$ for large t ($t \geq 1$ sec).

ii) Figures (2a) and (2b) illustrate that $\langle n_a|n_b\rangle'$ is appreciable for a broader range of values of n_a and n_b for small t ($t \geq 1$ sec) than that for large t.

iii) Figure (3) illustrates that the diagonal values of $\langle n_a|n_b\rangle'$ approach unity with increase of t.

iv) From figure (4) we infer that off diagonal terms of $\langle n_a|n_b\rangle'$ approach zero with increase of t.

Conclusions (iii) and (iv) are a consequence of the fact that the operator a and b become identical in the limit of large t in view of equation [20] and the behaviour of γ and G as a function of t discussed earlier.

The average photon number $\langle b^+b\rangle$ is given by the expression (60). This expression converges to a good degree of approximation for $0 \leq n_a$, n_a', $n_b' \leq 50$ and $t \geq 0.7$ sec. Thus in our numerical study in the present section we have evaluated $\langle b^+b\rangle_{n_b}$ for $t \geq 0.7$ sec.

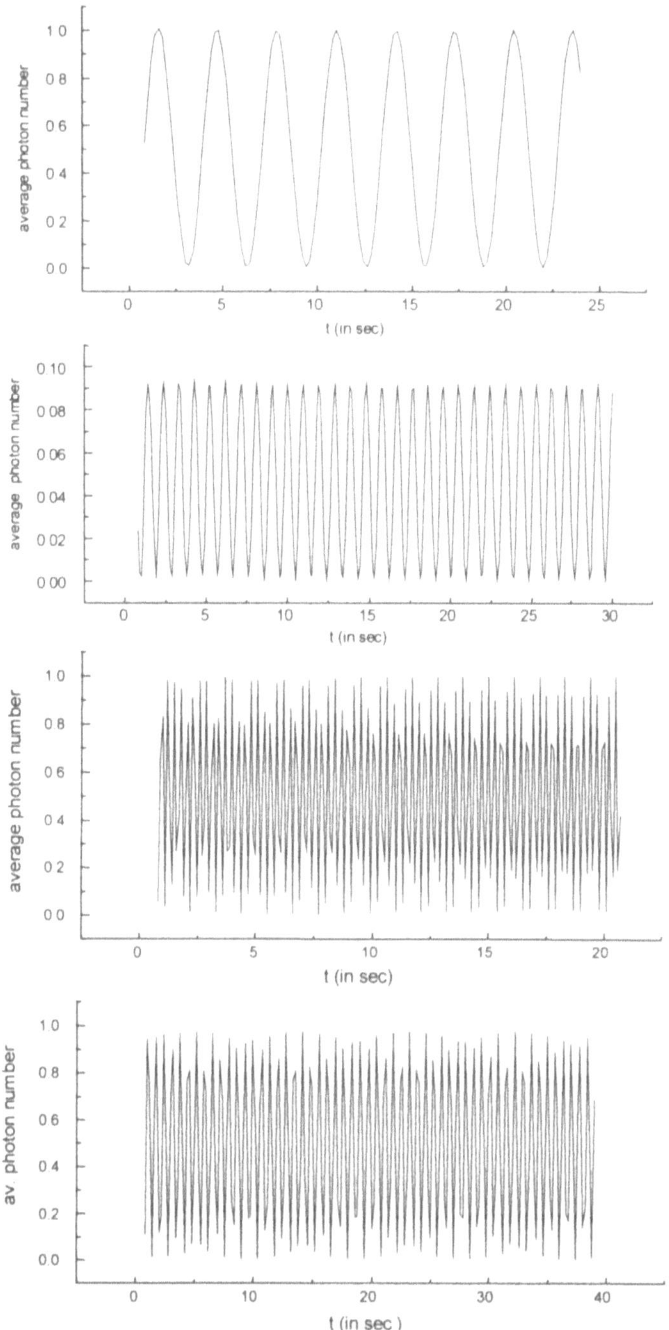

Figure 5. Evolution of average photon number for initial vacuum state field and the atom initially inverted with (a) $g = 1$, $\Delta\Omega = 0$, (b) $g = 1$, $\Delta\Omega = 6.273$, (c) $g = 20$, $\Delta\Omega = 0$ and (d) $g = 20$, $\Delta\Omega = 6.273$.

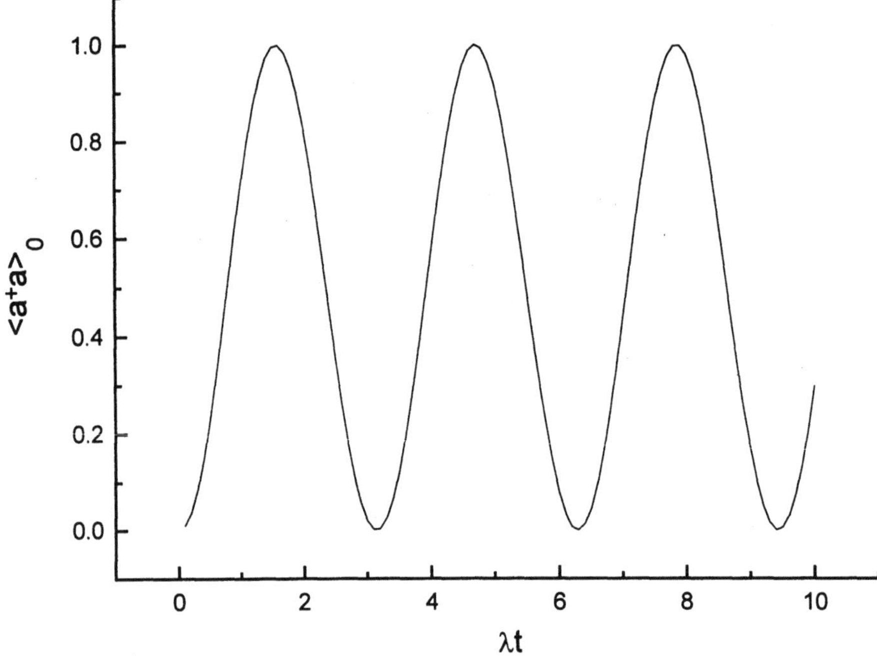

Figure 6. Evolution of average photon number for initial vacuum state field and the atom initially inverted in the RWA case with $\Omega = 2\pi$.

Figures 5(a) through (d) show the results of numerical evaluation from equation (60) of behaviour of $\langle b^+b \rangle_{n_b}$ with t for large and small g in the resonance and off-resonance case.

On comparing these graphs with that obtained in RWA by Phoenix et al (Figure 6), one observes the following points:

1. Figures 5(a-d) show that $\langle b^+b \rangle_{n_b}$ oscillates with time between n_b and $n_b + 1$. This behaviour is similar to that for the RWA case shown in figure (10). It is expected as at $t = 0$, one has initial number of photons n_b and the atom is in the excited state.

2. For small values of the coupling constant g, $\langle b^+b \rangle_{n_b}$ exhibits similar behaviour for the RWA and the non-RWA case both for resonance and off-resonance.

3. Figures5(c) and (d) show that for large values of the strength of perturbation g in the present case (non-RWA), the amplitude of variation exhibits large modulation which is very different from the behaviour obtained for RWA.

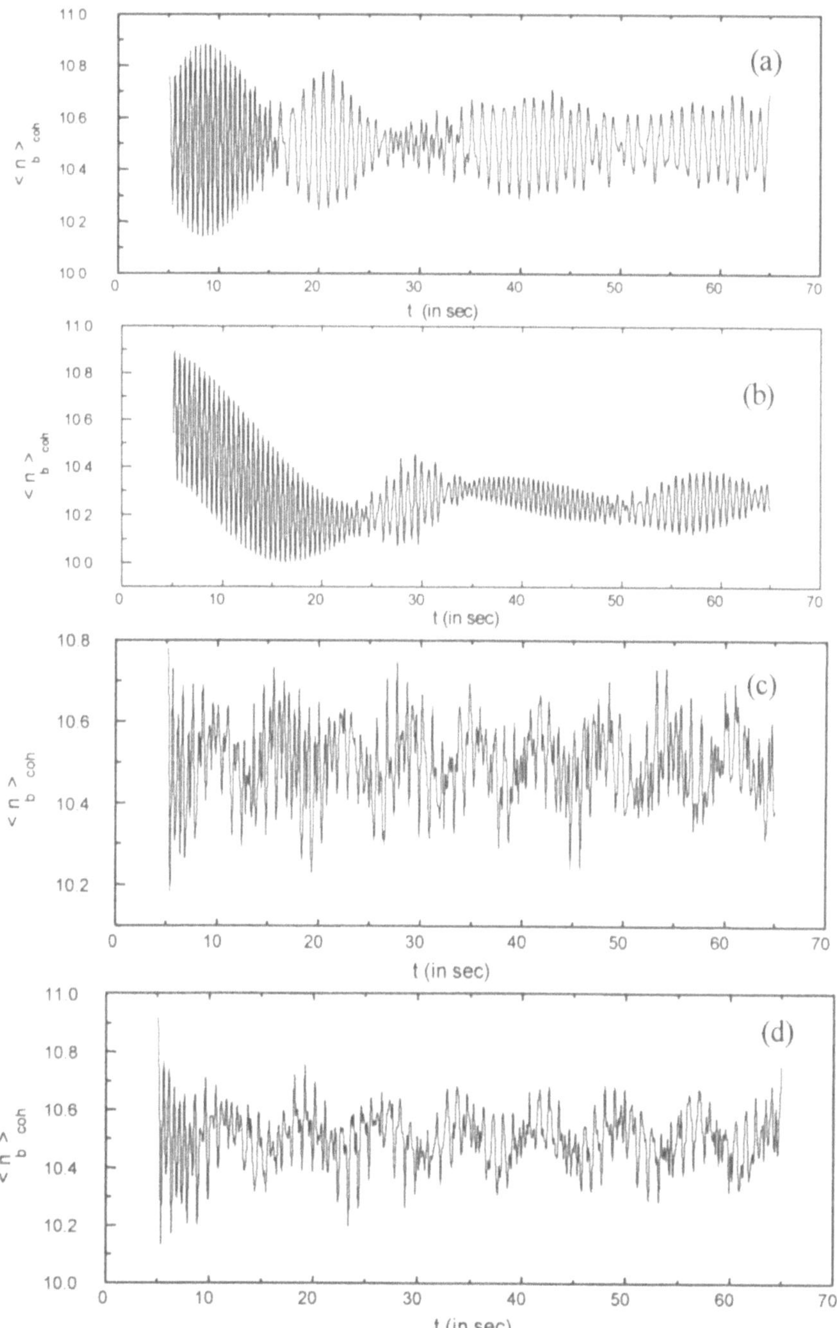

Figure 7. Evolution of average photon number for initial coherent state field and the atom initially inverted with $|\alpha|^2 = 10$ and (a) $g = 1$, $\Delta\Omega = 0$, (b) $g = 1$, $\Delta\Omega = 6.273$, (c) $g = 20$, $\Delta\Omega = 0$ and (d) $g = 20$, $\Delta\Omega = 6.273$.

214

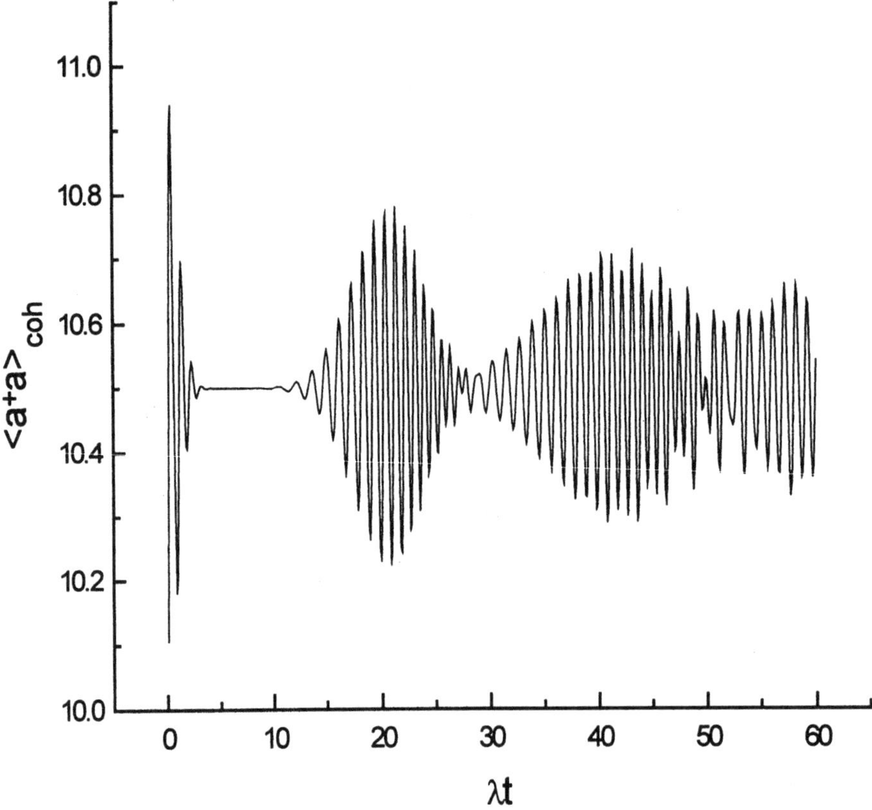

Figure 8. Evolution of the photon number for initial coherent state field and the atom initially inverted in the RWA case with $|\alpha|^2 = 10$ and $\Omega = 2\pi$.

The average photon number in the coherent state basis is given by expression (67). Figures 7(a-d) show the results of numerical evaluation for equation (67) for large and small g. Comparison of these plots with that obtained by Phoenix and Knight [9] (shown in Figure 8) shows that:

1. For small values of strength of perturbation, the behaviour of $\langle b^+ b \rangle_{coh}$ for non-RWA as shown in figures (11-14) and that for RWA (shown in Figure 8) are similar. In both the cases $\langle b^+ b \rangle_{coh}$ exhibits collapse and revival as a function of time but the overall time dependence of $\langle b^+ b \rangle_{coh}$ is entirely different for the off-resonance case.

2. For large g $\langle b^+ b \rangle_{coh}$ in the present study shows rapid fluctuations with small amplitude around the average value. This behaviour is very different from the RWA case.

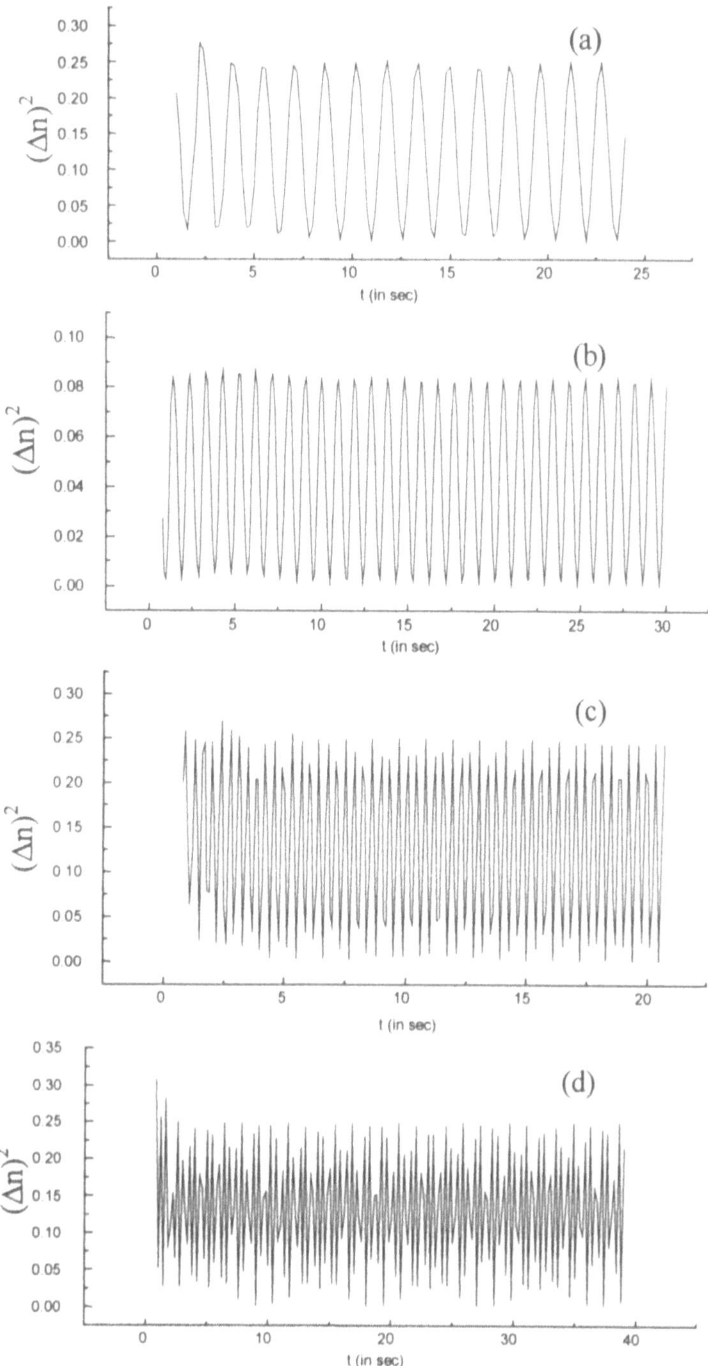

Figure 9. Evolution of the photon number variance with the field initially in the number state basis and the atom initially inverted with (a) $g = 1$, $\Delta\Omega = 0$, (b) $g = 1$, $\Delta\Omega = 6.273$, (c) $g = 20$, $\Delta\Omega = 0$ and (d) $g = 20$, $\Delta\Omega = 6.273$.

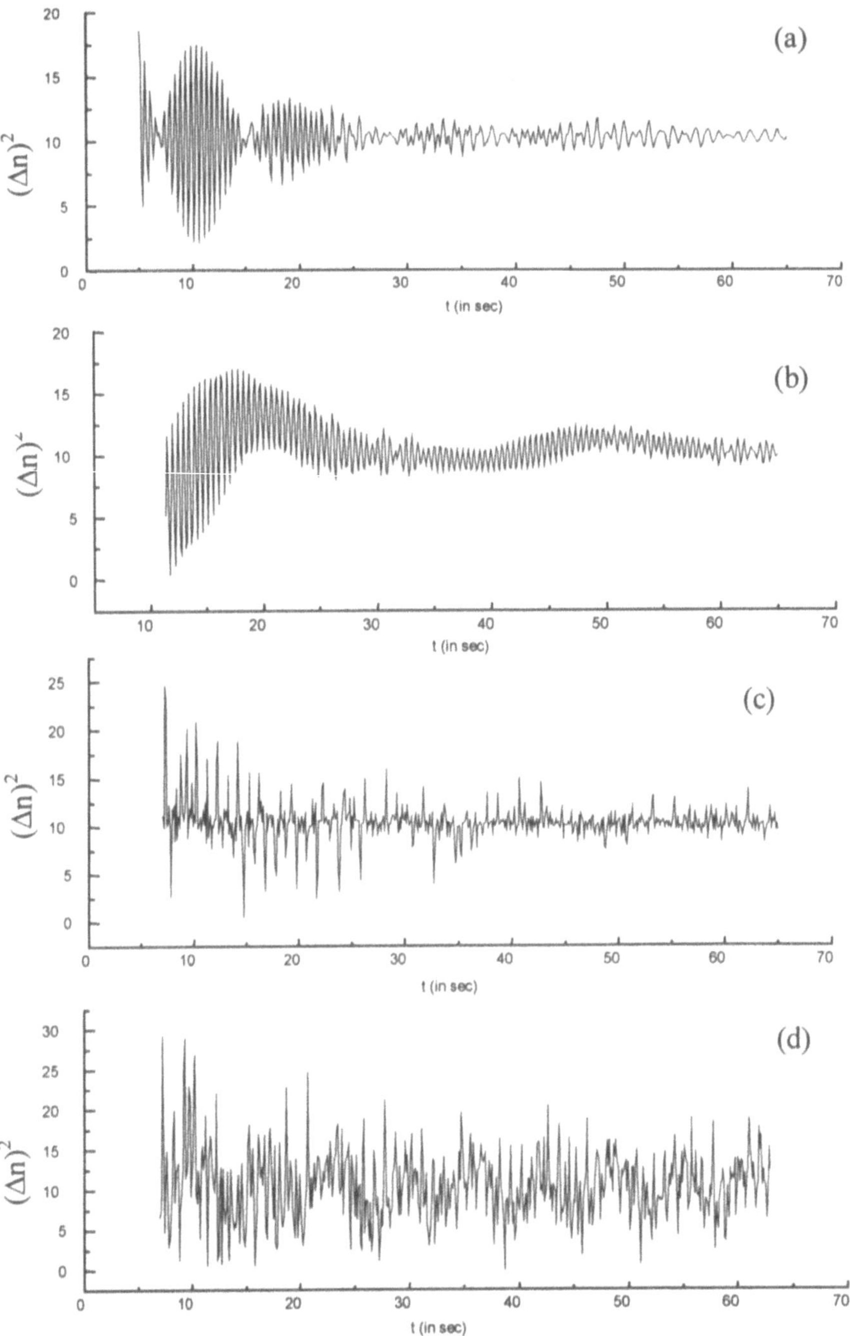

Figure 10. Evolution of the photon number variance for the initial coherent state field and the atom initially inverted with $|\alpha|^2 = 10$ and (a) $g = 1$, $\Delta\Omega = 0$, (b) $g = 1$, $\Delta\Omega = 6.273$, (c) $g = 20$, $\Delta\Omega = 0$ and (d) $g = 20$, $\Delta\Omega = 6.273$.

The fluctuations about $\langle b^+ b \rangle$ around the average value is given by equation (68). In the number state basis $(\Delta n)^2$ is obtained using (70), (60) and (68). It is plotted as a function of time in Figures 9(a-d) for large and small g.

1. The variance is a periodic function of time with frequency of the order of field frequency. The plot is similar to that obtained in RWA [9].
2. For the off-resonance case and small g the frequency of oscillation of $(\Delta n)^2$ is increased as shown in figure (17).
3. Figures (18) and (19) show the variation of $(\Delta n)^2$ with time for large g. The behaviour is seen to be different from RWA.

In the coherent state basis $(\Delta n)^2$ is obtained using (72), (67) and (68). It is plotted as a function of time in figures 10(a-d) for large and small g and for resonance and off-resonance condition. One observes from these figures that:

1. Fluctuation tends to a constant limit for the case of small coupling and for both resonance and off-resonance condition and for large coupling only for the resonance case.
2. For large coupling parameter and the off-resonance case the fluctuations exhibit undamped oscillation.

An examination of expressions for $\langle b^+ b \rangle_{n_b}$, $\langle b^+ b \rangle_{coh}$ and $(\Delta n)^2$ for the number and coherent states show that the major oscillatory time dependence of these quantities arises from terms like $\cos\left[t \left\{ ((\Delta\Omega)^2/4) + g^2(n_b + 1) \right\}^{1/2} \right]$. Thus one expects the various thermodynamic averages to vary with time with effective frequencies $\Omega_{eff} = [((\Delta\Omega)^2/4) + g^2(n_b + 1)]^{1/2}$. The time variations observed in various graphs are a consequence of the above behaviour. It is important to note that earlier studies [9] of the Hamiltonian (4) in the rotating wave approximation of time variation of thermodynamic quantities exhibit a frequency proportional to the interaction strength whereas without rotating wave approximation [13], the leading frequency is the field frequency Ω. Results obtained above show that in the resonance case, $\Omega = \omega$, $\Delta\Omega = 0$ and $\Omega_{eff} = g\sqrt{(n_b + 1)}$, one recovers the RWA result. For $\Delta\Omega \gg g\sqrt{(n_b + 1)}$ and $\Omega \gg \omega$, $\Omega_{eff} \sim \Omega$, the result obtained in [13]. Thus the result of the present investigations incorporate the earlier results as two limiting cases. However, the general result obtained in the present work shows that the thermodynamic averages vary with time with effective frequency depending on ω, Ω and g.

Thus one can conclude that the incorporation of non-RWA terms in the model have significant contribution in various physical cases.

REFERENCES

1. P. Meystre and M. Sargent III, *Elements of Quantum Optics*, Springer Verlag (1990)
2. E. T.Jaynes and F. W. Cummings, Proc. IEEE **51**, 89 (1963).
3. *Quantum Optics* edited by. R. J. Glauber, Academic Press, (1969).
4. S. Stenholm. Physics Reports **6C**, 1 (1973).
5. N. B. Narozhny, J. J. Sanchez–Mondragon and J. H. Eberly, Phys. Rev. A **23,** 23 (1981).
6. R. R. Puri and G. S. Agarwal, Phys. Rev. A **33**, 3610 (1986).
7. J. Gea-Banacloche, Phys. Rev. Lett. **65**, 3385 (1990).
8. S. M. Barnett and P. L. Knight, Phys. Rev. A **33**, 2444 (1986).
9. S. J. D. Phoenix and P. L. Knight, Ann. Phys. (N.Y.) **186**, 381 (1988).
10. C. J. Joachain, Quantum Collision Theory, North Holland, Amsterdam, (1983).
11. J. J. Sakurai, *Modern Quantum Mechanics*, The Benjamin/Cummings Publishing Company Inc., (1985).
12. G. H. Weiss and A. A. Maradudin, J. Math. Phys. **3**, 771 (1962).
13. L. Bonci, R. Roncaglia, B. J. West and P. Grigolini, Phys. Rev. Lett. **67**, 2593 (1991).

Collision Physics as a Tool for Environmental Physics

N J Mason, S K Pathak*, J M Gingell and N C Jones

Department of Physics & Astronomy,
University College London,
Gower Street, London, WC1E 6BT, UK

**Department of Physics,*
Christ Church College, Kanpur 208 001, India

Collision physics can play an important role in understanding the basic spectroscopy and dissociation dynamics of those molecular systems that are prevalent in the terrestrial atmosphere. In this brief review we will give several examples of how collision physics is providing fundamental information on the ozone molecule.

1.   INTRODUCTION

In the last few decades the possible detrimental impact mankind is having upon the planet Earth has caused increasing concern. The first evidence of industrially induced global warming, the widespread phenomenon of acid rain and the growing evidence of health problems caused by urban pollution has attracted world-wide attention from both social and political commentators. Substantial (and often heated) debates have taken place, both in the scientific and political communities, as to the actual evidence for such phenomena and what action should be taken to correct such impacts.

To understand and assess the possible dangers to the Earth caused by our exploitation of its resources and the development of industry, a new branch of science has evolved in the last forty years; that of 'Environmental Science'. Environmental science embraces many areas of 'traditional physics and chemistry', particularly through the adoption of tools developed

Trends in Atomic and Molecular Physics,
Edited by Sud and Upadhyaya. Kluwer Academic/Plenum Publishers, New York, 2000.

220

in well established fields that may be adapted for the study of specific environmental issues. Many mathematical modelling techniques developed for 'pure research' have been utilised to model environmental issues, e.g. models of fluid dynamics have been adapted to study water flow through the soil and cloud physics; atomic and molecular codes developed for physical chemistry have also been used to determine spectroscopic parameters for remote sensing within the Earth's atmosphere. Similarly experimental apparatus designed in the laboratory to study fundamental properties of matter have been adapted to make field measurements of atmospheric pollutants, the most common example being the design of high resolution laser systems for probing troposphere pollution [1, 2]. In this brief review we will discuss how techniques developed in the atomic, molecular and optical physics community are being used in environmental studies. In particular we will discuss how such techniques are being used in the study of global ozone depletion.

2. SPECTROSCOPIC TECHNIQUES

Spectroscopic studies remain the core of basic atomic, molecular and optical research and the refined techniques developed for studying the fundamental structure and reaction dynamics of atoms and molecules have been widely adapted by the environmental community, particularly for the remote study of the Earth's atmosphere. Indeed it is only through the adoption of spectroscopic techniques that we have been able to appreciate and understand the complex nature of the dynamic chemistry that underpins all of the major current environmental issues e.g. the formation and transport of acid rain; the emission of industrial pollutants and their world wide dispersal causing ozone depletion and enhanced global warming. Hence our knowledge of the spectroscopy of atomic and molecular compounds is essential to understanding and monitoring the terrestrial environment. However, there remain atmospheric molecules for which data is, at best, fragmentary and often is completely absent.

Two major techniques are currently being used to probe the spectroscopy of molecular systems important in the study of the terrestrial environment; (1) Photo-absorption spectroscopy and (2) Electron energy loss spectroscopy

2.1 Photo-absorption spectroscopy

Many of the experimental studies of photo-absorption have involved measurement of the attenuation of a beam of photons traversing a gas-filled absorption cell. Such an arrangement may be easily established in any laboratory using commercial UV-vis or FTIR instruments allowing spectra to be recorded from the infra-red to the near ultraviolet (200 nm). For studies in the far UV such a gas cell may be incorporated into one of the user beam lines at a synchrotron facility. Figure 1 shows a photo-absorption apparatus attached to a beam line at the UK Daresbury facility. Synchrotron radiation falls on a rotatable diffraction grating where it is dispersed. The dispersed radiation is then focused into the absorption cell with a single exit slit between the grating and the absorption cell selecting the resolution of the incident radiation.

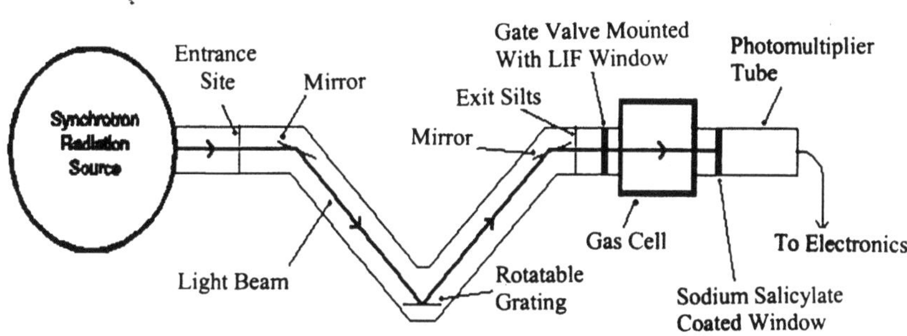

Figure 1. Photoabsorption apparatus as used at the UK Daresbury synchrotron facility.

The intensity of the incident radiation is then recorded by a photo-multiplier placed at the exit of the gas cell (to convert UV radiation into visible radiation a sodium salicylate window may be placed between the gas cell and the photo-multiplier). If light of flux I_0 is recorded with no gas in the gas cell the measured intensity falls to some lower value I when gas in introduced. If X is the path length in the cell and then these two intensities are related by the equation.

$$I = I_0 \exp(-\mu x)$$

where μ is the absorption coefficient for the particular wavelength in use. μ may then be related to the photo-absorption cross section σ_{pa} by $\mu = \sigma_{pa}N$ where N is the number density of the gas. There are however two conditions on the validity of this equation. First, the radiation must be so nearly monochromatic that there will be no change in effective absorption as the radiation progresses through the gas. Then μ will be independent of the value of x chosen, and Lambert's law will be obeyed. Secondly μ must also be independent of temperature and pressure, so that Beer's law is obeyed. If either of these conditions is broken then so-called 'line saturation' phenomena may occur resulting in erroneous absolute photoabsorption cross sections. It is therefore necessary to measure apparent σ_{pa} at different incident resolutions and gas pressures if absolute cross sections are to be derived.

2.2   Electron energy loss spectroscopy

Photo-absorption measurements are necessarily restricted to those transitions of an 'allowed' nature, forbidden transitions (those not obeying the simple dipole radiation $\Delta l = \pm 1$) will therefore only have weak absorption cross sections, often too low for detection with conventional light sources. Forbidden transitions produce long lived excited states which are amongst the most chemically reactive species in the atmosphere. Such states are produced by photo-dissociation, recombination and collisional excitation in the stratosphere and ionosphere and are also responsible for the emissions in airglow and the aurora. However, to date there have been few experimental studies on the spectroscopy of such forbidden states. Therefore another experimental technique has been applied to probe the spectroscopy of aeronomic molecules, this is the technique of Electron Energy Loss Spectroscopy (EELS).

If the incident electron energy T is sufficiently large (T > 100 eV) and the scattering angle small ($\theta \approx 0°$) the electric field induced at the site of the molecule by a passing electron is very similar to that which would be caused by an incident photon pulse. The electric field acts most strongly with the transition dipole of the molecule such that electric dipole (or optically allowed) transitions are predominantly excited, hence the inelastic electron scattering process simulates the photo-absorption process. An additional advantage of using EELS to measure optical absorption spectra is that a wide wavelength range may be recorded simultaneously without the need to change filters and detectors. It is possible to relate the energy loss spectrum obtained at $\theta \approx 0°$ and high incident energy (T $\gg E_L$, the electron energy loss) to the differential oscillator strength and thence the optical absorption cross section. Since EELS does not suffer from the effect of 'line saturation' such derived cross sections may provide a more accurate value of the absorption strength of particular electronic states than simple single photon absorption experiments.

At low impact angles and large scattering angles ($\theta > 90°$), i.e. the conditions of large momentum transfer within the collision event, structure observed in the energy loss spectra will be due to forbidden transitions. Hence by selective adjustment of scattering parameters towards large momentum transfer it is possible to probe those electronic states which are not able to be observed in one photon absorption experiments, but which may be excited in the Earth's atmosphere.

3.   OZONE: THE EARTH'S UV FILTER

3.1   Ozone chemistry

Ozone is restricted to a thin layer in the Earth's atmosphere, the ozonosphere, with a maximum concentration between 20 and 26 km above the Earth's surface. It is formed through the combination of atomic oxygen (O) and molecular oxygen (O_2). Atomic oxygen (O) is formed by the photo-dissociation of O_2 at around 100 Km by solar radiation with $\lambda < 175$ nm, the process for which may be summarised as follows:

$$h\nu\ (\lambda<175\ nm) + O_2 \rightarrow O + O.$$

The free oxygen atoms may then combine with oxygen molecules in a three body collision (to conserve energy) to form ozone:

$$O + O_2 + M \rightarrow O_3 + M$$

where M is any atom or molecule (e.g. O_2, N_2) capable of absorbing the excess energy liberated in the exothermic chemical reaction.

Once formed there are two natural destruction mechanisms for ozone; Photo-dissociation

$$h\nu + O_3 \rightarrow O_2 + O \tag{1}$$

and collisional dissociation:

$$O + O_3 \rightarrow 2O_2 \tag{2}$$

the latter being a net result of complex catalytic chemical reactions, either natural or, more recently, manmade. Natural catalytic destruction occurs through the presence of the OH radical formed by photo-dissociation of water vapour in the atmosphere.

$$OH + O_3 \rightarrow HO_2 + O_2$$
$$HO_2 + O \rightarrow OH + O_2$$

$$\text{Net } O + O_3 \rightarrow 2O_2 \tag{3}$$

in which the OH radical survives. These processes and reaction (3) ensure that there is little ozone formed naturally in the troposphere, while lack of

O_2 (restricting reaction (1)) prevents ozone formation in the upper atmosphere.

3.2 The 'Ozone' hole

In 1974, Molina and Rowland [3] suggested that the uncontrolled release of chlorofluorocarbons (CFC's) into the terrestrial atmosphere would lead to the catalytic destruction of ozone. Although chemically stable within the troposphere, in the stratosphere the CFC's can be broken down by solar radiation and release chlorine atoms which in turn form ClO catalytic species such that

$$ClO + O_3 \rightarrow ClO_2 + O_2$$
$$ClO_2 + O \rightarrow ClO + O_2$$
$$\text{------------------------------}$$
$$Net \ O + O_3 \rightarrow 2O_2$$

This mechanism is far more destructive than those naturally occurring mechanisms involving the OH radical as one ClO radical may destroy several hundred ozone molecules before it itself is removed form the ozone chemical cycle. Thus Molina and Rowland suggested even trace (ppm) amounts of any CFC in the stratosphere could have a dramatic effect upon stratospheric ozone concentrations. Subsequently their predictions were upheld and they were awarded the 1995 Nobel prize for chemistry, although the chemistry involved in ozone loss has in fact proved to be considerably more complex than this simple mechanism would suggest.

In 1985 Joe Farmer, Brian Gardiner, and Jonathan Shankin [4] of the British Antarctic Survey discovered a sharp reduction in ozone concentrations above Antarctica; this was soon termed the ozone 'hole'. Satellite monitoring of the ozone 'hole' has shown it is expanding and its depth increasing (Figure 2) and there is now evidence of similar but less marked changes over the Arctic and the more densely populated northern latitudes.

Such was the concern of the possible consequences of continued ozone depletion that global governmental meetings led to the agreement to phase out the production of all the CFC's (the Montreal Protocol (1987) and Rio treaty (1992)). However, due to long residence times of these compounds in the atmosphere it will be several decades before the negative influences of the CFC's are removed from the global atmospheric cycle. Attention is now being drawn to the effect of other halogen-containing species (e.g. bromine derivatives) and the increased emissions of NO due to a rise in the use of nitrates in agriculture, in turn leading to the increase in NO within the stratosphere, NO being another catalytic destructive agent for ozone. Other chemical pollutants may yet be found to play a role in ozone depletion. Continued industrial development and the introduction of new chemical production schemes make it likely that we will continue, unwittingly, to alter the atmosphere's natural balance. Thus research into the physics and chemistry of the ozonosphere will continue, as will monitoring of global ozone levels.

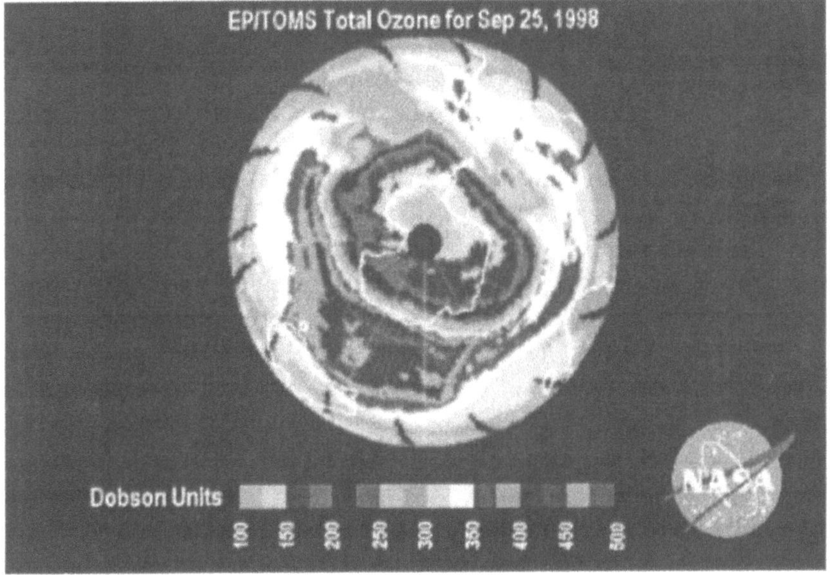

Figure 2. The 'ozone hole' as recorded by NASA, September 1998.

4.   OZONE SPECTROSCOPY

4.1   Ozone photo-absorption spectroscopy

It is the presence of ozone in the atmosphere that shields the Earth's surface from harmful solar UV radiation through its ability to absorb all solar radiation with wavelengths less than 295 nm. Figure 3 shows the absorption 'cross section' of ozone recorded using the apparatus illustrated in Figure 1 at the UK Daresbury synchrotron facility [5]. Uniquely amongst those molecules in the Earth's atmosphere ozone has a strong 'absorption band' (the so-called 'Hartley Band' named after its discover Hartley) between 210 and 300 nm. Hence ozone filters out the Sun's ultraviolet radiation below 300 nm preventing these wavelengths from reaching the Earth's surface.

226

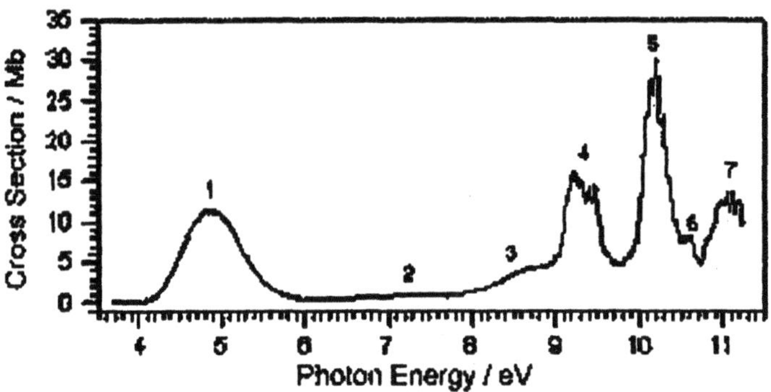

Figure 3. The Absorption spectrum of ozone.

Biological molecules have evolved under conditions of such filtering and thus while they do not absorb visible light they can safely absorb ultraviolet light since such radiation does not reach the Earth's surface. Figure 4 shows the solar emission spectrum reaching the Earth's surface for wavelengths below 340 nm together with the absorption spectra of two important biomolecules, DNA, the carrier of the genetic code, and α-crystallin the major protein of the mammalian eye lens. The absorption of light by both these biological molecules is essentially zero in the region $320 < \lambda < 400$ nm, the near-UV or UV-A region but it is intense in the region $200 < \lambda < 290$ nm, the far UV or UV-C region. The presence of ozone in the terrestrial atmosphere ensures that the absorption spectrum of these biological molecules only overlaps the solar spectrum at the Earth's surface in the wavelength region $290 < \lambda < 320$ nm, the mid-UV or UV-B region.

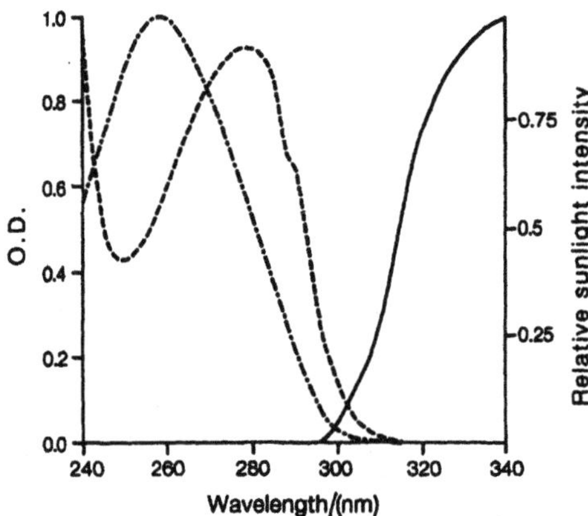

Figure 4. Absorption spectra of the biomolecules DNA ·······, the carrier of the genetic code, and α-crystallin ———— the major protein of the mammalian eye lens.

DNA and other biomolecules are therefore extremely sensitive to changes in the radiant solar flux an increase of which may therefore lead to severe damage of the genetic material (mutagenesis). It is in the mid-UV wavelength range that the reduction in the ozone shield will have the greatest influence upon biological systems. In particular it is now well established that the chromosphores of DNA are especially prone to UV damage, one consequence of which is the occurrence of erythema (sunburn) in human skin. Figure 5 shows the action spectrum (the damage done by a unit of irradiation of a certain wavelength) for the production of erythema. The probability of recurring erythema increases by five orders of magnitude between 350 and 280 nm, precisely that region where ozone is absorbing solar radiation, hence even small changes in ozone density and thence changes in the solar UV flux reaching the Earth's surface may lead to a dramatic increase in the number of cases of erythema. Similarly the mutagenesis of α-crystallin may lead to cataract formation in the human eye and, in the most severe cases, resulting blindness. Whilst humans may adapt to these changes (for example by wearing sun glasses and sun cream) other mammals are equally vulnerable and have less obvious methods of prevention. Nor are plants immune to the effects of changing solar UV levels. It is these potentially disastrous consequences of ozone depletion that have led to intense research upon the ozone layer in the past two decades and how it is being influenced by the Earth's increasing industrialisation.

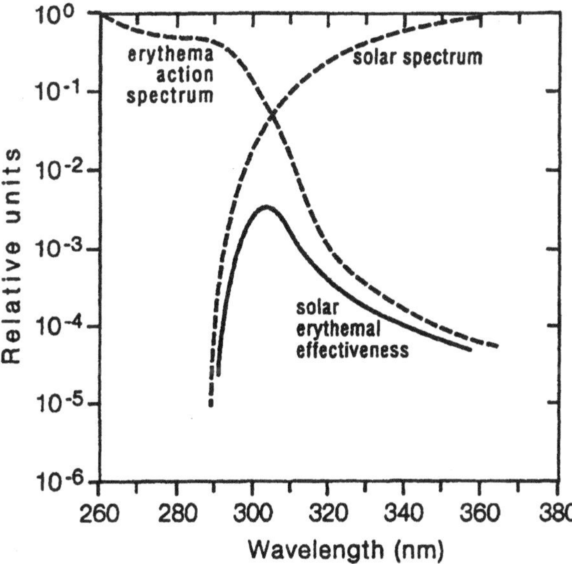

Figure 5. Erythema (sunburn) action spectrum plotted with solar spectrum and solar erythermal effectiveness (probability of damage to skin).

4.2   Ozone forbidden states

Considerable theoretical efforts have been made to calculate the excitation energies and dissociative properties of the electronically excited states of ozone. However until recently the results have been conflictory, and the ordering of the lowest lying states, and thence the assignment of observed absorption band structures, have remained ill-determined.

The method of electron-energy loss spectroscopy finally answered several of the remaining uncertainties on the order, nature and excitation energies of the lowest lying optically forbidden states of ozone [6]. Figure 6 shows an EELS spectrum under scattering conditions favouring triplet (forbidden) state excitation, while the Hartley band is still clearly evident between 4 and 6 eV, in the region of the Wulf and Chappuis bands new structure is observed comprising of a band with superimposed vibrational structure.

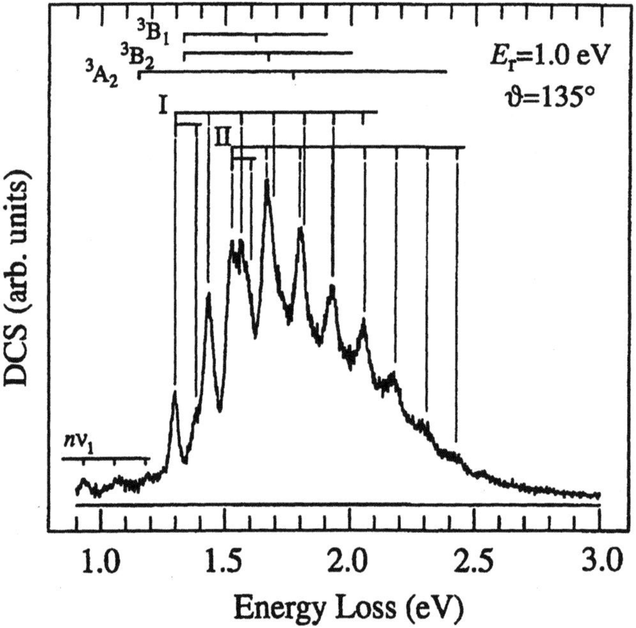

Figure 6. EELS spectrum of ozone showing forbidden transitions.

The vibrational structure can be plausibly, but not unambiguously divided into two progressions labelled I and II. The first has an origin at 1.29 ± 0.1 eV and can be ascribed to the $^3B_2$ state. The second progression may have an origin at 1.528 ± 0.01 eV or 1.45 eV the later value allowing it to be assigned to the $^3B_1$ state.

The recent EELS measurements therefore appear to have established the main spectroscopic features of the lowest lying excited states of ozone responsible for the Wulf and Chappuis bands. In particular they reveal that at least two of these states, $^3B_2$ and $^3B_1$ support vibrational excitation and therefore are likely to be 'metastable' in character. Thus if such states are excited in collisional processes within the atmosphere, or if they are populated during ozone formation processes, they will act as an energy 'sink', while providing the nascent ozone with sufficient internal energy to overcome certain chemical reaction barriers previously regarded as impenetrable. The role of such 'metastable states' in the ozone depletion mechanisms will therefore need to be evaluated by atmospheric modellers.

5. SPECTROSCOPY AND PHOTOLYSIS OF OZONE DEPLETING COMPOUNDS.

In understanding the ozone depletion processes discussed above it is obviously necessary to understand the formation mechanisms of the catalytic reactive species that destroy the nascent ozone within the stratosphere. The formation of such species is derived by simple one photon photolysis (dissociation) of the so-called 'reservoir compounds' e.g. $OClO$ and N_2O_5. It is therefore necessary to understand the absorption spectroscopy and dissociation dynamics of such species in as much details as the spectroscopy of ozone itself. Considerable effort has been expended in the compilation of a spectral atlas of those molecules believed to play the major role in global ozone depletion [7]. However our knowledge of the photoabsorption cross sections remains incomplete with relatively few temperature dependent cross sections having been measured. Most of the experimental measurements have been made using laboratory UV-vis instruments which are limited to a lower wavelength of 200 nm. Such a lower limit was felt to be satisfactory since the solar flux falls rapidly below 220 nm. However for several important stratospheric compounds the photoabsorption cross section rises rapidly below 200 nm such that the photolysis rate in this region may actually be comparable to that at longer wavelengths (where the cross section is lower but incident solar flux higher). This is clearly shown in Figure 7 for N_2O_5 an important reservoir compound in the lower terrestrial atmosphere. Such cross sections are easily measured using synchrotron radiation and form a major part of the authors' own research programme both in the UK and at other synchrotron sites world wide. The data collected will then be used in detailed computational models of to assess the role of such longer wavelength radiation in global ozone depletion.

230

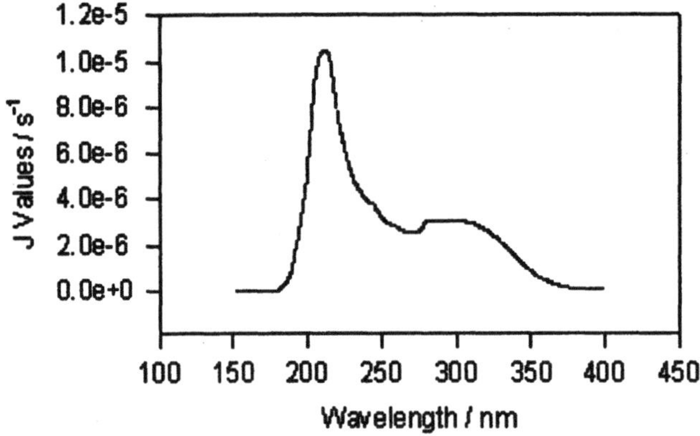

Figure 7 : Photolysis rate of N_2O_5 at 50 km.

6.   OZONE IN THE EARTH'S IONOSPHERE

Photo-dissociation and photo-ionisation processes dominate in the Earth's ionosphere, but a non-negligible proportion of the ionic species are formed by electron impact ionisation, the incident electron being formed in photo-ionisation processes. Indeed the local chemistry in the D-layer of the ionosphere is dominated by electron impact dissociation processes. Hence it is important to study the dissociation of stratospheric molecules by electron impact.

The chemistry of the D-layer of the ionosphere is dominated by the presence of negative ions. Such ions are produced by Dissociative Electron Attachment (DEA) of primary species followed by charge transfer between the products. Models of the chemical pathways in the D-layer therefore rely on accurate knowledge of the formation rate of O^- and O_2^- since this determines the formation of the CO_2^- and NO_3^- negative ions which dominate the local chemistry [8].

The possible DEA pathways in ozone may be summarised as follows

$$O_3 \; + \; e^- \longrightarrow \; O^- \; + \; O_2$$
$$O_2^- \; + \; O$$

DEA studies of reactive aeronomic molecules have only commenced in the last few years and there remain very few measurements of absolute cross sections for the formation of product anions. Recent experiments on DEA in ozone highlight the difficulty in measuring absolute cross sections for anion formation. The basic mechanisms of DEA in ozone were well characterised by the experiments of Allan et al [6]; Walker et al [9] and Skalny [10] but the last experiment was unable to detect fast anionic fragments. Skalny therefore failed to locate the DEA channel producing fast O^- ions above 2 eV and hence underestimated the DEA cross section for O^- at higher energies.

The recent experiments of Rangwala et al [11] and Senn et al [12] have measured the absolute DEA cross sections of both O^- and O_2^- from ozone. The two experiments agree both in the magnitude and the shape of the cross section for formation of O_2^- but Senn et al report, for the first time, (Figure 8) the presence of a low (essentially zero) energy O^- formation process. This low energy anion formation channel has important consequences for the chemistry of the D-layer.

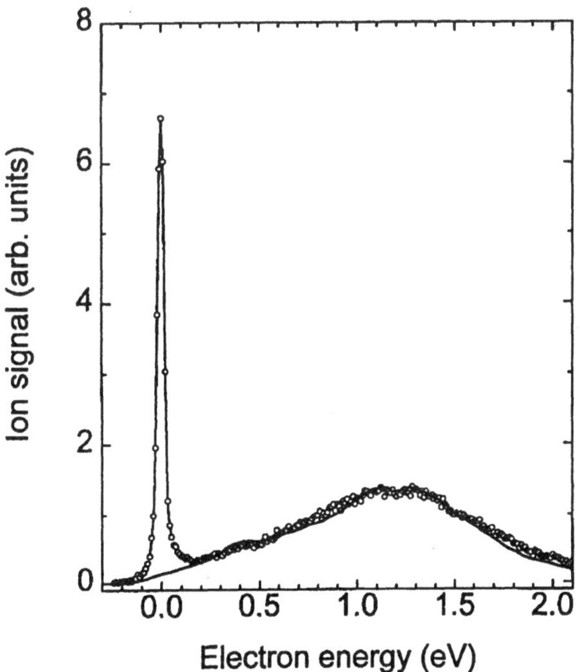

Figure 8. DEA production at low energies to ozone.

To date the major anion formation process in the D-layer has been attributed to the three body reaction

$$e^- + O_2 + O_2 \rightarrow O_2^- + O_2$$

however the discovery of a low energy dissociative electron attachment process in ozone suggests that an equal number of anions may be formed by the process

$$O_3 + e^- \longrightarrow O^- + O_2$$

Thus the subsequent ion-molecule chemistry in the D-layer will be as much due to nascent O^- as O^{2-}.

7. MOLECULAR CLUSTERS IN THE STRATOSPHERE

In the 1990s the role of heterogeneous chemistry within the terrestrial atmosphere became apparent [13].   Most of the key environmental regulating processes do not in fact occur in the gaseous phase but upon the surface of atmospheric aerosols or on cloud droplets.   Therefore it is important to study both the structure and composition of particulates within the terrestrial atmosphere, their solvation dynamics and hence the heterogeneous chemistry that may occur on the gas-surface interface.

During the Antarctica winter a vortex is formed over the continent. This develops a core of very cold air and it is these very low stratospheric temperatures below -80 °C which in turn lead to the formation of high altitude clouds commonly known as Polar Stratospheric Clouds (PSCs). The polar stratospheric clouds are formed due to condensation of nitric acid (HNO_3), sulfuric acid (H_2SO_4), and water (H_2O) in the stratospheric region. When freezing, the hydrates of HNO_3 become stable and are commonly known as nitric acid hydrate (NAD = $HNO_3.H_2O$) and nitric acid trihydrate (NAT = $HNO.3H_2O$).   The particles that contain appreciable HNO_3 are termed as Type I PSCs. Alternatively, the liquid aerosol can freeze to form sulfuric acid tetrahydrate (SAT) or other sulfate hydrates especially in the absence of HNO_3, these are known as Type II PSCs.

In considering such heterogeneous chemistry the simple photo-dissociation mechanisms described in Section 3 are no longer strictly true. A photolysed fragment may become trapped within a surface defect and recombine with other photolysed products to form a new isomer of the original compound e.g. photolysis of OClO may yield ClOO. Alternatively since as the gaseous molecules condense upon the surface they may form clusters either of the parent species or, since water is prevalent, solvated clusters.

The first spectroscopic studies and electron collision experiments involving clusters of atmospheric molecules have only been reported in the last decade. Most experiments have been concerned with water clusters [14] but more recently experiments on OClO [15] and ozone [16] have been reported. Electron impact studies have provided the first measurements of the ionisation energies of ozone and OClO dimers from which information on the structure of the dimer may be inferred. These experiments provide the first evidence that the dimer may form a cyclic structure. This leads to the possibility that cage like structures of ozone water may preferentially form on PSCs or in the higher pressure troposphere, both of which have profound implications for stratospheric ozone chemistry.

ACKNOWLEDGMENTS

The authors wish to thank all their colleagues Dr J A Davies, Dr H Zhao, Mr L Kaminski and Ms K Brown (University College London); Dr I C Walker (Heriot-Watt University, UK); Dr G Marston and Mr B Osbourne (Reading University,UK); Professor J Delwiche and Drs M J Hubin-Franskin and F Motte-Tollet (University of Liege, Belgium); Professor M Allan and Drs M Stepanovic and D Popovic (University of Fribourg, Switzerland); Professor T Mark, Mr G Senn, Mr G Denifl, Ms S Muigg (University of Innsbruck, Austria); Drs S V Kumar and E Krishnakumar and Mr S Rangwala (Tata Institute of Mumbai, India). We also thank the EPSRC, NERC, CLRC for funding this research and the Royal Society, British Council and NATO for funding collaborations. S K Pathak in particular wishes to thank the Royal Society for the award of a visiting fellowship during the preparation of this manuscript. Finally the authors thank the organisers of NCAMP XII for the invitation to present this talk and write this review.

REFERENCES

1. D. E. Heard, Atmospheric Enviroment **32**, 801 (1998).

2. *Spectroscopy in Environmental sciences* in Advances in Spectroscopy **24**, eds. R. J. C. Clark and R. E. Hester, Wiley (1995).

3. M. J. Molina and J. S. Rowland, Nature **249**, 810 (1974).

4. J. Farmer, B. Gardiner, and J. Shanklin, Nature **315**, 207 (1985).

5. N. J. Mason, J. M. Gingell, J. A. Davies, H. Zhao, I. C. Walker and M. R. F. Sigel, J. Phys. B **29**, 3075 (1996).

6. M. Allan, N. J. Mason and J. A. Davies, J. Chem. Phys., **105** 5665 (1996).

7. NASA Report No. 12.

8. L. Thomas and M. R. Bowman, Ann. Geophys. **4**, A3 219 (1986).

9. I. C. Walker, J. M. Gingell, N. J. Mason and G. Marston, J. Phys. B **29**, 4749 (1996).

10. J. D. Skalny, S. Matejcik, A. Kiendler, A. Stamatovic and T. D. Mark, Chem. Phys. Lett. **255**, 112 (1996).

11. S. A. Rangwala, S. V. K. Kumar, E. Krishnakumar and N. J. Mason, *6th EPS Conference on Atomic and Molecular Physics* (Siena) p4-124.

12. G. Senn, D. Muigg, G. Denifl, S. Matejcik, P. Cicman, J. Skalny, Y. Chu, N. J. Mason, E. Illenberger, A. Stamotovic, P. Scheier and T. D. Mark, *6th EPS Conference on Atomic and Molecular Physics* (Siena) p4-139.

13. R. P. Wayne et al, Halogen oxides; Radicals, sources and reservoirs in the laboratory and in the atmosphere. *Air Pollution Research Report 55* European Commission Directorate-General Science, Research and Development Environment Research Programme.

14. M. Ahmed, C. J. Apps, C. Hughes and J. C. Whitehead, Chem. Phys. Lett. **240**, 216 (1995).

15. K. Fenner, A. Furlan and J. R. Huber, J. Phys. Chem. A **101**, 5746 (1997).

16. S. Matejcik, P. Cicman, A. Kiendler, J. D. Skalny, E. Illenberger, A. Stamatovic and T. D. Mark, Chem. Phys. Lett. **261**, 437 (1996).

Confined Atoms: A New Path Towards Controlled Orbital Collapse

J.-P Connerade[1], P. Anantha Lakshmi[1,2] and V. K. Dolmatov[3]

[1]Physics Department
Imperial College, London SW72BZ, UK

[2]School of Physics
University of Hyderabad, Hyderabad 500 046, India

[3]Starodubtsev Institute
Tashkent 700084, Uzbekistan

1. INTRODUCTION

Atomic orbital collapse, which arises as a result of centrifugal barrier phenomena in many-electron atoms is a well-established effect [1], and has been invoked to account for anomalies in the filling of the d and f transition sequences [2], for the presence of giant resonances in many of their spectra [3], as well as for critical phenomena which appear as perturbations of Rydberg series [4].

Orbital collapse appears naturally as a function of atomic number, when one studies the filling of successive shells [5, 6]. It can also appear as a function of charge, if one considers the giant resonances of an element, of its singly-, doubly-, triply-, etc charged ions [7]. It can also be studied down a homologous sequence [8] (i.e. down columns of the Periodic Table). The possibility of controlling orbital collapse by changing the excitation state of the atom has also been suggested [9]. Finally, one can envisage another form of control in which either the chemical environment of the atom [10] or the solid state lattice surrounding an atom are modified [11].

In the present report, we consider another, related path, namely the influence of an external cavity on the atom. In most of our examples, it is

Trends in Atomic and Molecular Physics,
Edited by Sud and Upadhyaya. Kluwer Academic/Plenum Publishers, New York, 2000.

taken to be in its ground state. We treat the problem theoretically, and we show that the influence of the cavity is very similar to that of fractional ionisation (in which the ion carries an adjustable non-integral charge). We also consider the problem of whether the outermost electron is bound to the atom, or whether it is simply confined by the cavity.

The theoretical problem of a hydrogen atom in a cavity was considered by Sommerfeld and Welker [12] who explained its basic features and attempted to relate it to the problem of computing the atom in the solid. They showed how to construct wavefunctions for the hydrogen atom which satisfy the boundary condition of dropping to zero amplitude at the radius of the cavity by making use of the nodes of excited state functions, and they deduced an interpolation formula for the energy as a function of the radius of the cavity. However, their treatment was not applicable to many-electron atoms, and the role of angular momentum was not explored.

In later papers [13, 14] the properties of many-electron atoms in cavities were also considered. However, these papers treated the general problem, and did not tackle the issue of orbital collapse, which is perhaps the single most dramatic change within the atom which can be induced by changing the radius of the cavity.

We have developed simple Hartree-Fock models for an atom placed at the centre of a spherical cavity. By using such models we have shown that (a) the ground state of an atom or ion can be altered by inducing orbital collapse [15] (b) internal and external control of orbital collapse display remarkable similarities [16, 17] and (c) that there are general rules which determine the order of filling for atoms placed inside a spherical cavity, these rules being different from those of the conventional Bohr-Stoner *aufbau prinzip*.

A characteristic feature of confined transition elements is that their compressibility is strongly affected by *orbital collapse* [1]. This is readily understood in terms of the simple model of an atom with an internal short-range well of binding strength close to the critical condition for trapping an additional state [18]. Clearly, small changes in an externally applied potential can precipitate large changes in the wavefunction for this state. Taking the case where the *3d* function is close to critical binding because of the action of the centrifugal force, whereas the *4s* function is not, because it experiences no centrifugal force, we see that configurations involving *4s* electrons will be much less sensitive than those involving *3d* to the influence of a confining potential. This principle is exactly the same as the one which underpins the study of orbital collapse in homologous sequences [8]. We further deduce that confinement, by reinforcing the binding strength of the inner well, should favour *3d* filling over *4s*.

Recently, it was demonstrated [17] that the effect of confinement on the properties of *3d* and *4s* functions over the range where orbital collapse occurs is very similar to the effect of a (hypothetical) fractional increase in nuclear charge for a free atom. Since a fractional increase in charge can be used to subsume radial correlations [19], a spherical cavity may also be used in the same way. This result has implications for solid state physics, where better wavefunctions for localised states may be obtained by the device of an increase in the apparent charge of ionic centres. For compressed atoms, it suggests that *3d* filling will be favoured, because an increase in hydrogenicity (resulting from the increase in apparent charge) brings the order of filling closer to the one which would occur if the levels followed the strict hydrogenic ordering implied by the *aufbau* principle.

Self-consistent field calculations confirm these general ideas, with some restrictions. In the present paper, we provide examples to illustrate various aspects of orbital collapse and the filling of shells. We concentrate on the *d*-transition elements, which have the particular property that successive rows in the Periodic Table for free atoms do not fill in the same way, a property which, as will emerge in our paper, disappears for atoms within a certain confinement radius.

2. THEORETICAL METHOD

Our model is the same as previously used [15], viz. a Hartree-Fock configuration average scheme with a modified potential. We add a finite step in the potential of 10 atomic units, large enough for penetration of the barrier to be very small, outside a radius r_c. We implement a full Hartree-Fock procedure extending outwards beyond the radius of the sphere, to account for the very small but finite penetration depth.

Concerning convergence, several procedures were used in difficult cases. Solutions could be 'started' from a multiply charged ion, then the charge reduced in steps, based on the principle that convergence is more easily obtained as the hydrogenicity increases. This is a standard method for free atoms, and also works for atoms inside a cavity. Another option was to use a solution obtained for one value of the cavity radius as a starting function for solving for another cavity radius; this method was found to be very useful to explore ranges just above and just below the point of orbital collapse. Finally, there is the possibility of slowly turning up the height of the potential step, and retaining the self-consistent solution for one step height as the starting solution for another. Between them, these different methods allowed (i) solutions to be obtained for a large number of cases and (ii) solutions to be checked for numerical accuracy by approaching convergence through distinct paths.

238

We illustrate our results by a set of examples, showing the general principles of orbital collapse in confined atoms. The main result of our numerical study is to confirm the tendency for the filling of shells to favour d rather than s electrons as the system is confined, but one must also take account of the tendency of the system to 'ionise' as the cavity becomes smaller, i.e. for the electron to be bound not by the atom alone but by the cavity itself. This, as observed in ref 12, is similar to the emergence of a conduction band, because the electron is no longer localised on the atom. Rather, it is confined by the cavity, so that 'ionisation' can be taken as a primitive description of the birth of the conduction band in a Wigner-Seitz cell.

We now consider various aspects of the solutions for confined atoms.

2.1 Wavefunctions

Orbital collapse is known to occur as a result of competition between the inner short range well which develops in transition elements and rare earths and the outer, long range part of the effective radial potential for d and f states. The origin of this 'double-well' structure is discussed by Connerade [1, 18] and by Karaziya [20]. Key points to note for the present paper are (a) the orbitals involved are those with the highest angular momentum occupied in the system, because the double-well structure can only occur in the presence of a large centrifugal potential (b) that a positive barrier does not need to occur between the two wells for orbital collapse to manifest itself. It suffices that there be two regions, one with a short- and the other with a long-range well and (c) that s electrons, by virtue of the fact that $l = 0$, exhibit no centrifugal effects, which means that they can be used as 'markers', indicating the binding strength of the electrostatic potential within an atom

When a transition element or rare-earth is placed inside an external impenetrable cavity, the long-range outer well is replaced by a finite well which is less binding, and so the competition between the inner and outer wells is altered in favour of the former. As a result, orbital collapse into the inner well is induced by gradually reducing the radius of the cavity. The effect is similar to that of increasing the depth of the inner well, i.e. raising the effective charge of the nucleus. This similarity for excited states of Cr has been noted by Connerade and Dolmatov [17]. Here, the main emphasis is on the ground configurations of atoms and ions, since our concern is with the order of filling. We therefore present, in Fig. 1, an example demonstrating how the $3d$ valence electron of a rare-earth element can be controlled either by varying the cavity radius or by altering the charge state of the atom. As already noted, altering the charge state is not as artificial as might seem. Non-integral charge can be used to subsume certain radial correlations [19] whose effect is indeed to induce or modify orbital collapse.

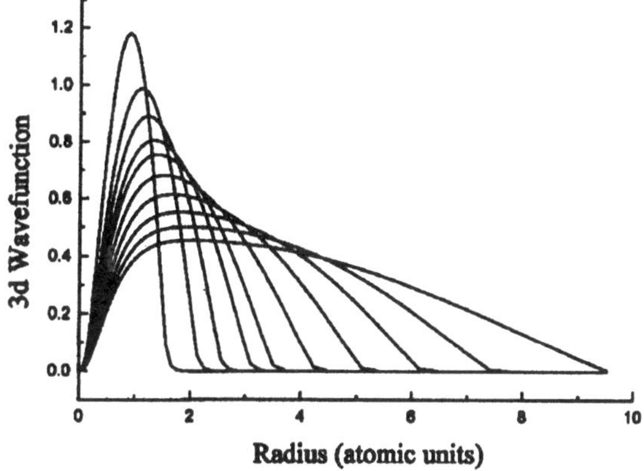

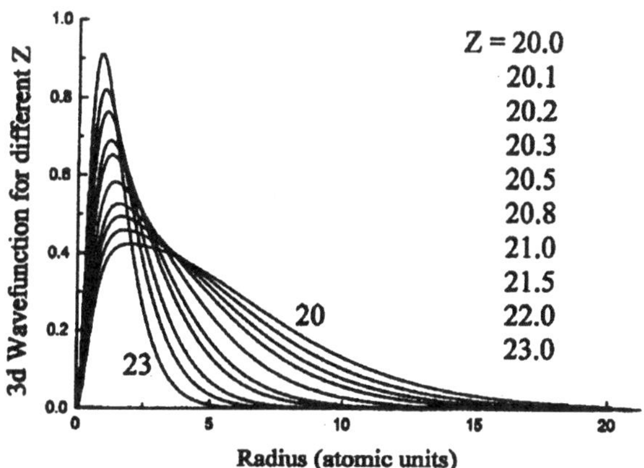

Figure 1. (a) The *3d* wavefunctions of Ca$^+$ inside a cavity. The wavefunctions fall to zero at a radius roughly equal to that of the cavity. The cavity radii are as follows (in atomic units): 1.5, 2.0, 2.5, 3.0, 3.5, 4.0,5.0,6.0,7.0,9.0; (b) the *3d* wavefunctions of a fictitious Ca$^+$ ion with a non-integral nuclear charge increasing above 20. The actual values are shown on the figure. Some of them correspond to real ions. Notice the similarity between the inner reaches of (a) and (b) commented on in the text.

From the standpoint of the *aufbau* principle, one should compare the *n = 3* wavefunctions among themselves, because anomalies of filling are due to differences of behaviour resulting in part from the angular momentum barrier. In Fig. 2, we show that the *3d* functions are 'soft' as compared with *3p* functions in the same element. This is demonstrated by plotting effective

240

radii as a function of the cavity radius, which brings out the large influence of the cavity on the *3d* electron.

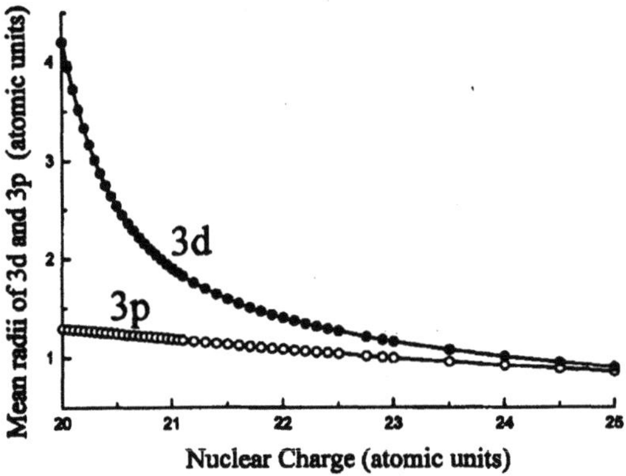

Figure 2. Mean radii of the *3p* and *3d* wavefunctions of Ca$^+$ as a function of the nuclear charge (the same situation as for Figure 1(b)). Notice that the *3p* core size changes rather slowly as the *3d* orbital collapses with increasing nuclear charge

Finally, in Fig. 3, we show the *3p* functions themselves for different cavity radii. This figure demonstrates that core wavefunctions are not much altered by the cavity over quite a wide range of sizes. Notice the rapid cut-off close to r_c, the radius of the cavity. In the calculations, the potential used had a high step, so that the penetration of the wall of the cavity is small.

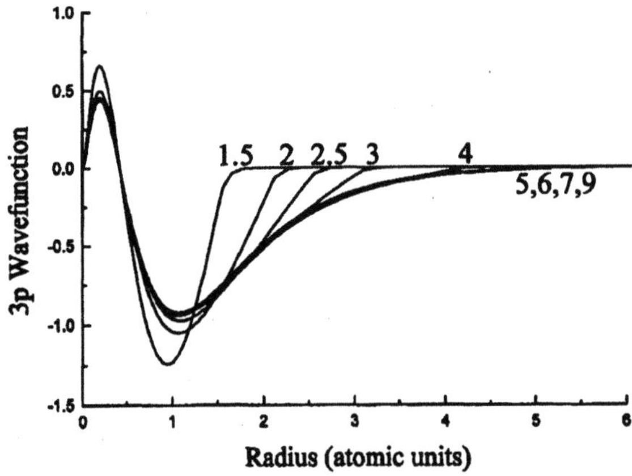

Figure 3. The change in the *3p* orbital for several different values of cavity radius (the same situation as for the *3d* orbital in Figure 1(a)). The values of r_c are indicated next to each curve. The inner lobe of the *3p* wavefunction and the first node change very little, even when the cavity radius becomes comparable with that of the core.

2.2 Shell filling within the same row

Inter-comparing elements within the same row of the Periodic Table is the basic method of identifying orbital collapse [5, 6]. As the atomic number increases, the inner well of the effective radial potential deepens. Orbital collapse occurs rather suddenly once the quantum condition for the inner well to hold a weakly-bound state is satisfied. Thereafter, filling can occur inside the atom rather than on its surface, and is responsible for the appearance of the long periods as the inner well continues to deepen. For the *(n-1)d*-transition elements, there is the added complication that the *ns* orbitals lie at roughly the same binding energy, with a radius comparable to the defining 'edge' of the d-inner well. Consequently, the filling of the two subshells occurs in competition: when an additional *ns* electron is introduced, it influences the properties of the *(n-1)d* electrons through the action of the self-consistent field. Since this is a very sensitive effect, and depends on the radial equation of the atom in a way which differs for different periods, *s-d* competition occurs differently for different rows. For example, Pd is the only element with a closed d^{10} outer subshell.

When atoms are confined, the situation is altered, and in particular the competition between *(n-1)d* and *ns* filling no longer appears in the same way. To understand why this happens, we consider first, in Fig. 4(a), the cases of K, Ca and Sc which lie at the beginning of the *3d* row. It is found that, with decreasing cavity radius, configurations involving a *3d* electron drop below corresponding configurations in which a *3d* electron is replaced by *4s*, so that the ordering begins to conform more closely to the ordering of principal quantum numbers. This behaviour is readily traced to the occurrence of orbital collapse: as the atom is compressed, the outer *(n-1)d* electrons find it easier to penetrate the core than the ns electrons, because the *d*-potential exhibits an inner well whereas the *ns*-potential does not. In effect, the influence of the cavity is to precipitate orbital collapse in the same way as an increase in the charge state, which would also lead to a change in the ordering.

It follows that, when the atoms are appropriately confined, the order of filling no longer conforms to the order for free atoms, but rather to that which would be expected if the *aufbau* principle did not break down at the long periods. For a confining radius of about 3 atomic units, orbital collapse is displaced towards lighter elements: *3d* filling begins not at Sc but at K, the lowest configuration of Ca is $3d^2$ rather than $4s^2$, and the lowest configuration of Sc is $3d^3$ rather than $3d4s^2$ (all within an independent particle, configuration average approximation). We will adopt the notation [Ca]4 for a Ca atom confined in a cavity of radius 4 atomic units.

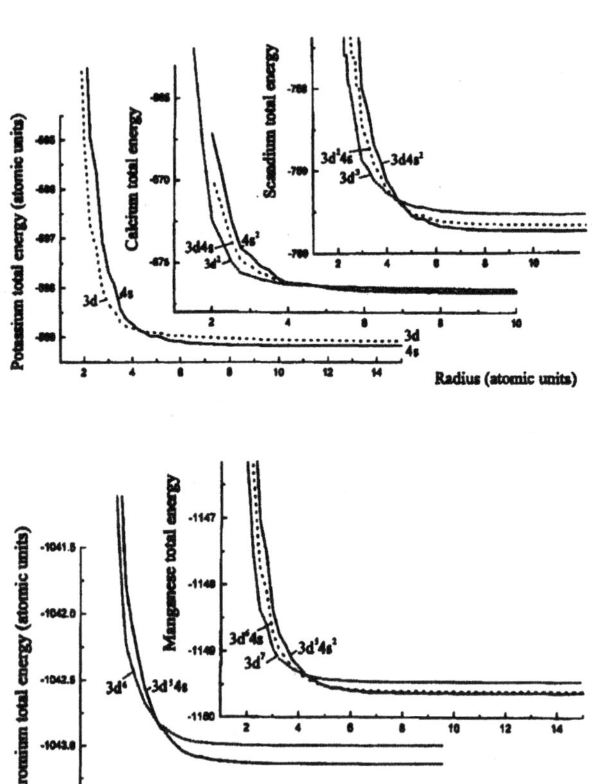

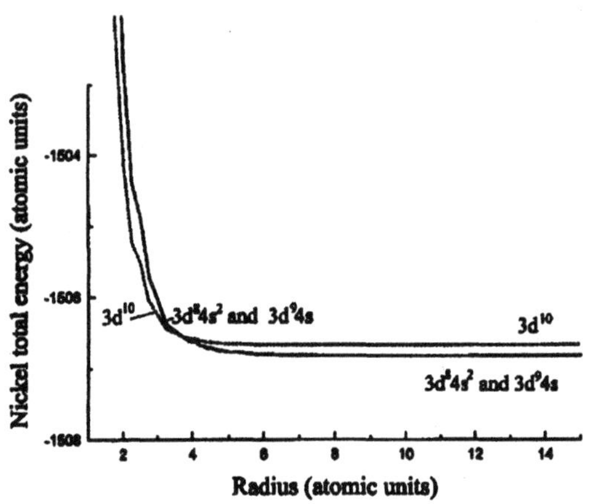

Figure 4. Total energies of the lowest configurations of: (a) Potassium, Calcium and Scandium; (b) Chromium and Manganese; and (c) Nickel; in spherical cavities, plotted as a function of the radius of the cavity. Notice that all the curves exhibit crossings, which are due to the higher compressibility of the *3d* functions as compared with *4s*.

Notice that the rare gases retain their closed shell configuration inside a cavity. Thus, the ground configuration of [Ar]3 remains $3p^6$, but there are also other consequences concerning excited states. Thus, for [Ar]3, the lowest excited configuration is not $3p^54s$ but $3p^53d$, with the $3d$ orbital collapsed, so that a $3p$-$3d$ giant resonance will occur as the dominant feature of the valence excitation spectrum.

Orbital collapse also has the consequence that $3d$-$4s$ competition is inhibited under compression. As we proceed down the row, all elements exhibit a similar tendency. Thus, half-way across the period (see Fig. 4(b)) [Cr]4 and [Mn]4 have lowest configurations $3d^6$ and $3d^7$. As we approach the end of the row, the $3d$ shell fills earlier than for free atoms, and thus [Ni]4 has the lowest configuration $4d^{10}$. It is only beyond Ni that the $4s$-subshell begins to fill under pressure.

It follows from this discussion that all the usual chemical properties of transition metals (in particular, their valence) are modified under pressure. This is further illustrated in section (iii) below, where we inter-compare elements from two different rows of the Periodic Table.

2.3　　Differences between successive rows

As explained in the previous section, s-d competition which characterises the order of filling of the free transition elements can be eliminated under pressure, so that the order of filling for confined atoms can proceed as predicted by the *aufbau* principle, that is to say: in the order implied by the principal quantum numbers.

One consequence of this is that differences between successive rows are eliminated. In other words, all the long periods will fill systematically for atoms in appropriate cavities. To emphasise why this happens, we show, in Fig. 5, the case of Sr. Notice that [Sr]4 has as lowest configuration $4d^2$, just as [Ca]3 has $3d^2$ in Fig. 4(a), and that the general behaviour of the configuration energies for [Sr] parallels the behaviour of the outermost configurations of [Ca].

This example raises the question of homologous sequences, i.e. groups of elements forming columns rather than rows in the Periodic Table. Orbital collapse down extended homologous sequences for free atoms was studied by Connerade [8]. For confined atoms, the situation is modified and, indeed, the sequences are formed of elements which acquire greater similarities from row to row when they are confined.

For simplicity, we discuss just one case. Pd is the only element of the Periodic Table for free atoms which possesses a closed d^{10} subshell. However, as noted in section (ii), the lowest configuration of [Ni]4 is $3d^{10}$. When Pd is placed inside a cavity, however, since the ground state is already

244

$4d^{10}$, no change of ordering can be induced by compression. Thus, the influence of the external cavity tends to restore the same order of filling for successive rows. A comparison between the energy curves for [Ni] and [Pd] is shown in Fig. 6. Note the absence of any crossing for Pd. In principle, these rules for filling extend also to heavier systems, and to the case of Pt. However, for heavy elements, a full treatment of relativistic effects would really be necessary, and is beyond our present code (see section (v) below for some further comments on heavy elements).

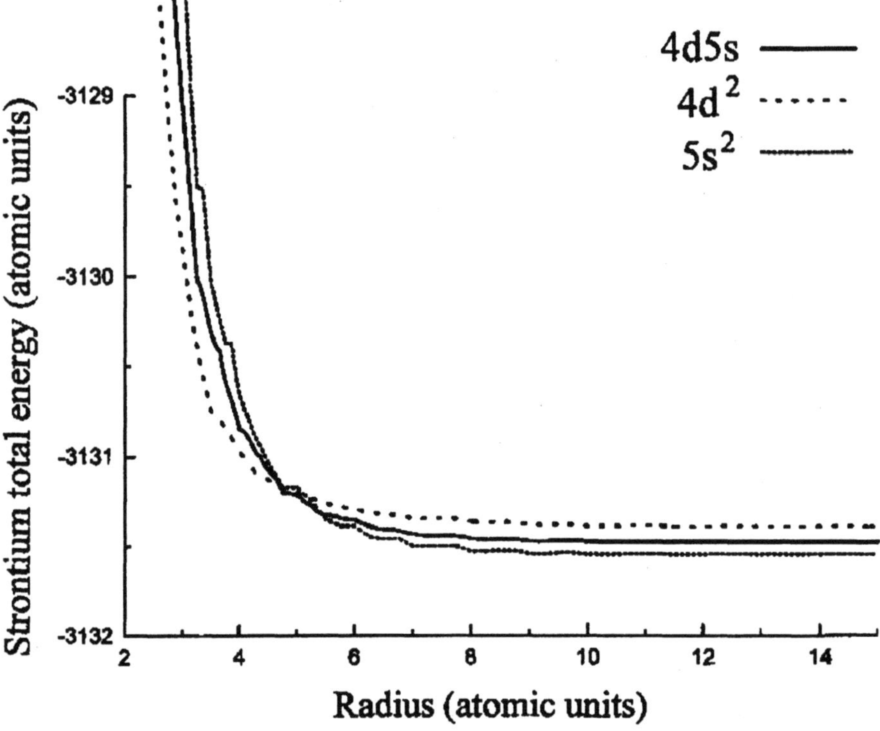

Figure 5. Total energies for the lowest configurations of Sr. This figure should be compared with Figure 1(a), and demonstrates that the behaviour of the *4d* period at the onset of filling is analogous to that of the *3d* period at its onset, and that the *4d* orbitals are more compressible than *5s*.

While on the subject of homologous comparisons, it is interesting to note that K, Rb and Cs in the metallic phase all have a higher resistivity than Na, despite the fact that the corresponding free atoms all have lower first ionisation potentials than Na, and that the element with the *lowest* first ionisation potential (Cs) gives the alkali metal with the *highest* resistivity. This is more readily understood if one notes that all these alkalis have np^6 cores, but that only Na has no corresponding *nd* orbital to fill. Thus, delocalisation of the *(n+1)s* valence electron alone does not suffice to account for the conductivity of the alkalis. The *nd* orbital must also be taken into account.

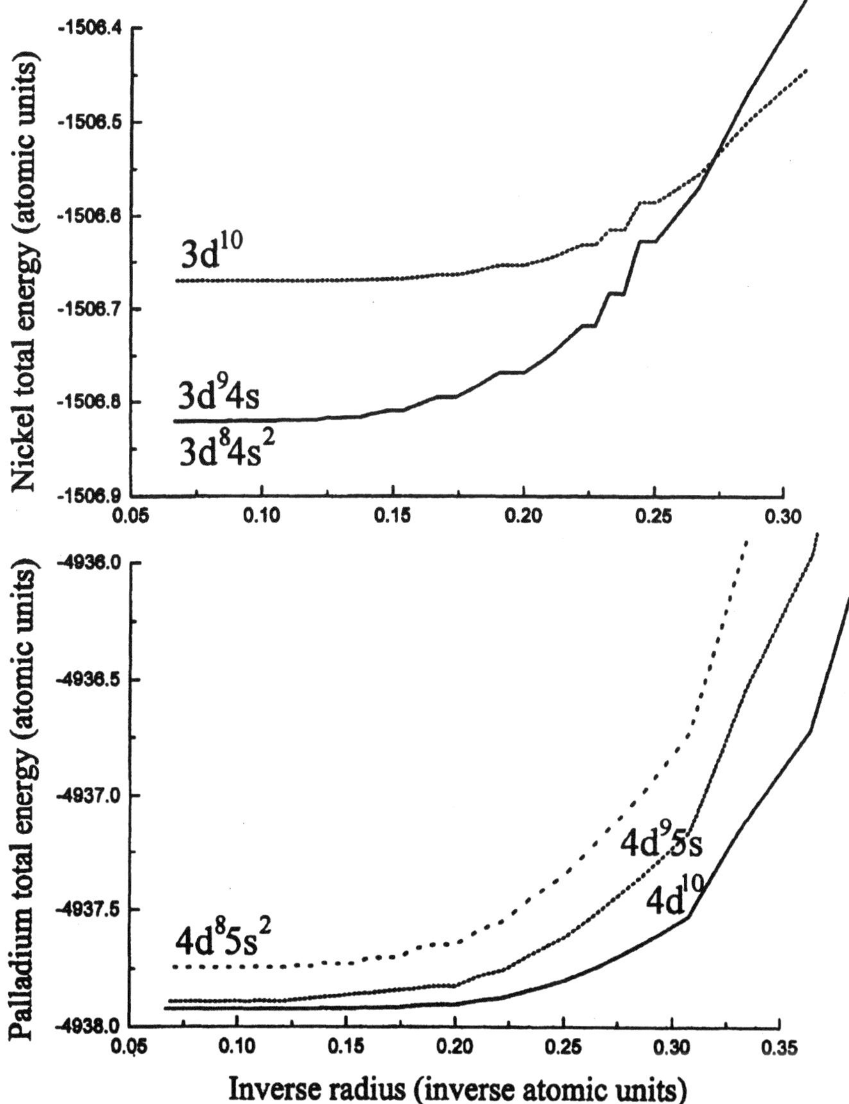

Figure 6. Comparison between the situation in Nickel (towards the end of the *3d* period) and Palladium (at the end of the *4d* period) showing that, for large cavities (free atom limit) the ordering of the configurations is not the same in the *3d* and *4d* periods, but that, as the cavity is made smaller *d^{10}* becomes the ground configuration in both cases, so that this anomaly of ordering is removed and both rows become identical under pressure.

246

2.4   'Ionisation' of confined atoms

A matter we have, for simplicity, held back until now is the question whether the outermost electron remains bound to the atom inside the cavity, or whether it is 'free' in the sense that it is confined only by the cavity.

This aspect of the problem was first considered by Sommerfeld and Welker [12] in their study of hydrogen inside a cavity. They showed that the electron remains bound to the proton so long as $r_c > 1.85\ a_0$. Below this cavity radius, it becomes an 'unbound' electron, in the sense that it is confined only by the cavity. In this respect, we can regard it as having 'ionised', or we can use the analogy pursued by Sommerfeld and Welker, who compared the spherical cavity with a Wigner-Seitz cell, and this 'ionised' electron with the birth of a conduction band.

This situation is quite clear-cut when dealing with hydrogen. With many-electron atoms, however, there is a richer variety of phenomena, owing to the presence of more than one ionisation threshold.

The lowest single-ionisation thresholds are of course those of the alkalis. Despite this, alkalis behave essentially like one-electron atoms, because removal of a second electron requires one to break open the closed p^6 subvalence shell, which is energetically much higher in energy. The alkaline-earth elements, on the other hand, also have very low double-ionisation thresholds, and are therefore the elements for which rearrangements due to 'ionisation' within the cavity produce the most striking effects. We therefore consider the case of calcium (Fig. 7).

Because of the low double ionisation threshold, calcium has extremely low-lying doubly-excited states. Indeed, the $3d^2$ configuration of Ca I lies below the single-ionisation threshold. The lowest energy doubly-excited configuration is $3d^2$. As can be seen on the right hand side of Fig. 7, for free atoms $3d^2$ is an excited state but lies below the ground state of the single ion. This rather unusual circumstance is actually due to orbital collapse, which explains why $3d^2$ is so sensitive to the presence of a cavity.

As the atom is compressed, the ordering of configurations reverts to the order predicted by the *aufbau* principle (see above), in which $3d^2$ becomes the ground state of the neutral atom. Before this occurs, however, it crosses above the first ionisation threshold and becomes a doubly-excited state situated above the $4s^2$ ground state, embedded in the ionisation continuum. This occurs in the range of r_c lying between Y and slightly to the left of X on Fig. 7. For free atoms, we would call such states autoionising states, and we recognise this situation as a normal one for doubly excited configurations: it is usual for them to lie in the autoionising range, i.e. above the first ionisation threshold, where lifetime broadening occurs.

Further compression takes the atom well past the point labelled X on Fig. 7. What now happens is that the *3d* ground state of the ion lies below the *3d²* ground state of the atom. Under these circumstances, by analogy with the situation when r_c lies between X and Y, we see that the ground state of the atom is no longer stable, but will 'autoionise' in the cavity. For sufficiently short periods of time, the atom will be in a superposition of states involving both the *3d²* discrete state and the *3d* ion plus one 'free' electron confined by the cavity.

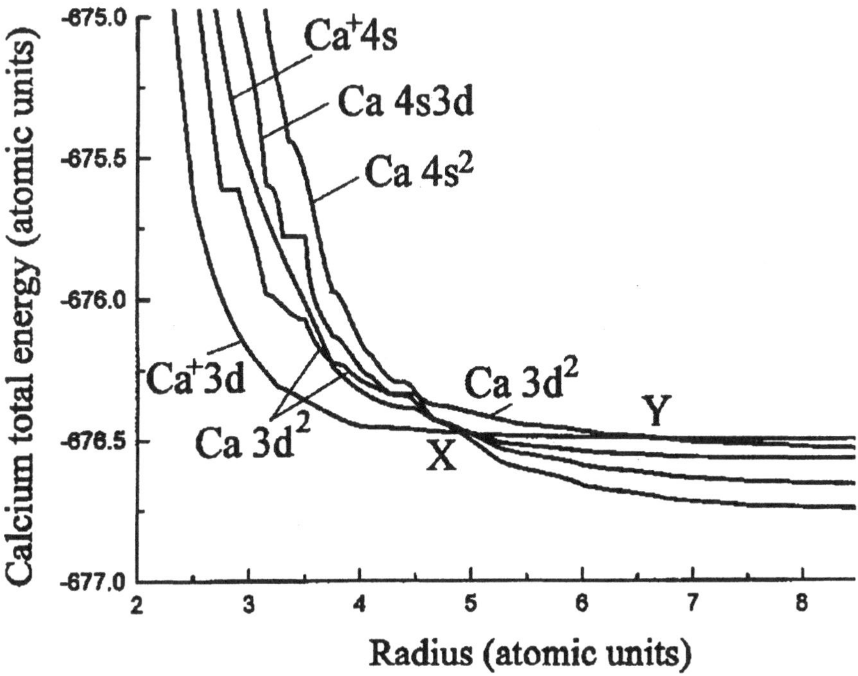

Figure 7. Total energies for neutral and ionised calcium in a cavity, plotted as a function of the cavity radius. Many of the curves cross in a point labelled X in the figure. Towards the left of this point, the configurations containing more *3d* electrons for a given charge state lie lowest while, towards the right of X, the configurations with more *4s* electrons and a given charge state are lowest in energy. The *3d²* configuration of Ca behaves differently. It is a bound state to the right of Y, rises above the first ionisation limit to the left of Y and is then an 'autoionising' excited state. Slightly to the left of X, *4s²* and *3d4s* rise above it, so that it becomes, in effect, an 'autoionising' ground state. Note that these curves are displayed as functions of inverse cavity radius rather than radius.

To pursue the analogy suggested by Sommerfeld and Welker [12], this situation resembles mixed valence in a solid: the atom then has the choice of how many electrons remain attached, and how many go into the conduction band. Both options may coexist on the same site.

Quite clearly, one can pursue the study of such crossings for other atoms. The alkaline-earths, however, have particularly low double-ionisation thresholds, and therefore provide the most interesting examples.

3. CONCLUSION

We have presented a discussion of how the Periodic table for atoms in the configuration average Hartree-Fock approximation is modified when the atoms are confined within a cavity. We note a tendency of shell-filling for confined atoms to follow the rules of the Bohr-Stoner *aufbau* principle beyond the point where it breaks down for free atoms, and for the long sequences of the Periodic Table to replicate from one row to the next. Thus, the chemistry of such elements in the solid and in compressed phases, and in particular such properties as valence differ from those of the corresponding free atoms.

Another application of confined atoms is in solid-state and atomic cluster physics: if we think of a solid as being made of atoms piled together, then they are of necessity confined, and the most appropriate starting wavefunction to use in their description is not that of the free atom. In particular, for transition metals and rare-earths, account must be taken of localised states deep inside the atom which can modify their valence properties and inhibit their conductivity.

Finally, there is the possibility that our conclusions may be relevant in discussions of dense matter produced under irradiation by high-power lasers.

To extend our work, we intend to perform the calculations in a Dirac-Fock scheme, with inclusion of spin-dependence and correlations for confined atoms, so that heavier species can be included, and the whole Periodic Table established within a single consistent model for confined atoms.

ACKNOWLEDGMENTS

The present research is supported financially under a Joint Project by the Royal Society of London. PAL is grateful for hospitality in the Laser Optics and Spectroscopy Group in the Physics Department at Imperial College, where the present work was performed while on leave from the School of Physics of the University of Hyderabad.

REFERENCES

1. J.-P. Connerade, Contemp. Phys. **19**, 415 (1978).

2. M. Goppert-Mayer, Phys. Rev. **60**, 184 (1941).

3. J.-P. Connerade and M. W. D. Mansfield, Proc Roy Soc **A341**, 267 (1974).

4. J.-P. Connerade and M. W. D. Mansfield, Proc Roy Soc **A346**, 565 (1975).

5. D. C. Griffin, K. L. Andrew and R. D. Cowan, Phys. Rev. **177**, 62 (1969).

6. D. C. Griffin, R.D. Cowan and K. L. Andrew, Phys. Rev. **A3**, 1233 (1971).

7. T. J. Lucatorto, T. McIllrath, J. Sugar and S. M. Younger, Phys. Rev. Lett. **47**, 1124 (1981).

8. J.-P. Connerade, J. Phys. B **24**, L109 (1991).

9. J.-P. Connerade, J. Phys. B **11**, L381; and L409 (1978).

10. A. A. Maiste et al, Sov. Phys. JETP **51,** 474 (1980); **52**, 844 (1980).

11. J.-P. Connerade, J. Phys. C (Solid State) **15**, L367 (1982).

12. A. Sommerfeld and H. Welker, Ann. der Phys. **32**, 56 (1938).

13. W. Jaskolski, Phys. Reports **271**, 1 (1996).

14. J. C. A. Boeyens, J. Chem. Soc. Faraday Trans. **90** (22), 3377 (1994).

15. J.-P. Connerade, J. of Alloys and Compounds **255**, 79 (1997).

16. J.-P. Connerade, V. K. Dolmatov and P. Anantha Lakshmi (submitted to Phys. Rev. A).

17. J.-P. Connerade and V. K. Dolmatov, J. Phys. B **31**, 3557 (1998).

18. J.-P. Connerade 1984 *New Trends in Atomic Physics* Vol II Les Houches, Lectures in Theoretical Physics Session XXXVIII Elsevier Amsterdam p645.

19. J.-P. Connerade, K. Dietz, M. W. D. Mansfield and G. Weymans, J. Phys.B **17**, 1211 (1984).

20. R. I. Karaziya, Sov. Phys. Usp. **24**, 775 (1981).

Ionization of Atoms Under Heavy Particle Impact

S. C. Mukherjee

*Indian Association for the Cultivation of Science
Jadavpur, Calcutta - 700 032, India*

1.   INTRODUCTION

When a charged heavy particle collides with an atom, a number of reactions can occur. In elastic collisions no energy is transferred. There may be excitations of either of the nucleus, charge transfer, ionization, transfer ionization and other reactions. If A and B are one electron atoms, the end products are like

$$A^+ + B \rightarrow A^+ + B \qquad \text{(a)}$$
$$\rightarrow A^+ + B^* \qquad \text{(b)}$$
$$\rightarrow A + B^+ \qquad \text{(c)}$$
$$\rightarrow A^* + B^+ \qquad \text{(d)}$$
$$\rightarrow A^+ + B^+ + e^- \qquad \text{(e)}$$

If the projectile is antiproton A^-, the reactions (c) and (d) are absent. In case of ionization (e) the final state contains three charged particles which mutually interact by long range Coulomb forces and as such the theoretical treatment of the problem is somewhat complicated. The investigations on ionization by heavy particle collision started with the calculation of total ionization cross sections (TICS) by Bates and Griffing [1] on $H^+ - H$ collision in first Born approximation. Calculated values in the higher energy region were in good agreement with the experimental data of Gilbody group

Trends in Atomic and Molecular Physics,
Edited by Sud and Upadhyaya. Kluwer Academic/Plenum Publishers, New York, 2000.

group at Belfast performed 10 years later. Below 200 keV incident proton, marked disagreement was noticed between theory and the experiment.

The main difficulty is the representation of the final electronic states where the emitted electron travels in the presence of two Coulomb potentials. Salin [2] and Mukherjee and Chowdhury [3] considered the distortion of the final states due to the presence of two Coulomb potentials. To study the effects of distortion in both channels Belkic [4] applied continuum distorted wave (CDW) approximation such that the initial bound wave function was distorted by a projectile continuum factor and the final wave function was taken as a product of a plane wave and two continuum factors corresponding to the projectile and residual target. Crothers and McCann [5] introduced CDW-EIS. In this model the final state is chosen as in the CDW but the initial state is represented as a bound state distorted by a projectile eikonal phase (EIS). To get deeper insight into the problem it is more useful to study the doubly differential cross section as a function of electron emission angle and energy. Rudd et al [6] measured doubly differential cross sections (DDCS) of H_2 in the energy range of 100 to 300 keV incident proton. Recently Kerby et al [7] measured the cross section in the energy region of 20 to 114 keV.

Classical trajectory Monte Carlo (CTMC) technique applied by Abrines and Percival [8] gives good results for electron removal by heavy particle impact. The method based on the simulation in ion atom collision in which a large ensemble of projectile – target configurations is sampled.

According to Bethe-Born approximation the ionization cross section of the charged particle impact depends on the collision velocity and square of the projectile charge.

$$\sigma_{ion} \approx Z_p^2 \log(v)/v^2,$$

Thus the cross sections for equivelocity proton and antiproton should be identical in this approximation. Recent experiments, however, have shown that for intermediate to small velocity the ionization by p and p⁻ impact have marked differences in both total and differential cross sections. Experiments [9] also show that the ratio of proton and antiproton cross sections in double ionization by helium tends to a factor of 2 ($\sigma_{p^-}^{++}/\sigma_p^{++} \approx 2$) at two to five MeV/amu. Clearly more detailed theoretical study seems necessary to explain this behaviour.

The production of fluxes of slow antiproton suitable for atomic collision experiment became possible after the availability of low energy antiproton ring (LEAR) facility at CERN since 1983. Antiprotons are produced by the bombardment of a thick target by a high energy proton beam from CERN proton synchrotron. The protons collide with nuclei in the target and proton/antiproton pairs are produced, along with other particles.

The high-energy antiprotons are then cooled and decelerated in a series of storage rings. Measurements comparing proton and antiproton single ionization cross section have been carried out by the Arhus group at Denmark with collaboration of CERN in the energy range of 0.5 to 5.0 MeV corresponding to velocity range 4.5 to 14 a.u.

In the case of heavy ion impact we can simplify the problem by applying the impact parameter formalism in which the projectile and target nuclei are assumed to move in a classical trajectory and only the electron is subjected to quantum mechanical laws. Unlike electron impact collisions there is no exchange interactions for heavy particle impact. However, for ionization in heavy particle collisions we should consider the coupling with other channels especially in the low energy region where the projectile and the orbital electron velocities are comparable. The role of two centred effects i.e. the combined influence of the target and the projectile Coulomb field is quite significant in the ionization by heavy particle impact.

For low and intermediate energy proton collision we applied a coupled channel semiclassical impact parameter calculation [10, 11] in which we have retained the effect of charge transfer channel in the initial state by solving time dependent Schrödinger equation by applying a variational method. The final state wave function considers the electron to be moving in the continuum of both the projectile and target nuclei.

2.   THEORY

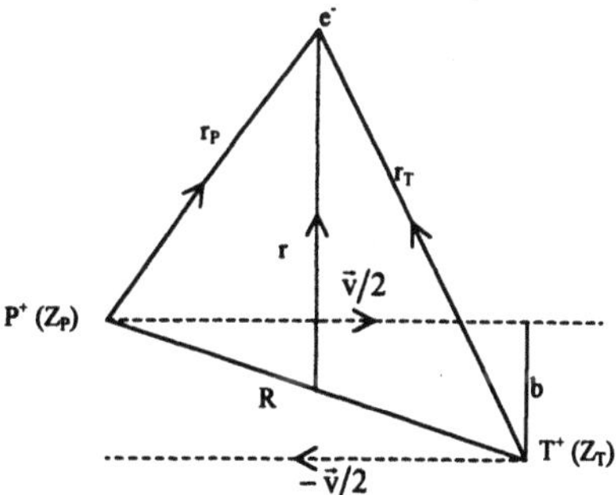

Let us denote the target T^+ and the projectile by P^+. The mid-point of the line joining P^+ and T^+ is taken to be at rest and is chosen as origin. P^+ and T^+ move with equal and opposite velocities $\bar{v}/2$ and $-\bar{v}/2$, $\bar{v}$ being relative velocity of the incident proton with respect to target.

254

Hamiltonian H_e for the electron moving in the field of the projectile and residual target is given by

$$H_e = -\frac{1}{2}\nabla^2 - \frac{Z_P}{r_P} - \frac{Z_T}{r_T} \tag{1}$$

(atomic units are considered throughout)
The nucleus-nucleus interaction term is omitted as this can be removed by a canonical transformation and as such it should not affect any transition probability for a particular impact parameter.
The development in time 't' of the electron wave function Ψ is given by the time dependent Schrodinger equation

$$\left(H - i\frac{\partial}{\partial t}\right)\Psi = 0 \tag{2}$$

with initial condition, at $t = -\infty$, the electron is attached to the target nucleus in the ground state. We write

$$\Psi = \int c_{k'}\Psi_{k_c'}^{-} \, d\vec{k}_c' + \sum_i a_i\Psi_{P_i} + \sum_j b_j\Psi_{T_j} \tag{3}$$

First term is integral over continuum states and the others are discrete states in direct and charge transfer channel. c_k, $\dot{a}_i$ and b_j are functions of time. On putting Ψ in the Schrodinger equation (2) we obtain

$$i\int c_{k'}\Psi_{k_c'}^{-} d\vec{k}_c' = \left(H - i\frac{\partial}{\partial t}\right)\left[\sum_i a_i\Psi_{Pi} + \sum_j b_j\Psi_{Tj}\right] + \int c_{k'}\left(H - i\frac{\partial}{\partial t}\right)\Psi_{k_c'}^{-} d\vec{k}_c' \tag{3a}$$

$\Psi_{k_c'}^{-}$ is an approximate solution of the equation,

$\left(H - i\dfrac{\partial}{\partial t}\right)\Psi_{k_c'}^{-} \approx 0$, the solution being exact asymptotically.

We neglect $\int c_{k'}\left(H - i\dfrac{\partial}{\partial t}\right)\Psi_{k_c'}^{-} d\vec{k}_c'$, considering $c_{k'}$ to be small.

Further,

$$\int \Psi_{k_c}^{-*} \, \Psi_{k_c'}^{-} \, d\vec{r} = \delta\left(\vec{k}_c - \vec{k}_c'\right)$$

Now multiplying both sides of the equation (3a) by $\Psi_{k_c}^{-*}$ and integrating over space we have

$$\frac{\partial}{\partial t}(c_k) = -i \int \Psi_{k_c}^{-*}\left[\left(H - i\frac{\partial}{\partial t}\right)\left(\sum_{ij}(a_i \Psi_{Pi} + b_j \Psi_{Tj})\right)\right] d\vec{r} \qquad (3b)$$

The ionization amplitude $c_k(t = +\infty)$ is obtained by time integration with initial condition $t = -\infty$, $c_k = 0$ and $|c_k(t = \infty)|^2$ corresponds to ionization probability.

If we denote the discrete parts of the wave function Ψ as Ψ_d, equation (3b) takes the form

$$\frac{\partial}{\partial t}(c_k) = -i \int \Psi_{k_c}^{-*}\left[\left(H - i\frac{\partial}{\partial t}\right)\Psi_d\right] d\vec{r} \qquad (4)$$

The continuum state wave function occurring in the final channel is represented by the product of two Coulomb wave functions around both the target and the projectile,

$$\Psi_{kc}^{-} = N_1 N_2 e^{i\vec{k}.\vec{r}} \, _1F_1\left(i\alpha_p, 1; -i\left(k_p r_p + \vec{k}_p.\vec{r}_p\right)\right) \times \, _1F_1\left(i\alpha_T, 1; -i\left(k_T r_T + \vec{k}_T.\vec{r}_T\right)\right) e^{-\frac{ik^2 t}{2}} \qquad (5)$$

where $\alpha_p = -\dfrac{Z_P}{k_p}$ and $\alpha_T = -\dfrac{Z_T}{k_T}$; $\vec{k}_p$ and $\vec{k}_T$ are the velocities of electron in the projectile and target frame of reference and $N_1 = \exp\left(-\pi\alpha_p/2\right)\Gamma\left(1 + i\alpha_p\right)$ and $N_2 = \exp\left(-\pi\alpha_T/2\right)\Gamma\left(1 + i\alpha_T\right)$.

The above wave function Ψ_{kc}^{-} asymptotically satisfies the Schrodinger equation

$$\left(-\frac{1}{2}\nabla_r^2 - \frac{Z_P}{r_p} - \frac{Z_T}{r_T} - i\frac{\partial}{\partial t}\right)\Psi_{kc}^{-} = 0. \qquad (6)$$

256

Now the discrete part Ψ_d is obtained by solving the time dependent Schrodinger equation

$$H\Psi_d = i\frac{\partial \Psi_d}{\partial t} \tag{7}$$

by a variational technique.

To solve the time dependent Schrödinger equation, we consider a trial wave function Ψ_T where the effect of the continuum state has been neglected and a number of lower most discrete states around the two moving nuclei, which seem important have been incorporated.

We solve the differential equation of $\Psi(\Psi_d)$ by making the integral

$$I = \int \frac{1}{2}\left[\Psi^*\left\{i\frac{\partial}{\partial t} - H\right\}\Psi + \Psi\left\{-i\frac{\partial}{\partial t} - H\right\}\Psi^*\right]d\vec{r}\,dt \tag{8}$$

stationary with respect to arbitrary variations of Ψ and its complex conjugate Ψ^*.

The trial wave function is a linear combination of two ground state wave functions around two protons

$$\Psi_T = a(t)\Psi_1 + b(t)\Psi_2 \tag{9}$$

where

$$\Psi_1 = (\pi)^{-1/2}\exp\left(-r_1 + i\varepsilon_0 t\right)\exp\left(-\frac{i}{2}\vec{v}.\vec{r} - \frac{i}{8}v^2 t\right) \tag{10}$$

and

$$\Psi_2 = (\pi)^{-1/2}\exp\left(-r_2 + i\varepsilon_0 t\right)\exp\left(\frac{i}{2}\vec{v}.\vec{r} - \frac{i}{8}v^2 t\right) \tag{11}$$

Integrating equation (4) over time we get corresponding ionization probability. The detailed calculation can be obtained from [10, 12].

The double differential cross section can be written as

$$\frac{d^2\sigma}{dE_e d\Omega_e} = k \int d^2b \, | \, c_k(b) \, |^2 \qquad \text{and the total cross section}$$

$$\sigma_{total} = 2\pi \int \frac{d^2\sigma}{dE_e \, d\Omega_e} \mathrm{Sin}\theta_e \; d\theta_e \, dE_e \; .$$

3. RESULTS AND DISCUSSION

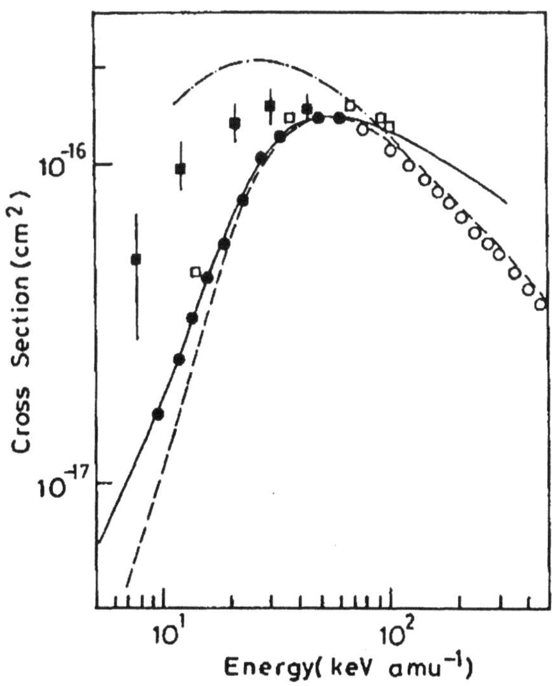

Figure 1. Total ionization cross section of atomic hydrogen by proton impact in the energy range of 5 to 300 keV/amu. The theoretical results used are as follows: Present results: (———), CDW EIS: (--------) and Born result: (-·-·-··-). The experimental results are as follows: Ref. [13]: (●), Ref. [14]: (○), Ref. [7]: (□) and Ref. [20]: (■).

In Figure 1 the present results for TICS of atomic hydrogen for proton impact energy of 5 to 300 keV are displayed and compared with the available experimental data and other theoretical findings. As evidenced from the figure the present calculation predicts cross section values especially in low and intermediate region up to 100 keV are found to be in better agreement with the data of Shah et al [13] and Shah and Gilbody [14].

However the present curve deviates from the experiment and other theoretical calculations above 100 keV impact energy. This may be due to non-inclusion of momentum transfer term in the calculation. Below 9.4 keV no experimental value is available for a proton –hydrogen system. The success of the present model at low and intermediate energies may be attributed to the fact that it takes into account the charge transfer channel the effect of which is very important particularly at low energies.

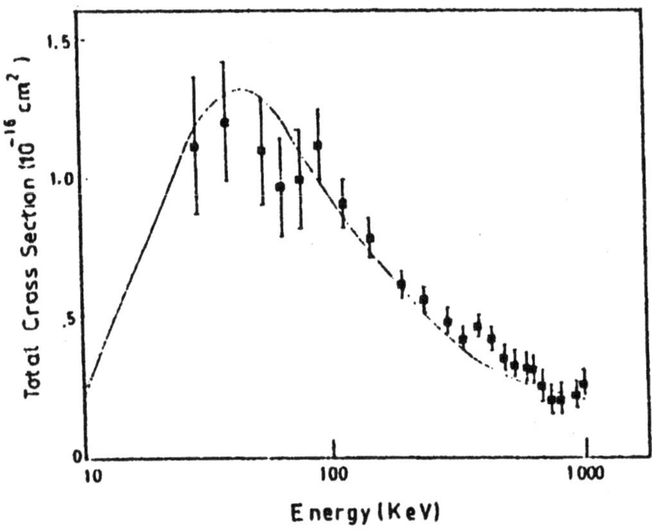

Figure 2. Total ionization cross section as a function of impact energy for antiproton impact on atomic hydrogen in the energy range of 10 to 1000 keV. Present theoretical results are represented by ———— and experimental measurements of Knudsen et al [9] are represented by ■.

In Figure2 we compare the present results of TICS for antiproton impact on atomic hydrogen in its ground state for incident energy ranging from 10 to 1000 keV. The present result agrees well with experiment in the energy range considered. In the calculated result we find the present curve drops off rapidly with decreasing energy which is in contradiction to the experiment. Fermi Teller [15] showed that for negatively charged heavy particle impact with positively charged particle in low energy region the electron moves under the influence of dipole like potential. In the limit when the internuclear separation R reaches a critical value .639/Z a.u. known as Fermi Teller radius, the bound state of the electron can not be supported [16, 17]. Thus the low energy ionization cross section will not drop off as the present curve does but would form plateau. The present perturbation approach may cause the unphysical drop off as this is basically a high energy approximation.

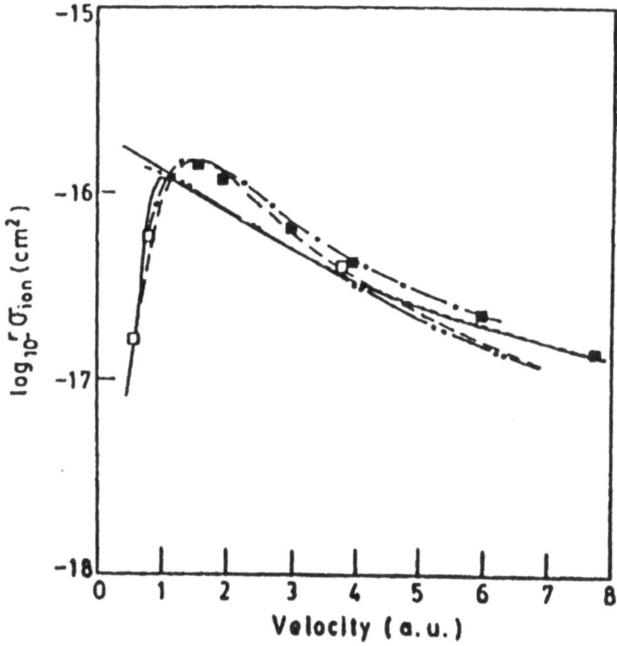

Figure 3. Total ionization cross section of atomic hydrogen by proton and antiproton impact in the energy range of 10 to 1566 keV/amu. Present theoretical results are represented by —— for proton and ------- for antiproton impact. CTMC results are represented by -··-··- for antiprotons and – – – for protons. CDW-EIS Crothers and McCann [5] results are represented by -··-··- for protons. Experimental results of Shah and Gilbody [14] are represented by ■ and Shah et al [13] by □.

In Figure 3 we display our results for total ionization cross section (TICS) for direct ionization in collision with protons and antiprotons with atomic hydrogen along with the available experimental measurements in the energy range of 10 to 1566 keV/amu. For the case of proton impact the present calculated result are in fairly good agreement with the classical trajectory Monte Carlo (CTMC) and CDW-EIS results as well as the experimental measurements at the intermediate and high energy region. As regards the maximum the CTMC and CDW-EIS results are slightly closer to the measurements of Shah et al than ours. In case of antiproton impact the present result almost coincides with the CTMC values in intermediate energy region. Figure 3 also reveals the difference in cross section between proton and antiproton impact. At small impact velocities there exists substantial difference on the basis of projectile charge. At low velocities the collision time is relatively lower and if a proton removes an electron from the target it has high probability of capturing it because of the attractive interaction with the electron. As a result the ionization cross section becomes smaller. At low velocities the effect of charge transfer process on

260

ionization is more significant. For antiproton impact on the other hand the electron after its removal from the target is repelled by the antiproton. Thus the ionization cross section values for antiproton impact definitely be higher than those for proton impact at low velocities.

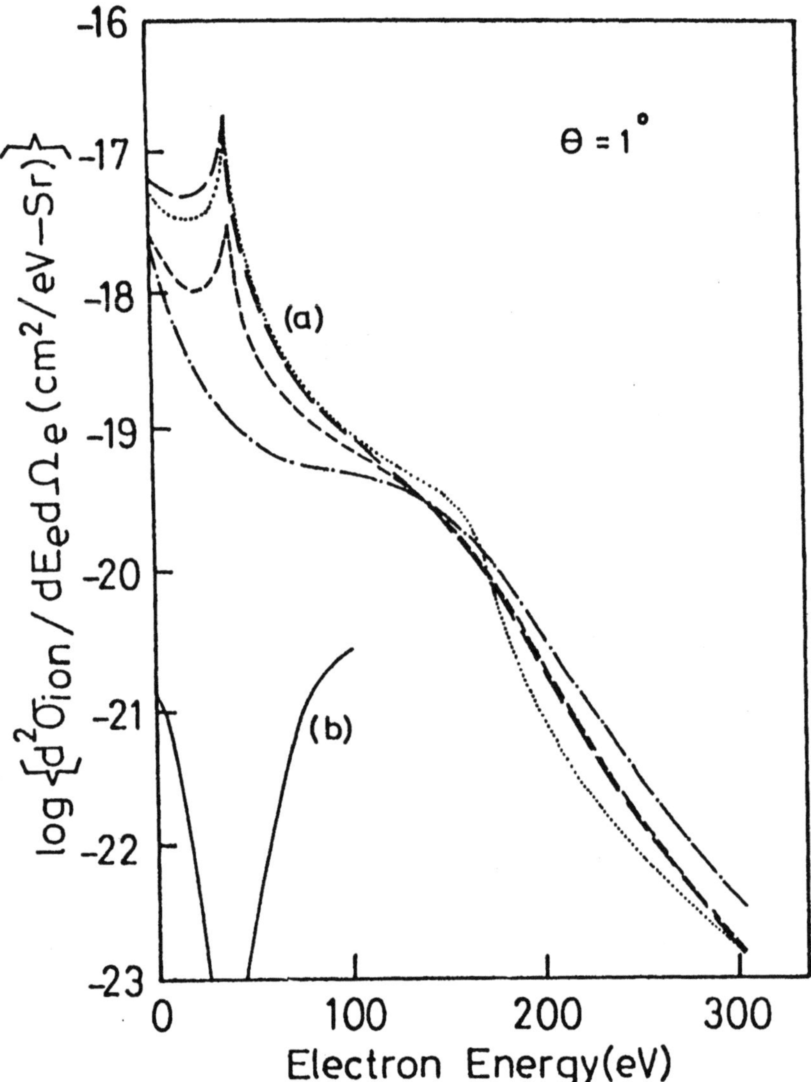

Figure 4. Doubly differential cross section as a function of electron energy for ejection angle of 1° for the ionization of atomic hydrogen in ground state for 70 keV proton impact (a) and antiproton impact (b). In figure (a) the results are represented as follows: Sahoo et al [12]:———, Born:-·-·-·-, CDW-EIS: ------ and CTMC: – – –. In figure (b) the present results are represented by ———.

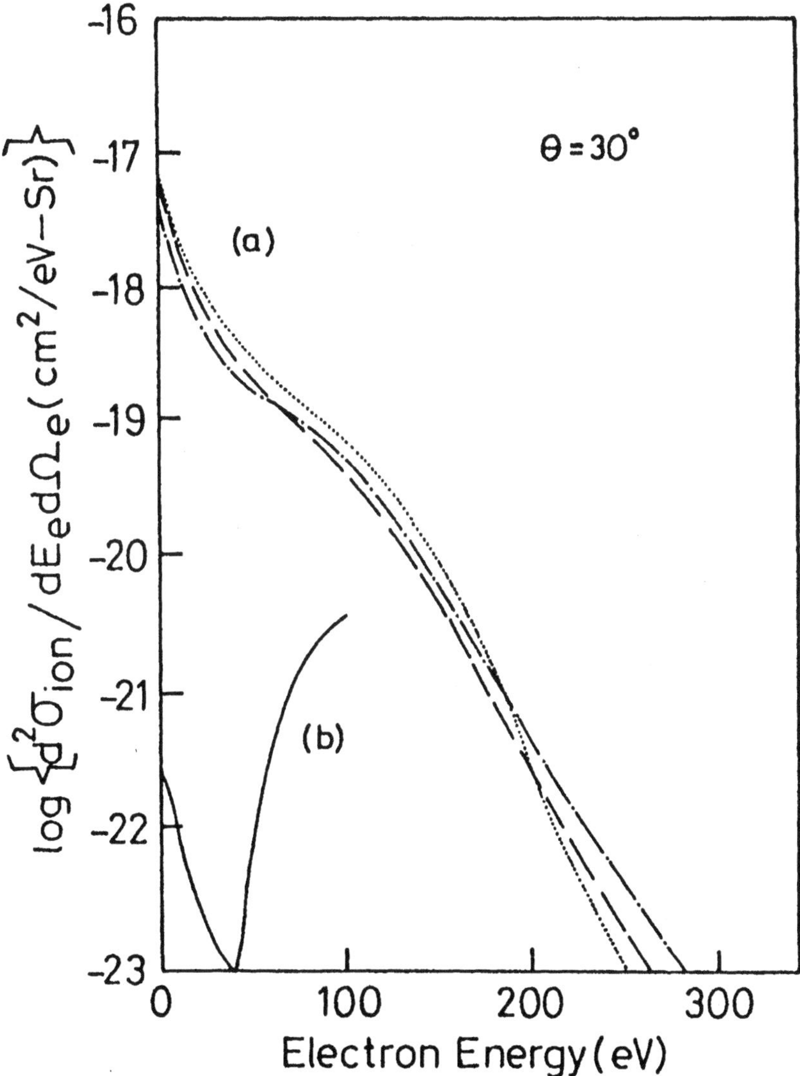

Figure 5. Doubly differential cross section as a function of electron energy for ejection angle of 30° for the ionization of atomic hydrogen in ground state for 70 keV proton impact (a) and antiproton impact (b). The group (a) of three curves represent results of Sahoo et al [12] (———), Born (-·-·-·-) and CDW-EIS (----------). The group (b) of only one curve depicts present results represented by ———— .

In figures 4 and 5 we compare the calculated results as a function of electron ejection energy for the ejection angles of 1° and 30° respectively for the ionization of atomic hydrogen for proton and antiproton of energy 70 keV with existing theoretical results. For proton impact the present result shows a maximum which corresponds to the well-known cusp due to electron capture to continuum and exponential dip for antiproton impact.

The cusp originates due to Coulomb focussing of electron ejected with nearly same velocity as that of projectile $\left(v_e(\text{ejected elecrton velocity}) \approx v_p(\text{projectile velocity})\right)$. Projectile energy of 70 keV corresponds to an equivelocity electron energy of 38 eV. The cusp is prominent at very small ejection angles. Our curve for proton impact at an ejection angle of 1° agrees well with the findings of CTMC and CDW-EIS approximations. In the Born calculation no such maxima are visible. These features in the differential cross section are related to the three-body character of the problem.

The electron ejection cross section depends on the density of states popularly known as Coulomb factor associated with the final state wave function.

$N_1 = \exp\left(-\pi\alpha_p/2\right)\Gamma\left(1+i\alpha_p\right)$; its square appears as a factor in the cross section. So

$$|N_1|^2 = \left|\exp\left(-\pi\alpha_p/2\right)\Gamma\left(1+i\alpha_p\right)\right|^2, \text{ where } \alpha_p = -\frac{Z_P}{k_p}$$

$$= \exp\left(-\pi\alpha_p\right)\frac{\pi\alpha_p}{\text{Sinh}\left(\pi\alpha_p\right)} = \frac{2\pi\alpha_p}{\exp\left(2\pi\alpha_p\right)-1}.$$

For a positively charged particle $Z_P > 0$, $\lim\limits_{k_{pe}\to 0}|N_1|^2 \to \infty$ gives rise to a sharp peak in the cross section and similarly for the case of negatively charged projectile ($Z_P < 0$) the factor $|N_1|^2 \to 0$ in the limit of $k_{pe} \to 0$. This gives rise to the singular structure that leads to an exponential dip in the cross section [18].

The sign of the projectile charge contained in the Coulomb factor is responsible to yield different structures in the DDCS. The cusp and anticusp in the DDCS can be considered as two-centred electron emission effect (TCEE). Due to long range character, Coulomb potential distorts the unperturbed wave function even when the two nuclei are far apart.

In Figure 6 we display the calculated results of Sahoo et al [19] for doubly differential cross sections of He^+ by 300 keV antiproton impact for fixed emission angle of 0° and10°. The doubly differential cross sections display the prominent feature in the electron ejected spectra (a sharp

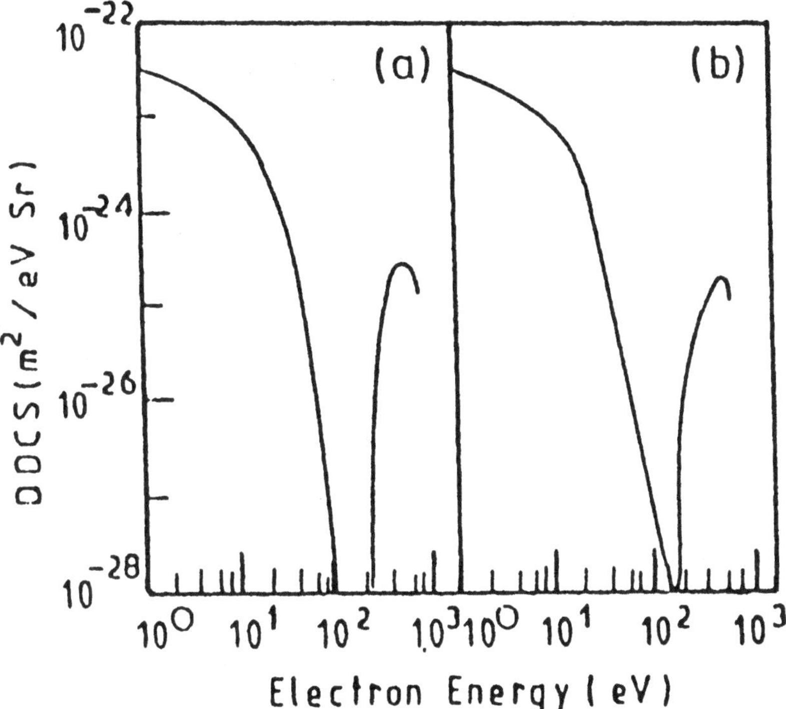

Figure 6. Doubly differential cross section of He$^+$ as a function of ejection energy by impact of antiproton of incident energy 300 keV for fixed emission angle of (a) 0° and (b) 10°. ———— represents the present results (Ref. [19]).

exponential dip in the region of electron capture to continuum (ECC) peak in the forward direction) and remains as an important feature even at10°. The exponential dip arises from ejected electron travelling with nearly same velocity as that of projectile as indicated above.

In Figures 7 and 8, we compare our calculated results as function of electron ejection energy for the angles of 1° and 30° respectively for the ionization of atomic hydrogen in the ground H (1s) and metastable H (2s) states by proton impact of energy 70 keV with existing theoretical results. The present calculated result, is identical with the other calculated results, shows a maximum which corresponds to the well known cusp due to electron capture to continuum peak (explained earlier). For the case of H (2s) the present results shows a very prominent binary encounter peak around ejected electron velocity $2v_p \cos\theta$ for both 1° and 30° similar to those obtained by other theoretical calculations.

264

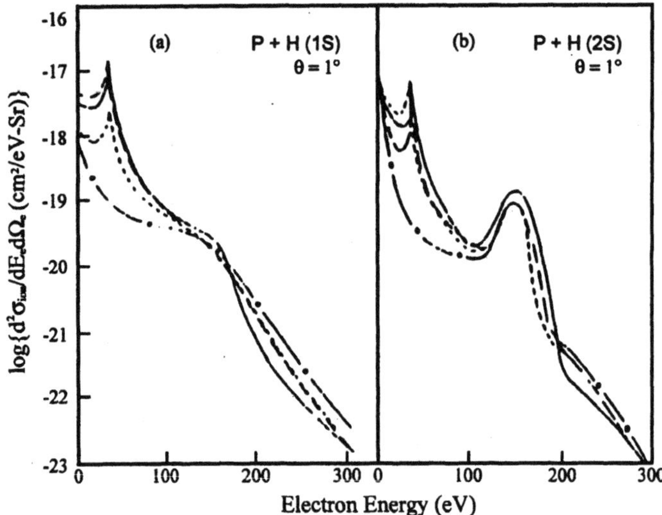

Figure 7. Doubly differential cross section as a function of electron energy for ejection angle of 1° for the ionization of atomic hydrogen in (a) ground state and (b) metastable state for 70 keV proton and antiproton impact. Both the windows (a) and (b) depict results of present calculations (————), Born (-·-·-··-), CDW-EIS (– – –) and CTMC (--------).

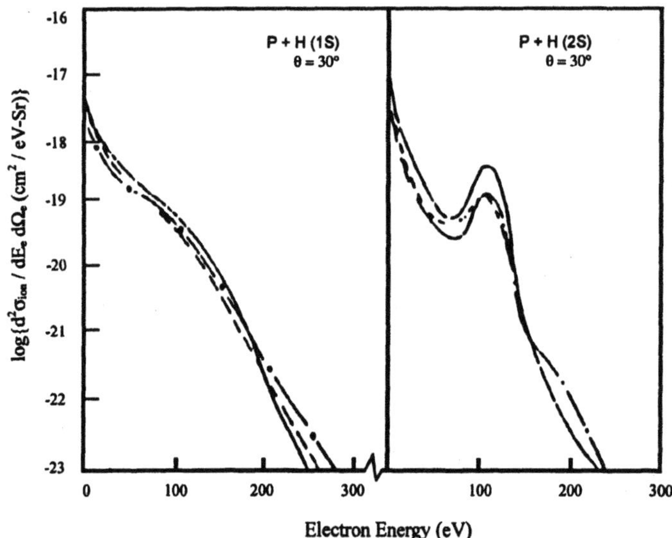

Figure 8. Doubly differential cross section as a function of electron energy for ejection angle of 30° for the ionization of atomic hydrogen in (a) ground state and (b) metastable state for 70 keV proton and antiproton impact. Both the windows (a) and (b) depict results of present calculations (————), Born (-·-·-··-), CDW-EIS (– – –) and CTMC (---------).

4. CONCLUSION

Total and Doubly differential cross sections for proton impact and total ionization cross section for antiproton impact on atomic hydrogen in both ground and metastable 2s states have been calculated by using a product of two Coulomb wave functions around the two nuclei in the final state. Comparisons have been made with the existing theoretical as well as experimental results. As expected, the present values for both proton and antiproton impact agree reasonably well with those of other available theoretical findings and experimental data in the intermediate and high energy region.

Total ionization cross sections for proton impact ionization of atomic hydrogen in ground state have been calculated by applying a coupled channel semiclassical impact parameter treatment. The results thus obtained are compared and found in very good accord with the experimental data and other theoretical findings in low energy region. It is interesting to note that the performance of the present coupled channel calculation, wherein the conservation of probability has been ensured, is very encouraging. This appears to point out that the enforcement of unitarity is rather more important than inclusion of a large number of terms in the expansion of the wave function.

ACKNOWLEDGMENTS

The author is grateful to S. Sahoo, K. Roy and N.C. Sil for their valuable comments and discussions. The author also likes to express his thanks to the International Atomic Energy Agency (IAEA), Vienna, for encouragement under Research Agreement No. 8003/CF. Financial support from Department of Science and Technology, Govt. of India, under Research Agreement No. SP/S2/K-01/93 is highly appreciated.

REFERENCES

1. D. R. Bates and G. Griffing, Proc. Phys. Soc A **66**, 961 (1953).
2. A. Salin, J. Phys. B **2**, 631 (1969a).
   A. Salin, J. Phys. B **5**, 979 (1972).
3. S. C. Mukherjee and S. M. M. R Chowdhury, Lett. Al. Nouvo Cimento **7**, 629 (1973).
4. Dz. Belkic, J. Phys. B **11**, 3529 (1978).

5. D. S. F. Crothers and J. F. McCann, J. Phys. B **16**, 3229 (1983).

6. M. E. Rudd, C. A. Sautter and C. L. Bailey, Phys. Rev. **151**, 20 (1966).

7. G. W. Kerby, M. W. Geally, Y. Y Hsu and M. E. Rudd, Phys. Rev A **51** 2256 (1995).

8. R. Abrines and I. C. Percival, Proc. Phys. Soc. **88**, 861 (1966).

9. L. H. Andersen, H. Hvelplund, H. Knudsen, S. P. Moller, K. Elsenser, K.-G. Rensfelt and E. Uggerhoj, Phys. Rev Lett **57**, 2147 (1986).

10. S. C. Mukherjee, S. Sahoo and K. Roy, Ind. J. Phys. B **69**, 396 (1995).

11. S. Sahoo, K. Roy, N. C. Sil and S. C. Mukherjee, Phys. Rev. A (in press).

12. S. Sahoo, K. Roy, N. C. Sil and S. C. Mukherjee, Physica Scripta **58** 126 (1998).

13. M. B. Shah, D. S. Elliot and H. B. Gilbody, J. Phys. B **20** 2481 (1987).

14. M. B. Shah and H. B. Gilbody, J. Phys. B **14**, 2361(1981).

15. E. Fermi and E. Teller, Phys. Rev **72**, 399 (1947).

16. I. Shimamura, Phys. Rev A **46** 3776 (1976).

17. M. Kimura and M. Inokuti, Phys.Rev A **38**, 3801 (1988).

18. S. Sahoo, Ind. J. Phys. (in Press).

19. S. Sahoo, K. Roy, N. C. Sil and S. C. Mukherjee, Fizzika A (in Press).

20. W. L. Fite, R. F. Stebbing, D. G. Hummer and R. T. Brackmann, Phys. Rev. **119**, 663 (1960).

Classical and Quantum Mechanical Investigations on Charge Transfer in Heavy Ion-Atom Interactions

C. R. Mandal

Department of Physics
Jadavpur University, Calcutta - 700 032, India

1. INTRODUCTION

Theoretical investigations on charge transfer in collisions of fully stripped ions with neutral atoms are exhaustive [1-4]. In contrast, a few theoretical studies [5-10] have so far been made to study charge transfer cross sections in collisions of partially stripped ions with neutral atoms at intermediate and high-energy region. In this respect, the scenario of experimental investigations on partially and fully stripped ions is well balanced. In this connection, works of Goffe et al [11] and Phaneuf et al [12] are worthy of mentioning. Olson and Salop [5] have calculated charge transfer and ionization cross sections by fully and partially stripped ions of B^{q+}, C^{q+}, N^{q+} and O^{q+} ($q \geq 3$) from atomic hydrogen in the framework of Classical Trajectory Monte Carlo (CTMC) simulation method in intermediate and high energy region. They have treated the interaction of the active electron with projectile ion as coulombic with an effective charge determined from spectroscopic data. Eichler et al [6] have calculated charge transfer cross sections for the same processes except for B^{q+} replaced by Li^{q+} ($q \geq 3$) in OBK- approximation. They have treated the projectile ions as bare ions with asymptotic charge (q). However, they have multiplied the calculated cross sections by a reduction factor obtained from eikonal approximation and successively by Pauli blocking factor. Under the circumstances, our present motivation is to have rigorous and exhaustive theoretical studies on charge transfer processes in collisions of partially

Trends in Atomic and Molecular Physics,
Edited by Sud and Upadhyaya. Kluwer Academic/Plenum Publishers, New York, 2000.

268

stripped heavy ions with neutral atoms both quantum mechanically as well as classically. Our quantum mechanical formalism has been confined to boundary corrected intermediate state (BCCIS) approximation originally proposed by Mandal et al [13]. Classical formalism is centred around Classical Trajectory Monte Carlo simulation method [14, 5]. Our formulations will be confined to their applications on $B^{q+} + H$ (q = 1-4) interaction only as test cases. This system has been chosen due to the fact that, very recently boron has been identified as the plasma facing material replacing high-Z surface materials such as Ni, Fe and Mo in the next step fusion reactor such as ITER.

Atomic units will be used throughout the entire text.

2. THEORY

2.1 Quantum Mechanical Formalism

A collision diagram for charge transfer is shown in Fig. 1. The total Hamiltonian of the whole system may be written as

$$H = H_0 + V_{Te}\left(\vec{r}_T\right) + V_{Pe}\left(\vec{r}_P\right) + V_{TP}\left(\vec{R}\right) \tag{1}$$

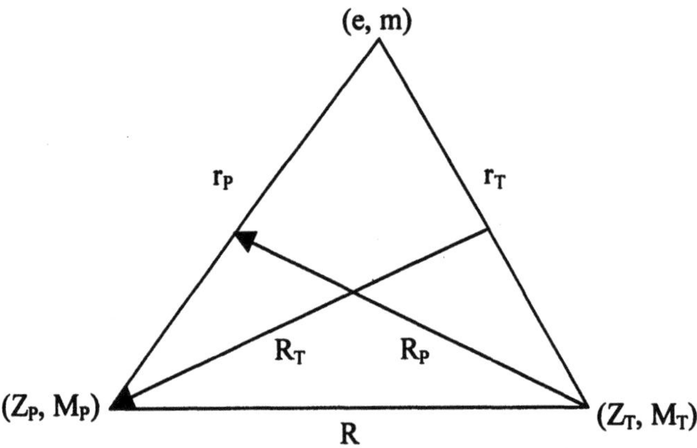

Figure 1. Collision diagram for charge transfer reaction. m, M_T and M_P are the masses of active electron, target ion and projectile ions respectively.

where,

$$H_0 = -\frac{1}{2\mu_T}\nabla^2_{R_T} - \frac{1}{2a}\nabla^2_{r_T} \quad \text{(entrance channel)} \tag{2a}$$

$$= -\frac{1}{2\mu_P}\nabla^2_{R_P} - \frac{1}{2b}\nabla^2_{r_P} \quad \text{(exit channel)} \tag{2b}$$

and

$$\mu_T = \frac{M_P(M_T + m)}{m + M_P + M_T}, \quad \mu_P = \frac{M_T(M_P + m)}{m + M_P + M_T}$$

$$a = \frac{m M_T}{m + M_T}, \quad b = \frac{m M_T}{m + M_T} \tag{3}$$

Here, two body potentials are identified by any pair, out of three subscripts (e, T, P) representing electron, target and projectile ions respectively. m, M_T and M_P are the masses of the active electron, target ion and projectile ion respectively. Total Hamiltonian may be separated in terms of channel Hamiltonian and interactions as

$$H = -\frac{1}{2\mu_T}\nabla^2_{R_T} - \frac{1}{2a}\nabla^2_{r_T} + V_{Te}(\vec{r}_T) + V_{Pe}(\vec{r}_P) + V_{TP}(\vec{R})$$

$$\leftarrow \qquad H_i \qquad \rightarrow\leftarrow \qquad V_i \qquad \rightarrow$$

$$H = -\frac{1}{2\mu_P}\nabla^2_{R_P} - \frac{1}{2b}\nabla^2_{r_P} + V_{Pe}(\vec{r}_{Pe}) + V_{Te}(\vec{r}_T) + V_{TP}(\vec{R})$$

$$\leftarrow \qquad H_f \qquad \rightarrow\leftarrow \qquad V_f \qquad \rightarrow \tag{4}$$

Channel wavefunctions satisfy the equations

$$(E - H_i)\Psi_i = 0 \text{ and } (E - H_f)\Psi_f = 0, \tag{5}$$

270

where, $\Psi_i = \varphi_i(\vec{r}_T)e^{i\vec{K}_i \cdot \vec{R}_T}$, $\Psi_f = \varphi_f(\vec{r}_P)e^{i\vec{K}_f \cdot \vec{R}_P}$, $\qquad$ (6a)

and

$$E = \frac{K_i^2}{2\mu_T} + \in_{Te} \quad \text{(entrance channel)} \qquad (6b)$$

$$= \frac{K_f^2}{2\mu_P} + \in_{Pe} \quad \text{(exit channel)} \qquad (6c)$$

where $\varphi_i(\varphi_f)$, $\in_{Te}(\in_{Pe})$ and $K_i(K_f)$ are respectively the initial (final) bound state wavefunction, the initial (final) binding energy and the initial (final) momentum of relative motion of colliding systems.

The prior form of the transition amplitude (with first term only) in the formalism of Dodd and Grieder [15] may be written as

$$T_{if}^{(-)} = \left\langle \Psi_f \left| \omega_f^{-*} \left[1 + g_x^- (V_f - W_f) \right]^* (V_i - W_i) \, \omega_i^+ \right| \Psi_i \right\rangle \qquad (7)$$

where the Moller operators ω_i^+ and ω_f^- and the propagator g_x^- are defined as

$$\omega_i^+ = 1 + \frac{1}{E - H_i - W_i + i\eta} W_i, \qquad (8a)$$

$$\omega_f^- = 1 + \frac{1}{E - H_f - W_f - i\eta} W_f, \qquad (8b)$$

$$g_x^- = \frac{1}{E - H + v_x - i\eta}. \qquad (8c)$$

W_i (W_f) is the distorting potential in the initial (final) channel and v_x is a distorting potential in an intermediate channel, whose choice is restricted by the condition that it should contain no two-body potential occurring in V_i.

Now, if we substitute

$$\omega_f^-\left|\Psi_f\right\rangle = \left|X_f^-\right\rangle \tag{9}$$

and

$$\left[1 + g_x^-\left(V_f - W_f\right)\right]\left|X_f^-\right\rangle = \left|\xi_f^-\right\rangle, \tag{10}$$

then $\left|\xi_f^-\right\rangle$ satisfies the equation (in the limit $\eta \to 0$)

$$\left(E - H + v_x\right)\left|\xi_f^-\right\rangle = 0, \tag{11}$$

provided, $v_x\left|X_f^-\right\rangle = 0.$ \tag{12}

Now writing $\left|\xi_f^-\right\rangle = \left|\phi_f(r_P)G_f^-\right\rangle$ and on substituting it in equation (11), we obtain,

$$\phi_f(r_p)\left[E - \varepsilon_{pe} - H_0 - V(R_{TP}) - V_{Te}(r_{Te})\right]G_f^- + \frac{1}{b}\vec{\nabla}_{r_p}\varphi_f(\vec{r}_p)\cdot\vec{\nabla}_{r_p}G_f^- + v_x\xi_f^- = 0. \tag{13}$$

Choosing, $v_x\xi_f^- = -\dfrac{1}{b}\vec{\nabla}_{r_p}\varphi_f(\vec{r}_p)\cdot\vec{\nabla}_{r_p}G_f^-$ and on substituting it in equation (13), we find that G_f^- satisfies the differential equation

$$\left(\frac{K_f^2}{2\mu_P} + \frac{1}{2\mu_T}\nabla_{R_T}^2 + \frac{1}{2a}\nabla_{r_T}^2 - V_{Te}(\vec{r}_T) - V_{TP}(\vec{R})\right)G_f^- = 0. \tag{14}$$

2.1.1   Construction of the interacting two-body potentials and final bound-state wave function

There is no ambiguity in the construction of $V_{Te}(r_T)$ and is uniquely determined by the Coulomb potential. The interaction of the active electron with the projectile ion may be estimated by a model potential as

$$V_{Pe}(\vec{r}_P) = -\frac{q}{r_P} - \frac{e^{-\lambda\,r_P}}{r_P}\left\{(Z - q) + br_P\right\}, \tag{15}$$

where Z and q are respectively the nuclear and asymptotic charge of the projectile ion. Such choice of the potential has the advantage that it satisfy the limiting conditions i.e.,

$$\lim_{r_P \to 0} V_{Pe}(\vec{r}_P) \sim -\frac{Z}{r_P} \quad \text{and} \quad \lim_{r_P \to \infty} V_{Pe}(\vec{r}_P) \sim \frac{q}{r_P}. \tag{16}$$

Here b and λ are two arbitrary parameters chosen variationally with respect to a Slater basis set in such a way that the corresponding Hamiltonian of the active electron in the final state is diagonalised to reproduce correct binding energies. These binding energies of the active electron on projectile ions of different charge state has been taken from the tables of Clementi and Roctti [16] and works of Clark and Abdallah [17]. Potential parameters for projectile ions of different charge state are given in Table I. It has been observed that a unique set of parameters for the potential reproduce binding energy of the captured electron in different subshells with better accuracy for

Table 1. Model potential parameters, λ and b in equation (15) for B^{q+} ions.

Ion	λ	b
B^{+}	2.1435	4.4119
B^{2+}	3.4012	5.8235
B^{3+}	6.2512	5.8616
B^{4+}	7.0512	8.2616

a particular shell of the projectile ion, which has initially a closed shell or subshell structure. For the capture to the open shell of the projectile ion, potential parameters have to change a little to find out the binding energies of the active electron in each different subshell to which the capture occurs. However, to check the accuracy of the wave function, the virial theorem has been tested and has been found to be accurate within 0.01% in all cases. We have treated the interaction of the projectile ion and the target nucleus as coulombic between two charges of magnitude q and 1 respectively. This is well justified as even if some short-range part exists, that will not affect the charge transfer cross section.

2.1.2 Construction of the transition amplitude and its evaluation

With all these consideration of the potentials, eq. (14) may be written as

$$\left[\frac{K_f^2}{2\mu_P}+\frac{1}{2\mu_T}\nabla_{R_T}^2+\frac{1}{2a}\nabla_{r_T}^2+\frac{1}{r_T}-\frac{q}{R_T}\right]G_f^-=0. \tag{17}$$

Here $\dfrac{1}{R}$ has been replaced by $\dfrac{1}{R_T}$. Solution for G_f^- may be written as

$$G_f^-=\left(b\mu_P\right)^{iv_1}e^{\pi\left(v_1-v_2\right)/2}\,\Gamma\left(1+iv_1\right)\Gamma\left(1-iv_2\right)e^{i\vec{K}_f\cdot\vec{R}_P}$$
$${}_1F_1\left(-iv_1;\,1;\,-ia\left(v_f r_T+\vec{v}_f\cdot\vec{r}_T\right)\right){}_1F_1\left(iv_2;\,1;\,-ib\left(K_f R_T+\vec{K}_f\cdot\vec{R}_T\right)\right) \tag{18}$$

where $v_1=\dfrac{1}{v_f},\ v_2=\dfrac{q}{v_f}.$

Since the incoming projectile ion in the entrance channel interacts with a neutral atom, we choose $W_i=0$ (asymptotically) and we may find

$$\left(V_i-W_i\right)\omega_i^+\left|\psi_i\right\rangle=V_i\left|\psi_i\right\rangle. \tag{19}$$

So the prior form of the transition amplitude in the framework of BCCIS - approximation may be written as

$$T_{if}^{(-)}=\int d\vec{r}_T\,d\vec{R}_T\,\varphi_f^*(\vec{r}_P)G_f^{(-)r}\left[\frac{q}{R}-\frac{q}{r_P}-\frac{e^{-\lambda r_P}}{r_P}\{(Z-q)+br_P\}\right]\Psi_i. \tag{20}$$

This six-dimensional integral may be reduced to a one-dimensional integral in compact form following an earlier investigation [18] by our group. Finally, cross sections are obtained by integration over scattering angles. These integrals are performed numerically in a 24-point and a 48-point Gauss Legendre quadrature method with an accuracy of 0.1%. Excited states are generated by parametric differentiations which have been obtained analytically.

2.2　　Classical Formalism (CTMC)

Classical Hamiltonian of the whole system in the initial channel may be written as

$$H=\frac{1}{2a}\left(p_1^2+p_2^2+p_3^2\right)+\frac{1}{2\mu_T}\left(p_4^2+p_5^2+p_6^2\right)+V_{Te}(\vec{r}_T)+V_{Pe}(\vec{r}_P)+V_{TP}(\vec{R}), \tag{21}$$

274

where,

$$r_p^2 = (aq_1 - q_4)^2 + (aq_2 - q_5)^2 + (aq_3 - q_6)^2,$$
$$r_T^2 = q_1^2 + q_2^2 + q_3^2. \tag{22}$$

Here q_i, $p_i (i = 1 - 3)$ be rectangular co-ordinates and conjugate momenta of the electron relative to the centre of mass of the target system respectively. The corresponding quantities of the projectile ion are q_i, $p_i (i = 4 - 6)$ respectively in the centre of mass co-ordinates.

Hamiltonian's equation of motion may be written as

$$p_i = a\dot{q}_i, \ i = 1, 2, 3$$
$$p_i = \mu_T \dot{q}_i, \ i = 4, 5, 6 \tag{23}$$

and

$$\dot{p}_i = -\frac{1}{r_p}\frac{\partial V}{\partial r_p}a(aq_i - q_{i+3}) - \frac{1}{r_T}\frac{\partial V}{\partial r_T}q_i - \frac{1}{R}\frac{\partial V}{\partial R}\frac{a}{\mu_T}\left(\frac{a}{\mu_T}q_i + q_{i+3}\right), \ i = 1, 2, 3$$

$$\dot{p}_i = -\frac{1}{r_p}\frac{\partial V}{\partial r_p}(q_i - aq_{i-3}) - \frac{1}{R}\frac{\partial V}{\partial R}\left(q_i + \frac{a}{\mu_T}q_{i-3}\right), \ i = 4, 5, 6 \tag{24}$$

where, $V = V_{Te}(\vec{r}_T) + V_{Pe}(\vec{r}_P) + V_{TP}(\vec{R}).$

These coupled twelve equations in two sets given by equations (23) and (24) completely describe the motion of the whole system in centre of mass frame.

These coupled equations are integrated numerically step by step from $t = -\infty$ to $t = +\infty$. At $t = -\infty$, the target system is unperturbed. So the initial values for q_i and $p_i (i = 1 - 6)$ may be assigned in terms of six random numbers [19] from a sequence of random numbers. At $t = +\infty$, q_i and $p_i (i = 1 - 3)$ are determined. From these values of q_i and $p_i (i = 1 - 3)$, energies ϵ_{Te} and ϵ_{Pe} of the electron in the sub-systems i.e. (electron, target ion) and (electron, projectile ion) respectively, are determined. Then the processes are identified as

(a) charge transfer, $\in_{Pe} < 0, \in_{Te} > 0$

(b) direct ionization, $\in_{Pe} > 0, \in_{Te} > 0, r_T < r_P$

(c) transfer ionization, $\in_{Pe} > 0, \in_{Te} > 0, r_T > r_P$

(d) elastic scattering and excitation, $\in_{Pe} > 0, \in_{Te} < 0$. $\qquad(25)$

Total ionization comes from the sum of direct and transfer ionization. Calculations are repeated for several thousand trajectories.

If N_T is the total number of trajectories calculated and N_R is the number of trajectories satisfies any of the criteria for a specific reaction (given by (a)-(d) as mentioned above), corresponding cross sections for a particular channel is given by

$$\sigma_R = \left(N_R / N_T\right)\pi\, b_{max}^2 \qquad (26)$$

where b_{max} is the maximum impact parameter beyond which no reaction takes place. The standard error(s) is estimated by

$$S^2 = \sigma_R\left[\frac{\pi\, b_{max}^2 - \sigma_R}{N_T - 1}\right]. \qquad (27)$$

3.   RESULTS AND DISCUSSIONS

As check to our computer program for CTMC calculation with model potential, we have reproduced the CTMC results of Olson and Salop [5] within 8% with supplied values of q and b by Z and zero respectively. To reduce the statistical error within 15% and to achieve a convergence within 1%, we have computed each of the results for six thousand trajectories.

Computed results for total charge transfer cross sections in collisions of B^{q+} (q=1-4) ion with atomic hydrogen are displayed in Figs. 2-5 in comparison with other existing theoretical results only. Unfortunately no experimental results are available for such processes. Reported data have been obtained by multiplication of the calculated data by Pauli blocking factor [6] given by

$$Q_{nl} = \left[1 - N_{nl}/2(2l+1)\right] Q_{nl}^C \qquad (28)$$

276

where N_{nl} is the number of electrons occupying the sub-shell of the incoming projectile and Q_{nl}^C is the calculated capture cross section.

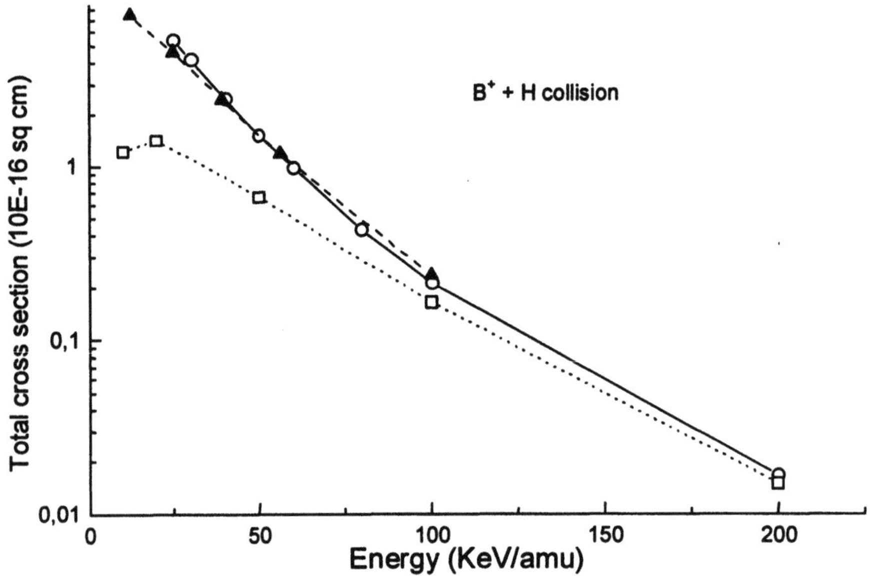

Figure 2. Comparison of total charge transfer cross section in B⁺ + H collision. ——— present BCCIS results; ········ coupled state results of Hansen and Dubois [8];----- present CTMC results.

Charge transfer cross sections for collisions of B⁺ with atomic hydrogen have been displayed in Fig 2. From the figure, we find that our results in BCCIS - approximation agree very well with coupled state results of Hansen and Dubois [8]. Our calculated results in CTMC method differ by a factor of two at lower energy side and agreement is increasingly good at higher energies.

In case of B²⁺ + H interaction (shown in Fig. 3), our CTMC results overestimates our findings in BCCIS approximation by a factor of two over the entire energy region. Fig. 4 shows the charge transfer cross section for B³⁺ + H collision. Here we find that BCCIS -results are in very good agreement with the coupled state results of Hansen and Dubois [8]. However, in this case, CTMC results of Olson and Salop [5] are better than our computed results in CTMC method with model potential. In case of collision of B⁴⁺ with atomic hydrogen, our computed results in CTMC method have favourable agreement in comparison to the CTMC -results of Olson and Salop [5]. However, it may be mentioned that Olson and Salop have given no data tables for charge transfer cross sections in their paper. We have to read it from their graphical presentations.

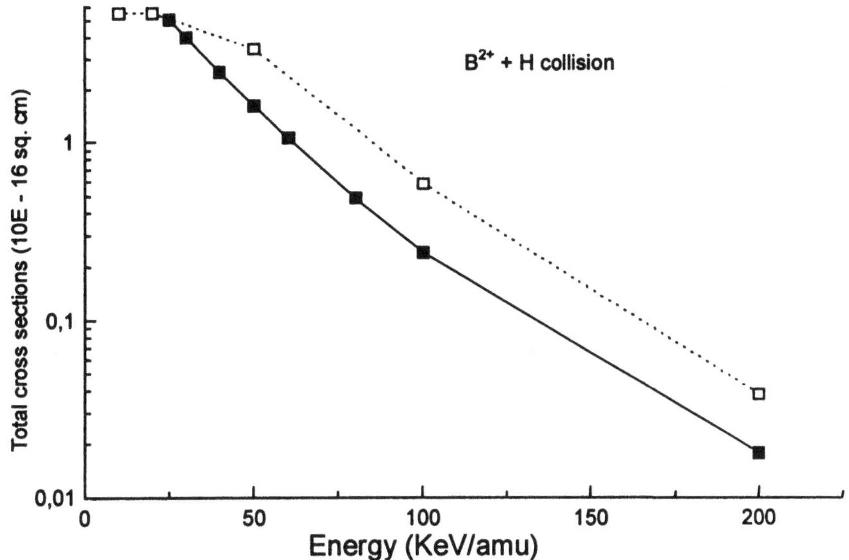

Figure 3. Comparison of total charge transfer cross section in B^{2+} + H collision. ———— present BCCIS results; ········ present CTMC results.

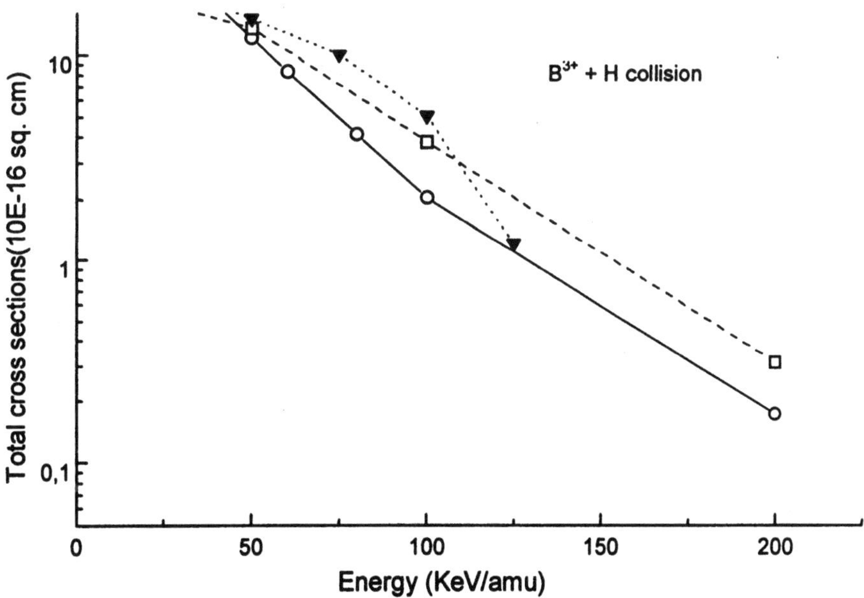

Figure 4. Comparison of total charge transfer cross section in B^{3+} + H collision. ———— present BCCIS results; —————— present CTMC results; ····· coupled state results of Hansen and Dubois [8].

From the above discussions, it is evident that CTMC method provides fair estimate of total charge transfer cross sections in collisions of partially stripped heavy ions with neutral atoms. Obvious question arises, why such classical predictions become competitive to its quantum mechanical counterpart in the description of a microscopic event like charge transfer.

One of the possible answers may be that ensemble interpretation (EI) is one of the possible result in finding the classical limit of quantum mechanical formalism. On the other way round, one may expect that, if we project classical equations of motion in an ensemble environment, meaningful results may be obtained at least to a certain degree of accuracy. However, this does not assure us that all features of quantum mechanics may be reproduced in this way. Under such circumstances, it may be mentioned that, when we attempted to find out the state selective capture cross sections by the method prescribed by Becker and Macjeller [20], results are not very encouraging in comparison to our computed results in BCCIS-approximation even if the number of trajectories goes up to eight thousands.

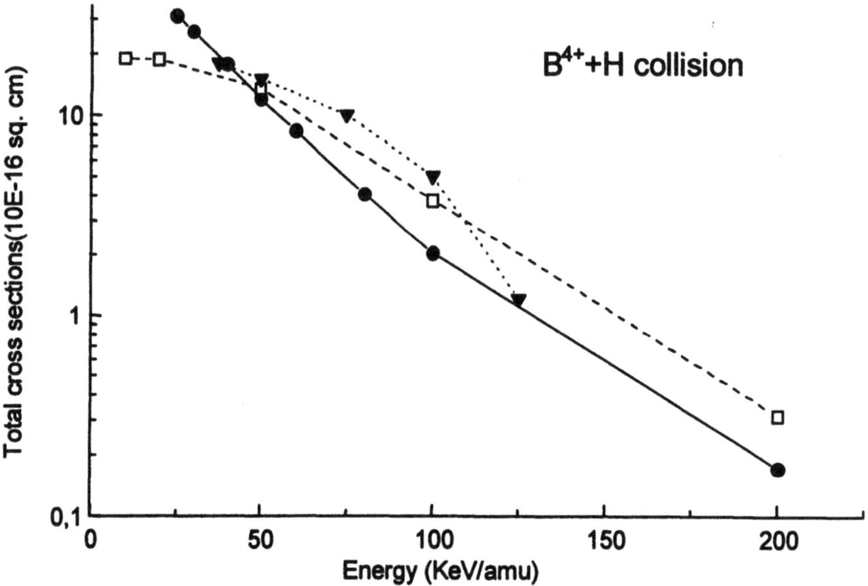

Figure 5. Comparison of total charge transfer cross section in B^{4+} + H collision. ——— present BCCIS results; ------ present CTMC results, ········ CTMC results of Olson and Salop [5].

4.   CONCLUDING REMARKS

Close agreement between our computed results in BCCIS-approximation and the coupled state results, indicate that former approximation may be employed with high degree of accuracy to find out charge transfer cross sections at intermediate and high energies for other systems as well. The success may be due to the fact that, (i) continuum intermediate states have been incorporated into the formalism, (ii) proper boundary condition for the scattering wave function is well satisfied and (iii) formulation may be extended to non - coulombic interactions as well. From our observations on CTMC-results, it is evident that this method may be employed to have fair estimate of cross sections of heavy ion-atom collisions which are very complex in nature. However, it is very difficult to explain why CTMC methodology works only in intermediate and lower part of high-energy region. Extensive research works may be carried out both theoretically as well as experimentally on studies of partially stripped heavy ions with neutral atoms to have answers to such questions.

ACKNOWLEDGMENTS

The author gratefully acknowledges the financial support provided by Department of Science and Technology (Govt. of India), New Delhi through grant no. SP/S2/LO3/95. The author also takes the privilege to thank Mr. M. Das, Mr. M. Purkait, Mr. A. Dhara and Mr. S. Sounda for their active support in carrying out such investigations. Thanks are due to Mr. Indrajit Sadhukhan for preparation of the manuscript.

REFERENCES

1.   J. S. Briggs and J. H. Macek, Adv. At. Mol. Opt. Phys. **28**, 1 (1990).

2.   W. Fritsch and C. D. Lin, Phys. Rep. **207**, 1 (1991).

3.   Dz Belkic, R. Gayet and A. Salin, At. Data Nucl. Data Tables **51**, 59 (1992).

4.   D. P. Dewangan and J. Eichler, Phys. Rep. **247**, 59 (1994).

5.   R. E. Olson and A. Salop, Phys. Rev. A **16**, 531 (1977).

6.   J. K. M. Eichler, A. Tsuji and T. Ishihara, Phys. Rev. A **23**, 2833 (1981).

7.   Y. D. Wang, N. Toshima and C. D. Lin, Phys. Scr. **T62**, 63 (1996).

8. J. P. Hansen and A. Dubois, Phys. Scr. **T62**, 55 (1996).

9. D. R. Schultz, P. S. Krstic and C. O. Reinhold, Phys. Scr. **T62**, 69 (1996).

10. M. Das, M. Purkait and C. R. Mandal, Phys. Rev. A **57**, 3573 (1998).

11. T. V. Goffe, M. B. Shah and H. B. Gilbody, J. Phys. B: Atom. Mol. Phys. **12**, 3763 (1979).

12. R. A. Phaneuf, F. W. Meyer and R. H. Mcknight, Phys. Rev. A **17**, 534 (1978).

13. C. R. Mandal, Mita Mandal and S. C. Mukherjee, Phys. Rev. A **44**, 2968 (1991).

14. R. Abrines and I. C. Percival, Proc. Phys. Soc. A **88**, 873 (1966).

15. L. D. Dodd and K. R. Greider, Phys. Rev **146**, 675 (1966).

16. E. Clementi and C. Roetti, At. Data Nucl. Data Tables, **14**, 237 (1974).

17. R. E. H. Clark and J. Abdallah Jr., Phys. Scr. **T62**, 7 (1996).

18. M. Ghosh, C. R. Mandal and S. C. Mukherjee, Phys. Rev. A **35**, 2815 (1987).

19. G. Peach, S. L. Willis and M. R. C. Mcdowell, J. Phys. B **18**, 3921 (1985).

20. R. L. Becker and A. D. Mackeller, J. Phys. B **17**, 3923 (1984).

Low Energy Electron Emission in Fast Ion-Atom Collisions

Lokesh C. Tribedi

Tata Institute of Fundamental Research
Mumbai (Bombay) – 400 005, India

We review here some of the recent studies on low energy electron emission in highly charged heavy-ion atom collisions involving atomic hydrogen as target. The double differential distributions of the low energy electrons, in terms of emission angle and energy, can provide a detailed understanding of the ionization process. We present the recent measurements on the energy and angular distributions of electrons emitted in ionization of atomic and molecular hydrogen and helium in collision with fast bare ions. It is shown that two centre effect plays a major role in heavy ion-atom ionization. A large forward-backward asymmetry in the angular distribution of low energy electrons has been observed which arises due to two centre effect and is discussed. A comparison with different theoretical models based on B1 (first Born) and CDW-EIS (continuum distorted wave eikonal initial state) approximations is presented.

1. INTRODUCTION

Ionization is one of the most important inelastic processes in fast ion atom collision in which an electron initially in a bound state of the target atom (or projectile) is ejected in the continuum. The total cross section for single ionization has been a topic of immense study for several decades in the past. It may be mentioned that the total cross sections are obtained by integrating over the emission angles and energies of the emitted electrons and over the projectile scattering angles and therefore a stringent test of the theoretical models often may not be possible. It has also been demonstrated

Trends in Atomic and Molecular Physics,
Edited by Sud and Upadhyaya. Kluwer Academic/Plenum Publishers, New York, 2000.

that the double differential cross sections (DDCS) of electron emission in terms of emission angle (θ) and electron energies (ε) (i.e. $d^2\sigma/d\Omega\,d\varepsilon$) can be very useful tool to investigate some of the salient features of the ion-atom collisions and can provide a stringent test to the theoretical methods. The main features of such electron DDCS spectrum, measured at zero degree, are soft electrons (SE), electron capture into continuum (ECC) cusp and binary encounter (BE) peak. The soft collision electrons are the lowest energy electrons $(\varepsilon = v_e^2/2 \rightarrow 0)$ which are mainly produced in large impact parameter collisions. Although the low energy electrons compose the bulk of the cross sections, they are difficult to detect. The ECC cusp, on the other hand, is observed at zero degree [1] and is composed of electrons emitted with a velocity equal to the projectile velocity (v) and is understood to arise due to the projectile-ejected electron interaction. The ECC peak disappears for $\theta \geq 10$ degree. The first Born (B1) calculations can not reproduce this peak since in this model such interaction is treated only as a perturbation which produces the transition of the active electron from a bound state to a continuum state. This model of ionization is not valid in case where the ionized electron moves in the presence of two moving sources of Coulomb potentials, which due to their long range nature, can distort the initial state wave function even when the two centres are far away. The BE electrons are produced due to the hard collisions between the projectile and the electrons and show up in the DDCS spectrum as a broad peak centred around electron velocity $v_e \cong 2v\cos\theta$. The width of the BE peak arises due to the Compton profile of the electrons in the initial state.

1.1 Background: Theoretical and Experimental

There have been several measurements on the angular distribution of electron DDCS in light ion and electron induced ionization [1-10]. In case of ionization by highly charged heavy ions the experimental data are scarce and theoretical methods often find difficulties in describing the final electronic state where the emitted electrons travel in presence of two moving Coulomb potentials centred at the residual target and projectile. The richness of the field originates from the fact that one can study the dynamics of the ionized electron in presence of two such strong centres in heavy ion atom collision. In B1 approximation it is assumed that the projectile potential acts as a perturbation to the initial electronic state of the target causing the transition from a bound state to a continuum state. This is also termed as single centre approximation. Madison [11] had employed the HFS wavefunction for the initial state and a numerical continuum wavefunction for the final state obtained as a solution of the same HS target potential in order to use the B1 formalism. The initial and final states, therefore, are described by the eigenfunctions of the same undistorted Hamiltonian and hence orthogonal. This approach provided a good agreement with the electron angular distribution data of Stolterfoht and co-workers [5] for p + He.

The ECC cusp and the direct ionization of target atom can be described in terms of Coulomb waves centred at the projectile and the target, respectively. In order to understand the shape of the ECC cusp one needs to invoke the simultaneous influences of the both the target and projectile nuclei on the ionized electron, in·its final state. Theoretical models based on continuum distorted wave (CDW) has been developed [12, 13] in order to explain such two centre electron emission (TCEE). In CDW the initial bound wavefunction is distorted by a projectile continuum factor and the final continuum wavefunction was taken as a product of a plane wave and two continuum factors corresponding to the projectile and residual target ion.

To improve the agreement with experimental data of Shah and Gilbody [14], for ionization of atomic hydrogen by light ions, Crothers and McCann [15] had developed the CDW-EIS (eikonal initial state) model in which final state is chosen as in the CDW but the initial state is represented as a bound state distorted by a projectile eikonal phase. This is first order in distorted wave series and has been shown to be adequate to describe the two centre electron emission in intermediate energy ion-atom collision. The model of Crothers and McCann for ionization from H(1s) were further generalized by Fainstein et al [16] for arbitrary initial state and for multielectronic targets [17]. In an independent particle picture the initial state of the active electron was represented by Roothian-Hartree-Fock [18]. The final continuum state was approximated by a screened hydrogen-like wavefunction with an effective nuclear charge $Z_{eff} = (2\varepsilon_b)^{1/2}$, derived from the initial binding energy, ε_b (in a.u.). It may be noted that the initial and final wavefunctions are not orthogonal in this model. To employ the orthogonal wavefunctions one needs to have numerical calculations of such wavefunctions as used by Madison [11] and Manson et al [5]. Such an improvement of the model was introduced only recently by Gulyas et al [19]. In this improved model HFS wave functions for the initial and final states were used for the electron to be ionized. So this model now combines two basic requirements for describing the electron emission in ion-atom collision: the two centre effect and realistic (numerical) target wavefunction. The predictions by different theoretical models are needed to be checked by experimental results.

The double differential cross sections of the low energy electrons have been measured in the past using low charged and light ion projectiles like proton, helium and electrons following the first measurements by Rudd and co-workers [1]. In case of highly charged ions as projectiles the ionization mechanism becomes more complicated since the two centre effect (TCE) being strong compared to that due to light ions. The ionized electron which moves in the Coulomb fields of the residual target and the projectile ions are highly influenced by the two centre effect and post collision interaction. Therefore a detailed study of electron emission in HCI induced ionization is required to have a better understanding of the TCE and hence for the ionization dynamics. Following the first measurement of DDCS using heavy ions (25 MeV/u Mo^{40+} + He) by Stolterfoht et al [20], there now exist a few measurements on the angular distributions of the low energy electrons emitted in ionization using varieties of HCIs as projectiles. Two centre electron emission has been studied via the measurements of angular distributions of e-DDCS in ionization of He and H_2 in collision with highly charged carbon, oxygen and neon ions having different energies between 1 and 5 MeV/u [21, 22, 23, 24, 25]. An excellent review on the topic can be found in Ref. [26].

However, most of the previous experiments on electron DDCS measurements are carried out by using multi-electron (atomic or molecular) targets such as H_2, He, Ne etc. for which various processes such as e-e correlation, transfer ionization, double ionization and dissociative ionization can also interfere with single ionization process. In case of atomic hydrogen target and bare projectile ions these processes are absent and hence one can study the pure three-body ionization. Due to experimental difficulties such measurements are very limited. The total ionization cross sections in p + H collisions have been measured in the past [27, 28, 14]. The DDCS, on the other hand, can provide much more important information on ionization dynamics compared to single differential or total cross section. However, the electron DDCS measurements using atomic hydrogen target have been carried out only for low charge projectiles such as electrons [29], protons [9] and helium ions [10]. The only measurement on e-DDCS using highly charged heavy-ions has been reported only recently [30]. It may be pointed out that as far as theoretical models are concerned no ambiguity exists for describing the ionization of atomic hydrogen since the wave function is known exactly.

In this paper we will review some of our recent studies on the energy and angular distributions of the low energy electron emission in fast ion-atom collisions using H, H_2 and He as targets and bare carbon ions as the projectiles. A comparative study will be presented for electron emission in ionization of H and He in collision with bare carbon ions.

2. MEASUREMENT TECHNIQUE

The experimental set up and measurement techniques are standard and can be found in several publications [22, 23, 9, 30]. We will describe, in brief, the set up used in recent e-spectroscopic studies at Kansas State University (KSU) using atomic hydrogen target. A Slevin [36] type commercially available RF (~35 MHz) discharge tube was used for producing atomic hydrogen target for collision experiments. The mixture of atomic and molecular hydrogen was allowed to the scattering chamber through a capillary tube of diameter 1 mm. The jet could be aligned with respect to the projectile ion beams by adjusting the x, y and z positions. Bare C ions at different energies ($v = 6$-10 a.u.) were obtained from the van-de-Graaff accelerator at J.R. Macdonald laboratory, KSU. The energy and charge state selected beam was collimated by a four-jaw-slits arrangement and was made to pass through another aperture of diameter 2 mm before it collides with the target gas. This second aperture was also used to prevent the scattered beam and the secondary electrons from entering the chamber. The current on the aperture was read separately and made negligible by reducing the beam dimension by the four-jaw-slits. This was necessary in order to reduce the background electron counts arising from slit scattering. A μ-metal shield inside the chamber and a current carrying coil placed in the horizontal plane around the chamber were enough to reduce the stray magnetic field below 5 mG in the region where the electrons travel before entering the analyzer.

The electrons emitted in ionization of target were energy analyzed with the help of a hemispherical electrostatic analyzer before they were detected by a channel electron multiplier (CEM). The analyzer is made of two hemispheres of diameter 50 and 70 mm made from oxygen free high conductivity copper. The spherical surfaces were coated with carbon soot to reduce the secondary electron production from the copper surface due to the electron bombardment. Before entering the analyzer the electrons had to pass through a collimator made of a copper tube with two rectangular grounded apertures one on each end. These two apertures of widths 4 and 3 mm mainly define the effective path-length solid-angle integral. Additional apertures at entrance and the exit of the analyzer were biased with a small voltage V_0 in order to pre-accelerate the electrons entering the analyzer. It was found that $V_0 = +5$ V was enough to improve the collection efficiency of the low energy electrons. The resolution of the spectrometer was about 5%. The energy-analyzed electrons were detected by a channel electron multiplier mounted on the exit of the analyzer. The cone of the CEM was biased at $+100$ V to help the low energy electrons reach the detector. Earlier measurements [32] have shown that with this bias the efficiency of the CEM is constant within 4% in the present energy range. The spectrometer could be rotated between 15° and 160° and the electrons were detected at ten to thirteen different angles different angles at an interval of about 15°. The data was collected in fine energy steps between 0.1 to 300 eV.

2.1   He and H_2 Targets

The chamber was flooded with the H_2 or He gas and the gas pressure was kept low (0.1 mT) in order to minimize the scattering of the low energy electrons emitted in the ionization of the target. In case of He gas target and for higher electron energies (30-300 eV) the gas pressure was about 0.3 mT. The data was corrected for the scattering of electrons in the gas target. The correction factor, in case of H_2 gas, was found to be about 9.3% at 1 and was less than 6% above 10 eV. This factor was even lower for He gas. These were estimated from the present geometry using the electron scattering cross section data from Golden et al [33,34]. For absolute normalization we have measured the e-DDCS spectra at several angles for 1.5 MeV p + He for which the absolute cross sections are known from earlier measurements [6].

2.2   Atomic Hydrogen Target

In case of atomic hydrogen experiment the data collection and normalization procedures are quite involved and need to be discussed. The RF source was fed with highly pure H_2 gas. The first step is to determine the dissociation fraction (D_f) i.e. the fraction of H_2 molecules which are dissociated in the RF and interacted with the projectile. This quantity is measured *in situ* using nine-eV ion method for each scattering angle under study. The spectrum of recoil ions produced in the collision of projectile ions with molecular hydrogen contains a broad peak around 9 eV energy. The hydrogen molecule gets doubly ionized in collision with the passing projectile and produces two protons of energy 9 eV as a result of Coulomb explosion [35]. These protons can also be produced as a result of dissociation of the $2p\sigma_u$ and some other nearby states of H_2^+ [36]. These protons are detected by the same spectrometer which is used to detect the electrons. In this case the polarities of the potentials on the two hemispheres are reversed compared to those used for electron detection. The front cone on the CEM was biased to a high positive voltage (+2 kV) and the rear was grounded. The dissociation fraction is given by $D_f = 1 - Y^{on}/Y^{off}$ where Y^{on} and Y^{off} are the yields of 9 eV protons for RF power *on* and *off* respectively, and are obtained from the area under the broad peak around 9 eV energy. A typical spectrum is shown in Fig. 1. The reduction in the yields of 9 eV protons with RF power *on* signifies the degree of dissociation of the molecules in the RF cavity. We have obtained about 80-85% dissociation fraction in these experiments and D_f was found to be almost independent of emission angles between 30 and 120 degrees. For extreme forward and backward angles the D_f was found to be smaller. Typical jet pressure was about 0.3-0.4 Torr and beam interacted with the jet 2 mm below the nozzle. The x, y and z positions of the jet were adjusted in order to get maximum value of D_f. To achieve the high degree of dissociation fraction extreme care was taken for the cleanliness of the vacuum chamber, RF cavity and the purity of the input H_2 gas.

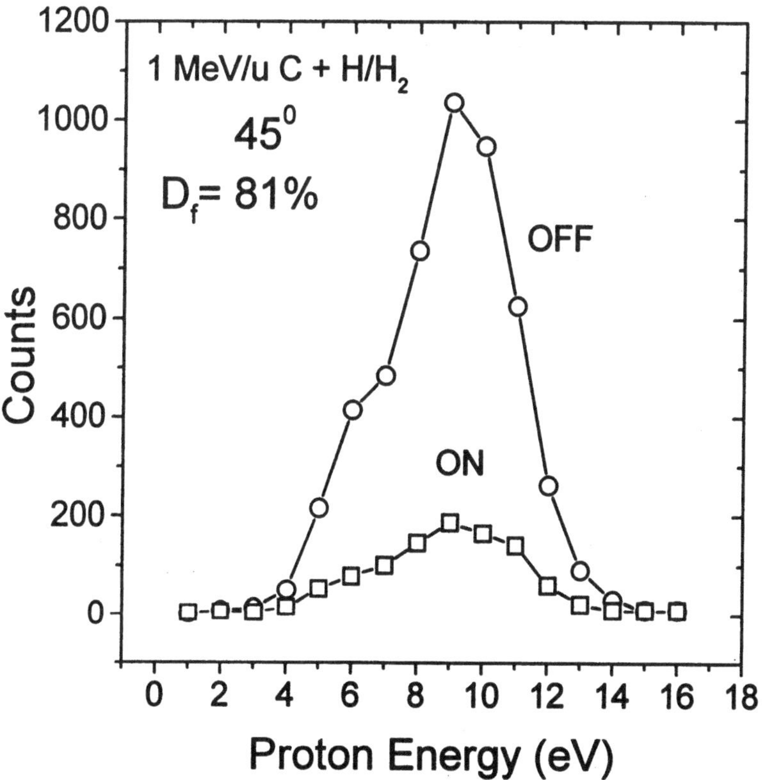

Figure 1. The spectrum of recoil-ions showing the broad peak of H⁺ ions around 9 eV in the collision of C⁶⁺ + H/H₂ jet with RF power *off* (O) and *on* (□) as indicated.

3. RESULTS AND DISCUSSIONS

The electron spectrum between 1 and 300 eV was taken for different angles at small intervals with for RF power *on* and RF power *off* at the same pressure of the jet. For a few forward angles higher energy electrons were also detected in order to measure the cross sections of BE process. The ratio (R_H) of DDCS for H and H_2 can be determined from these measurements and from the measured D_f at the same angle, using the following equation [9, 29], $R_H = \sigma_1/\sigma_2 = 1/(D_f \sqrt{2})[S^{on}(\varepsilon, \theta)/S^{off}(\varepsilon, \theta) - 1 + D_f]$. The quantities $S^{on}(\varepsilon, \theta)$ and $S^{off}(\varepsilon, \theta)$ are the background subtracted electron yields for energy ε and angle θ with RF *on* and *off*, respectively. The absolute DDCS (σ_2) for H_2 (as well as for He) was measured by flooding the chamber, as discussed above. The absolute DDCS (σ_1) for atomic hydrogen is then given by $\sigma_1 = R_H \sigma_2$.

3.1 Electron DDCS for He and H

Now we present a few examples of the electron DDCS spectrum

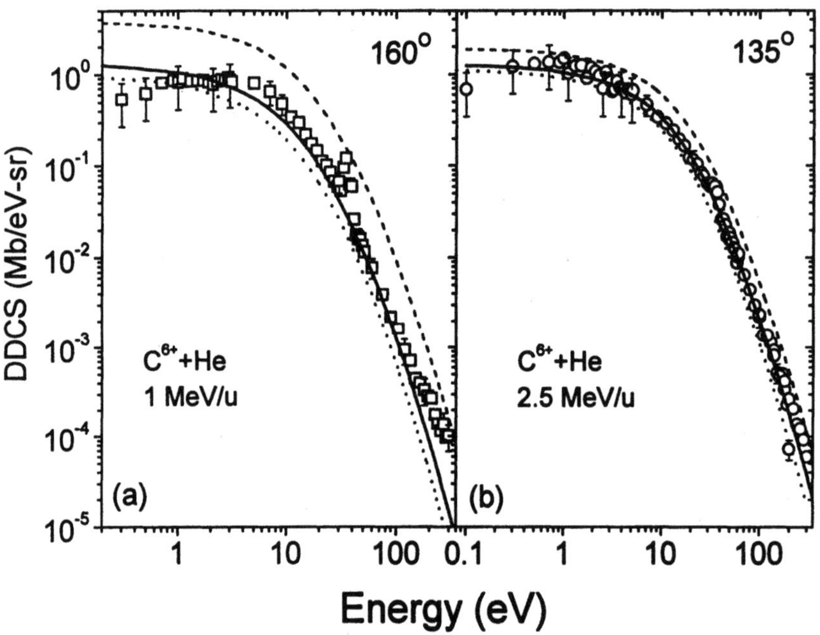

Figure 2. Measured electron DDCS spectrum for C^{6+}+He for two different projectile energies and two different emission angles in the backward directions. (a) Beam Energy = 1 MeV/u, $\theta = 160°$ [37] and (b) Beam Energy = 2.5 MeV/u, $\theta = 135°$ [23]. The solid and dotted lines are CDW-EIS calculations with HFS and H-like wave functions, respectively. The dashed lines are for B1 calculations using HFS wave functions.

measured at different emission angles for different collision systems. In Fig. 2 we have plotted the e-DDCS observed in large backward angles i.e. for

$\theta = 160^{\circ}$ (Fig. 2(a)) and 135° (Fig. 2(b)) for C^{6+}+He for collision energies 1 MeV/u (Fig. 2(a)) and 2.5 MeV/u (Fig. 2(b)), respectively. Measurements of DDCS at different angles have been reported earlier [23]. Here we emphasis on the comparison of the data with the different theoretical models. It can be seen easily that the CDW-EIS calculations (using HFS wave function) show a much better agreement with the data for both the collision energies compared to the CDW-EIS(H) calculations which employ H-like wave functions. A substantial difference between the two models can be noticed, the CDW-EIS(H) results being smaller than the CDW-EIS. The discrepancy between the two calculations is more for the lower beam energy (i.e. at 1 MeV/u) and the difference is found to be as large as a factor of 1.5 to 2.0. The B1 calculations overestimate the data at both the energies, although the difference is smaller at 2.5 MeV/u.

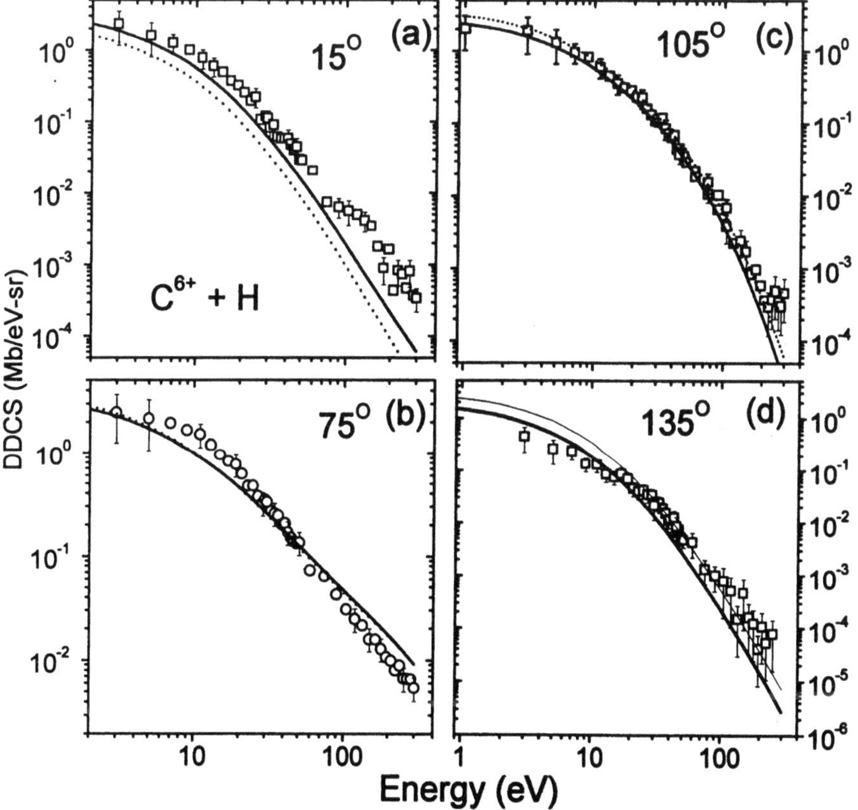

Figure 3. The electron DDCS for C^{6+}+H (v=10 a.u.) for four different angles (15°, 75°, 105°, 135°). The solid and dotted lines are the CDW-EIS and B1 calculations, respectively.

The peak around 35 eV is due to the doubly excited states of He target. So for multi-electron target (e.g. He in present case) the calculations are quite sensitive to the choice of initial state wave function. However, ionization of atomic hydrogen by bare ions is a pure three-body process and wave function being exactly known for the initial state the ambiguities in the calculations are removed. In Fig. 3 we show some examples [30] of the low energy e-DDCS spectrum for atomic hydrogen target in collision with 2.5 MeV/u (v = 10 a.u.) bare carbon ions. The data are shown for forward (15° and 75°) as well as backward (105° and 135°) angles. It may be seen that for extreme forward angle (i.e. 15°) the B1 calculations underestimate the data throughout the whole energy range (1-300 eV). The CDW-EIS calculation however gives a better agreement though it also underestimates the data somewhat above 5-10 eV. At 75° both the theories predict almost same cross sections and provide a good agreement between 30 and 100 eV. In case of 90° [not shown] and 105° both the theories provide a very good agreement with the data over entire energy range. For large backward angle, 135°, the CDW-EIS provides a better agreement with the data below 100 eV. However, a detail comparison of the angular distribution data and the theories will be provided in section 3.4.

3.2　　Comparison between the DDCS for H and He

The DDCS spectra measured for 60° are shown in Fig. 4(a) for C^{6+} + H and C^{6+} + He, along with the CDW-EIS calculations. The binary encounter peak is clearly visible at the higher energy part of both the spectra. The width of the BE peak in case of He is wider compared to that for H since the Compton profile is broader for the former case. It may be seen that the measured spectra for H target crosses that for He and theory reproduces this trend quite correctly. We show in Fig. 4(b) the ratio $R_{HeH} = \sigma_{He} / \sigma_H$ of the DDCSs of H and He targets as a function of electron energy. In Ist order approximation the expected value of R_{HeH} is estimated to be approximately 0.61 [i.e. $2 \times (0.5/.903)^2$] based on the initial binding energy consideration and the number of electrons in each atom. The binding energies for H and He targets are assumed to be those for the unperturbed states i.e., 0.5 and 0.903 a.u., respectively. The observed ratio is not a constant, rather it shows a definite structure. It may be seen that for the lowest energy electrons the experimental data fall almost on the dashed line denoted by $R_{HeH} = 0.61$. This is consistent with the fact that the lowest energy electrons are mostly emitted in the distant or large impact parameter collisions in which case the initial state of the target atom can still be considered as unperturbed and hence the assumption of the initial unperturbed binding energies holds good. As the energy increases, the ratio increases to a large value ~ 6.0. At the BE peak the ratio decreases again. At the binary encounter the projectile suffers a very close collision with the electron and in such a violent collision a large momentum is transferred to the electron. Therefore the cross section for electron emission at the BE peak is almost independent of the binding energy and the value of R_{HeH} should be close to two. Beyond the BE region the ratio increases again. The observed structure is also reproduced by the CDW-EIS calculations although there are some noticeable differences near the BE peak. The structure could also be related to the difference in the Compton profile of the two atoms.

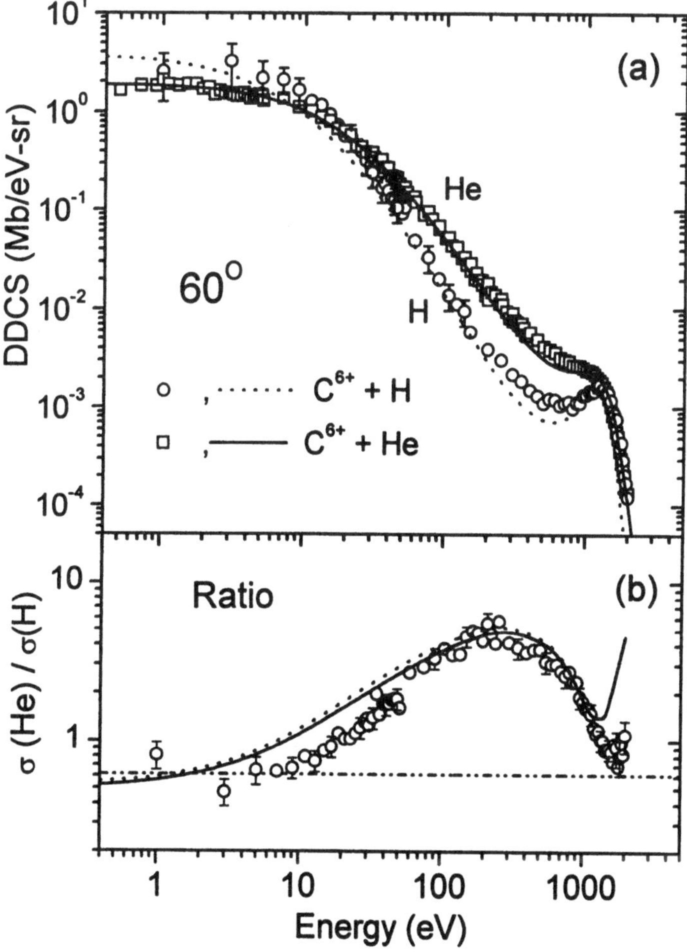

Figure 4 (a) The measured e-DDCS for C^{6+}+H and C^{6+}+He (v=10 a.u.). The dotted and solid lines represent the CDW-EIS calculations for H and He targets, respectively. (b) The ratio of the DDCS for He to that for H target along with the CDW-EIS (solid line) and B1 (dotted line) calculations. The horizontal line (dashed) represents the expected ratio = 0.61 (see text).

292

Since the B1 calculation (dotted line in Fig. 4(b)) reproduce the structure quite well it may be concluded that the two-centre effect may not be responsible for such structure. Similar structure was also observed in the similar DDCS ratio for H_2 and H targets [30].

3.3   The DDCS Ratio and TCE

It may be noticed that for the extreme forward angles the DDCSs are enhanced largely over the B1 prediction, an observation commonly found earlier also [21,23,24,30]. This is due to the TCE for which the moving ion has more influence on the ionized electron moving in the forward direction. Stolterfoht et al. [21] have shown that the TCEE can be visualised in a better way by plotting the DDCS ratio (R) i.e. the ratio of measured DDCS to that predicted by the B1 calculations as a function of electron energy. We show in Fig. 5 such a plot from our recent data on C^{6+} + He (v = 10 a.u.)[23].

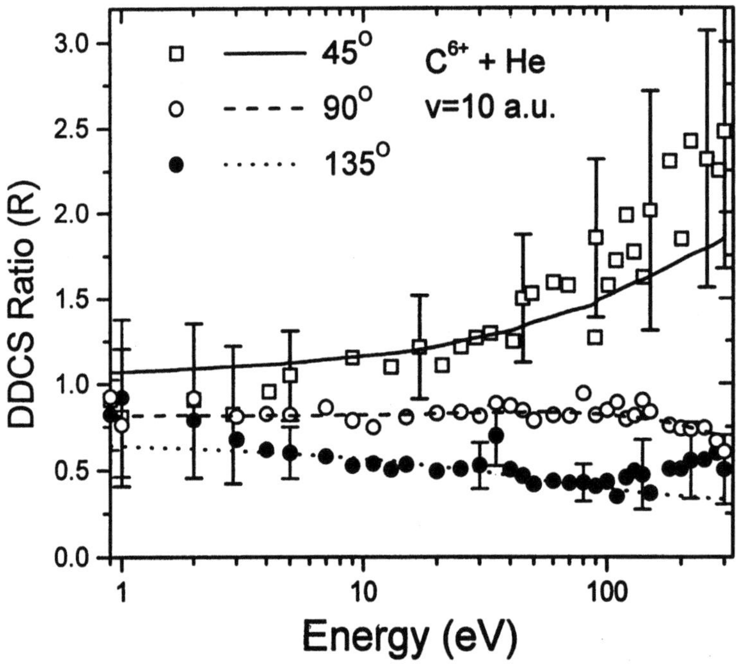

Figure 5. The ratios of measured e-DDCS to the B1 calculations for 45°, 90° and 135°. The solid (45°), dashed (90°) and dotted lines (135°) are for similar ratios for CDW-EIS calculations i.e. the ratio between the CDW-EIS and B1 calculations.

It can easily be seen that for forward angle the ratio R is larger than 1.0 and for backward angle it is less than 1.0 and therefore indicating the enhancement of the cross section in the forward angle and depletion in the backward direction with respect to the B1 prediction. For 90° the ratios are in between these two extremes and are slightly less than 1.0. However at lowest energies the value of R remains close to 1.0 for all three angles indicating that such low energy electrons could be effected less by the TCE since these electrons could be produced in large impact parameter collisions. The dynamics of these low energy electrons can be described as target-centre phenomenon and one centre electron emission can anyway be explained by the B1 approximation if realistic wave functions are used as in the present case. Higher energy electrons are produced in relatively close collisions in which the TCE plays a major role causing a large deviation in the value of R from 1.0. The CDW-EIS calculation (i.e. the ratio of CDW-EIS to B1 calculation) reproduce the DDCS ratio quite well for all the three angles and for all the energies. This demonstrates that the CDW-EIS calculations can in general describe the two centre electron emission quite well in fast ion-atom collision.

3.4 Angular distributions

A stringent test to the theories can be provided from the angular distribution plots of the DDCS for a given electron energy and such a plot can demonstrate the TCE more clearly. In Fig. 6 we show plots of such angular distributions for bare carbon ions colliding with atomic hydrogen (Fig. 6(a)), and helium (Fig. 6(b)) targets. It may be seen that for lowest energy electrons (11 eV in Fig. 6(a) and 3 eV in Fig. 6(b)) the B1 calculations predict almost a symmetric distribution about the peak at 75°. These calculations underestimate the 11 eV data for $C^{6+} + H$ at forward angles and overestimate for backward angles i.e. the observed asymmetry is not reproduced by the theory. At higher electron energies the deviation at small forward angles increases and the data are larger compared to the theory by a factor of ~ 3 (for 30 eV) to ~ 7 (for 180 eV) indicating a large TCE causing such discrepancies. However, at backward angles the theory gives a better agreement with the data at higher energies (30 and 180 eV) compared to that at low energy (11 eV). The CDW-EIS, on the other hand, provides a much better expianation with the data. For example, the theory almost reproduces the 11 eV data over the entire angular range with a small deviation about 30-40% at extreme forward and backward angles. At higher energies although the theory explains the data much better than B1 but it still underestimates the data for extreme forward and backward angles by about 50% for 30 eV. The deviations are even larger in case of 180 eV data at

294

small angles in case of $C^{6+} + H$. Interestingly, CDW-EIS provides an excellent agreement with the data for all angles between 60° and 120° and for all energies. In case of He target (Fig. 6(b)) the CDW-EIS provides an excellent agreement with the data at 3 eV, 30 eV and 240 eV over the whole angular range. For example, the theory almost passes through all the data points for 30 eV. A small deviation started being visible at the small angles for 240 eV data. The B1 calculations underestimate the data at forward angles and overestimate for the backward angles, especially at higher energies. The deviation of B1 calculations from the H data at forward angles is larger than that for He and similarly the CDW-EIS calculations provide a better agreement with the He data at forward angles compared to that for H. The reason for such behaviour is not fully understood at this point but these observations could provide crucial information for the development of the theory.

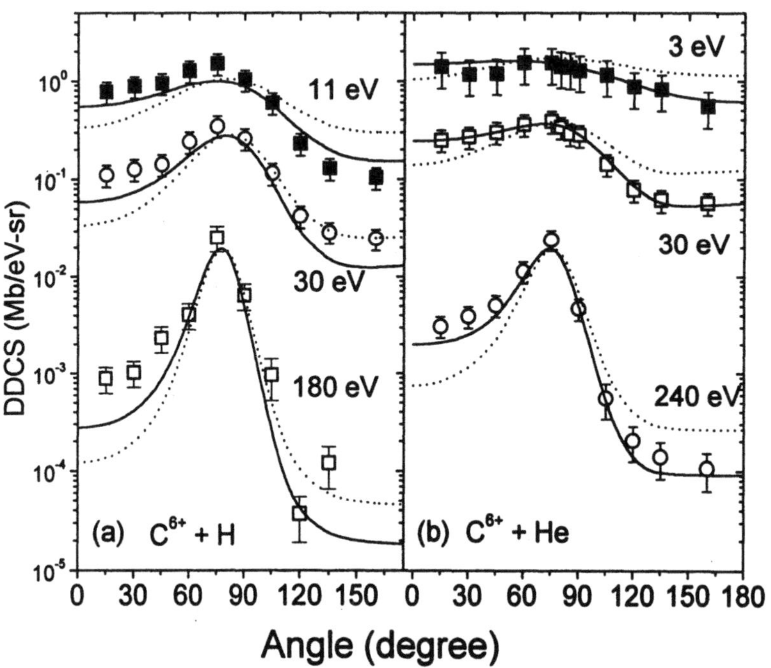

Figure 6 Angular distributions of electron DDCS for different electron energies for two different collision systems: (a) 2.5 MeV/u C^{6+}+H [30], (b) 2.5 MeV/u C^{6+}+He [23].

4. CONCLUSIONS

We have discussed some of our recent studies on the low energy electron emission in fast ion-atom collision with special emphasis on the experiments using atomic hydrogen target. A comparative study has been presented for the electron DDCS in collisions with He and H targets. From the study of angular distribution, and DDCS ratio it has been shown that two centre effect plays a major role in fast ion-atom ionization. Many of the observations can not be explained by B1 approximation. The CDW-EIS calculations using the HFS wave functions for the active electron in its initial and final states can provide an excellent agreement with the He data, although some discrepancies exist for electron emission in extreme forward angles. A better agreement with CDW-EIS is found for He data compared to that for H. The ratio of the electron DDCS for He to that for H shows a prominent structure. It is believed that these observations can provide valuable inputs for the further improvement of theoretical models.

ACKNOWLEDGMENTS

This work has been carried out in collaboration with P. Richard, L. Gulyas, M. E. Rudd, C. D. Lin and supported by the Division of Chemical Sciences, Office of Basic Energy Sciences, Office of Energy Research, U. S. Department of Energy.

REFERENCES

1. M. E. Rudd and T. Jorgensen Jr., Phys. Rev **131**, 666 (1963); M. E. Rudd, C. A. Sautter and C. L. Bailey, Phys. Rev **151**, 20 (1966).

2. G. B. Crooks and M. E. Rudd, Phys. Rev. Lett. **25**, 1599 (1970).

3. K. G. Harrison and M. Lucas, Phys. Lett. A **33**, 149 (1970).

4. C. B. Opal, E. C. Beaty and W. K. Peterson, At. Data Nucl. Data Tables **4**, 209 (1972).

5. Steven T. Manson, L. H. Toburen, N. Stolterfoht, Phys. Rev. A **12**, 60 (1975).

6. M. E. Rudd, L. H. Toburen and N. Stolterfoht, At. Data Nucl. Data Tables **18**, 413 (1976); *ibid.* **23**, 405 (1979).

7. M. E. Rudd, Y. Kim, D. Madison and T. Gay, Rev. Mod. Phys. **64**, 441 (1992).

8.   S. Suarez, C. Garibotti, W. Meckbach, and G. Bernardi, Phys. Rev. Lett. **70**, 418 (1993).

9.   M. W. Gealy, G. W. Kerby III, Y.-Y. Hsu, and M. E. Rudd, Phys. Rev. A **51**, 2247 (1995); G. W. Kerby III, M. W. Gealy, Y.-Y. Hsu, and M. E. Rudd, Phys. Rev. A **51**, 2256 (1995).

10. Y.-Y. Hsu, M. W. Gealy, G. W. Kerby III, and M. E. Rudd, Phys. Rev. A **53**, 297 (1996); Y.-Y. Hsu, M. W. Gealy, G. W.Kerby III, M. E. Rudd, D. R. Schultz and C. O. Reinhold, Phys. Rev. A **53**, 303 (1996).

11. D. H. Madison, Phys. Rev A **8**, 2449 (1973).

12. I. M. Chesire, Proc. Phys. Soc. **84**, 89 (1964).

13. Dz Belkic, J. Phys. B **11**, 3529 (1978).

14. M. B. Shah and H. B. Gilbody, J. Phys. B **14**, 2361 (1981); *ibid* J. Phys. B **14**, 2831 (1981); *ibid* J. Phys. B **15**, 413 (1982); *ibid* J. Phys. B **16**, L449 (1983).

15. D. S. F. Crothers and J. F. McCann, J. Phys. B **16**, 3229 (1983).

16. P. D. Fainstein, V. H. Ponce and R. D. Rivarola, J. Phys. B **23**, 1481 (1990).

17. P. D. Fainstein, V. H. Ponce and R. D. Rivarola, J. Phys. B **21**, 287 (1988).

18. E. Clementi and C. Roetti, At. Data Nucl. Data Tables **14**, 177 (1974).

19. L. Gulyas, P. D. Fainstein and A. Salin, J. Phys. B 28, 245 (1995).

20. N. Stolterfoht, D. Schneider, J. Tanis, H. Altevogt, A. Salin, P. D. Fainstein, R. Rivarola, J. P. Grandin, J. M. Scheurer, S. Andriamonje, D. Bertault and J. F. Chemin, Europhys. Lett. **4**, 899 (1987).

21. N. Stolterfoht, H. Platten, G. Schiwietz, D. Schneider, L. Gulyas, P. D. Fainstein and A. Salin, Phys. Rev. A **52**, 3796 (1995).

22. Lokesh C. Tribedi, P. Richard, D. Ling, Y. D. Wang, C. D. Lin, R. Moshammer, G. W. Kerby III, M. W. Gealy and M. E. Rudd, Phys. Rev. A **54**, 2154 (1996).

23. Lokesh C Tribedi, P. Richard, Y. D. Wang, C. D. Lin, L. Gulyas and M. E. Rudd, Phys. Rev A **58**, 3619 (1998).

24. J. O. P. Pedersen, P. Hvelplund, A. Petersen and P. Fainstein, J. Phys. B **24**, 4001 (1991).

25. R. D. DuBois, Phys. Rev. A **50**, 364 (1994).

26. *Electron emission in heavy ion-atom collision*, N. Stolterfoht, R.D. DuBois, R.D. Riverolla, Springer series on Atoms and Plasmas (1997).

27. W. L. Fite, R. Stebbings, D. G. Hummer, and R. T. Brackmann, Phys. Rev. **119**, 663 (1960).

28. M. B. Shah, D. S. Elliot and H. B. Gilbody, J. Phys. B **20**, 2481 (1987).

29. T. W. Shyn, Phys. Rev. A **45**, 2951 (1992).

30. Lokesh C Tribedi, P. Richard, W. DeHaven, L. Gulyas, M. W. Gealy and M. E. Rudd, J. Phys. B **31**, L369 (1998).

31. J. Slevin and W. Sterling, Rev. Sci. Instr. **52**, 1780 (1981).

32. M. A. Bolorizadeh and M. E. Rudd, Phys. Rev. A **33**, 882 (1986).

33. D. E. Golden and H. W. Bandel, Phys. Rev. **138**, A14 (1965).

34. D. E. Golden, H. W. Bandel and J. A. Salerno, Phys. Rev. **140**, 40 (1966).

35. V. V. Afrosimov, G. A. Leiko, Y. A. Mamaev, M. N. Panov, Soviet Phys. JETP **29**, 649 (1969).

36. R. M. Wood, A.K. Edwards and M. F. Steuer, Phys. Rev. A **15**, 1433 (1977).

37. Lokesh C Tribedi, P. Richard, L. Gulyas et al. (1999) (To be published).

Neutral Atom Traps for Bose-Einstein Condensation

B. N. Jagatap[*], A. P. Marathe[*], K. G. Manohar[*], R. C. Sethi[+], S. A. Ahmad[#]

[*]*Laser and Plasma Technology Division*
[+]*Accelerator and Pulse Power Division*
[#]*Spectroscopy Division*
Bhabha Atomic Research Centre,
Trombay, Mumbai - 400 085, India

1. INTRODUCTION

Optical and magnetic trapping of neutral atoms, to produce ultracold and dense samples of atomic vapours, is a new phenomenon in physics that has potential for use in many areas of research. Observation of Bose-Einstein condensation (BEC) in dilute vapours of alkali atoms [1-4] is one of the fascinating applications of these atomic trapping and cooling techniques. For BEC, one must produce a sample of bosonic particles whose thermal de Broglie wavelength exceeds the mean inter-atomic separation. Under this situation the Bose statistics favours the condensation of all the atoms into a single quantum state of the system [5]. Clearly such a phase transition can be observed only at ultra low temperatures and relatively high densities of bosonic particles. Fuelled by the search for high densities and very low temperatures of atomic vapours, last few years have seen a flurry of activities in demonstrating a variety of neutral atom traps [6, 7]. These have made it possible to obtain dense samples of atomic systems at unprecedented low temperatures for the observation of the collective quantum effects. Other applications of these neutral atom traps include very high resolution spectroscopy, metrology, nonlinear optics, atom optics and "non-accelerator" particle physics [8].

The experimental scheme, known as the "hybrid" scheme [1-3], which combines strengths of laser cooling and evaporative cooling for the observation of BEC is shown schematically in Fig. 1. The neutral atom

Trends in Atomic and Molecular Physics,
Edited by Sud and Upadhyaya. Kluwer Academic/Plenum Publishers, New York, 2000.

traps, both magnetic and optical, involved in these experiments are shown explicitly in this figure. While the confining force in magnetic traps originates from the interaction of the magnetic moment and a non-uniform magnetic field [6], the optical traps work on the radiation pressure force; either spontaneous or induced [7]. Magneto-optic trap (MOT) which works on the spontaneous radiation force in an inhomogeneous magnetic field is the fore runner of the neutral atom traps and is a vital component in all the laser cooling experiments [9-18]. Dark Spontaneous force optical trap (Dark SPOT) is a variation of MOT and is capable of achieving densities as high as 10^{12} atoms/cc at temperatures down to a few μK [17]. Although these temperatures and densities are not adequate to achieve BEC, they provide a good starting point for crossing the phase boundaries for BEC [1-3]. In the BEC demonstration experiments, MOT (or Dark SPOT) is used to load a magnetic trap, e.g. a spherical quadrupole magnetic trap (SQMT), for further cooling of the laser cooled sample by evaporative cooling. Time orbiting potential (TOP) trap [1] and SQMT with optical plug [2] are further innovations which overcome the difficulties posed by the Majorana spin-flip transitions at zero field in the trap centre of SQMT. These ideas have been successful in demonstrating BEC in dilute vapours of alkali atoms [1-3] and studying the properties of such condensates [4]. Clover-leaf trap (CLT) [19], which is based on the Ioffe-Pritchard trap (IPT) [6], is a much recent development in this area. It is now possible to obtain double BEC condensates which offer an opportunity to study the dynamics of the distinguishable and interpenetrating bosonic quantum systems [4, 20]. Very recently BEC confinement has been shown in optical dipole traps which can stably trap atoms in arbitrary hyperfine states [21].

In this article we discuss the ideas underlying these recent developments, describe the basic trap configurations and provide an insight into the experiments demonstrating BEC in weakly interacting bosonic systems. In BARC, we have started an intensive programme on laser cooling and trapping of neutral atoms [22]. Our own experience in the design of some of these traps will be discussed somewhat in details.

2. SPHERICAL QUADRUPOLE MAGNETIC TRAP (SQMT)

An atom of magnetic moment μ and placed in a magnetic field $\mathbf{B}$, experiences a force given by $F = -\nabla(\mu \cdot \mathbf{B}) = -|\mu| |\nabla| \mathbf{B} |\cos(\mu, \mathbf{B})$. In an inhomogeneous magnetic field this force can be used to confine the neutral atoms [6, 7]. The simplest of the magnetic traps is the SQMT, originally suggested by W. Paul for neutron trapping, where the field $|\mathbf{B}|$

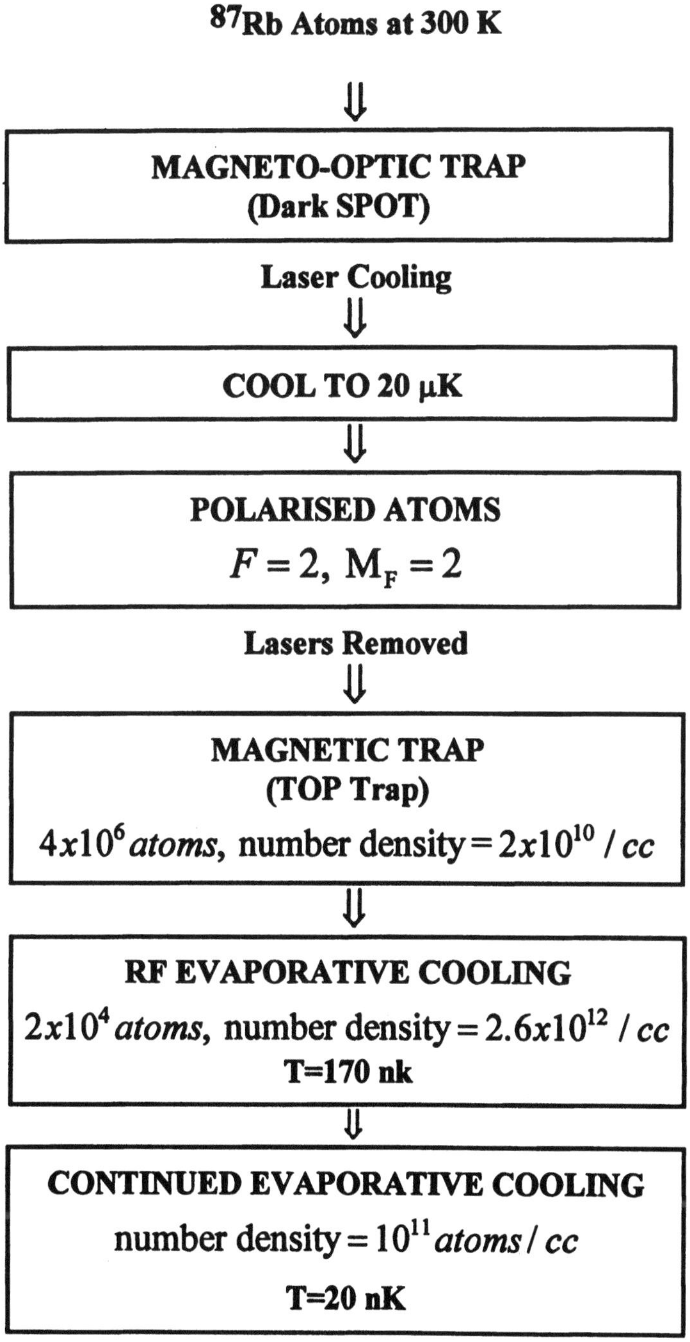

^{87}Rb Atoms at 300 K

⇓

MAGNETO-OPTIC TRAP
(Dark SPOT)

Laser Cooling

⇓

COOL TO 20 μK

⇓

POLARISED ATOMS

$F = 2,\ \mathrm{M_F} = 2$

Lasers Removed

⇓

MAGNETIC TRAP
(TOP Trap)

$4 x 10^6\,atoms$, number density $= 2 x 10^{10}\,/\,cc$

⇓

RF EVAPORATIVE COOLING

$2 x 10^4\,atoms$, number density $= 2.6 x 10^{12}\,/\,cc$
T=170 nk

⇓

CONTINUED EVAPORATIVE COOLING

number density $= 10^{11}\,atoms\,/\,cc$

T=20 nK

Figure 1. "Hybrid" approach for demonstration of Bose-Einstein condensation by Anderson et al. [1]. For evaporative cooling Time Orbiting Potential (TOP) trap is used. See text for details.

varies linearly in space. The trap consists of two identical coils of radius R, separated by distance S = 1.25 R and carrying opposite currents. In Fig. 2 we show schematically how a two coil configuration gives rise to a quadrupole field and a trapping potential.

The spherical quadrupole trapping field has cylindrical symmetry which makes it convenient to describe the fields in terms of the radial coordinate $\rho = \sqrt{x^2 + y^2}$ and axial coordinate z (along the coil axis). Multipole polynomial expansion [6] of the field reveals that the magnetic potential is given by $V \propto \sqrt{\rho^2 + 4z^2}$. Clearly the trap has a single centre ($\rho = z = 0$) where the field is zero and the field increases in all directions. Near the trap centre the field gradient is constant along any line through the origin but has different values in different polar directions. Specifically the field gradient in z direction is twice as much in the radial direction. The confining force is thus neither harmonic nor central and as a consequence the angular momentum of the trapped particles is not conserved in this trap. However the component of the angular momentum along the axis of the trap is a conserved quantity. In our experiments [18], we use two identical coils of R = 30 mm and S = 37.5 mm. Isopotential ($|\mathbf{B}|$) lines simulated for this two coil configuration is shown in Fig. 4. These simulations are performed for a spherical quadrupole excited by 100 A of current.

Because of its experimental simplicity and easy optical access, a SQMT is widely used to trap laser cooled atoms. To this end, let us recall that the trapping force is dependent on the angle between the field and the atomic moment. Therefore, in order to trap the atoms the atomic magnetic moment must be oriented so that they are repelled from the regions of strong magnetic field. This means that the atoms have to be produced in a low field seeking state prior to trapping in a SQMT. Such an orientation of the atomic spins is produced by optical pumping process. As described in Fig. 1, the neutral atoms are cooled in a MOT and subsequently spin polarised by optical pumping to load a SQMT. It must also be noted here that the orientation of the atomic moments must be preserved while the atoms are in the trap even though the trapping field changes directions in a complicated way. This requires atomic velocities slow enough so that the interaction between μ and $\mathbf{B}$ is adiabatic. In other words as long as the atoms move very slowly, the magnetic fields change slowly enough so that the magnetic moments precessing about the field at the Larmor frequency can follow it adiabatically.

The transfer of laser cooled atoms from a MOT to SQMT is affected in a number of controlled experimental steps. The sequence of operation as described by Wieman et al [12] is discussed here. Firstly the MOT is turned

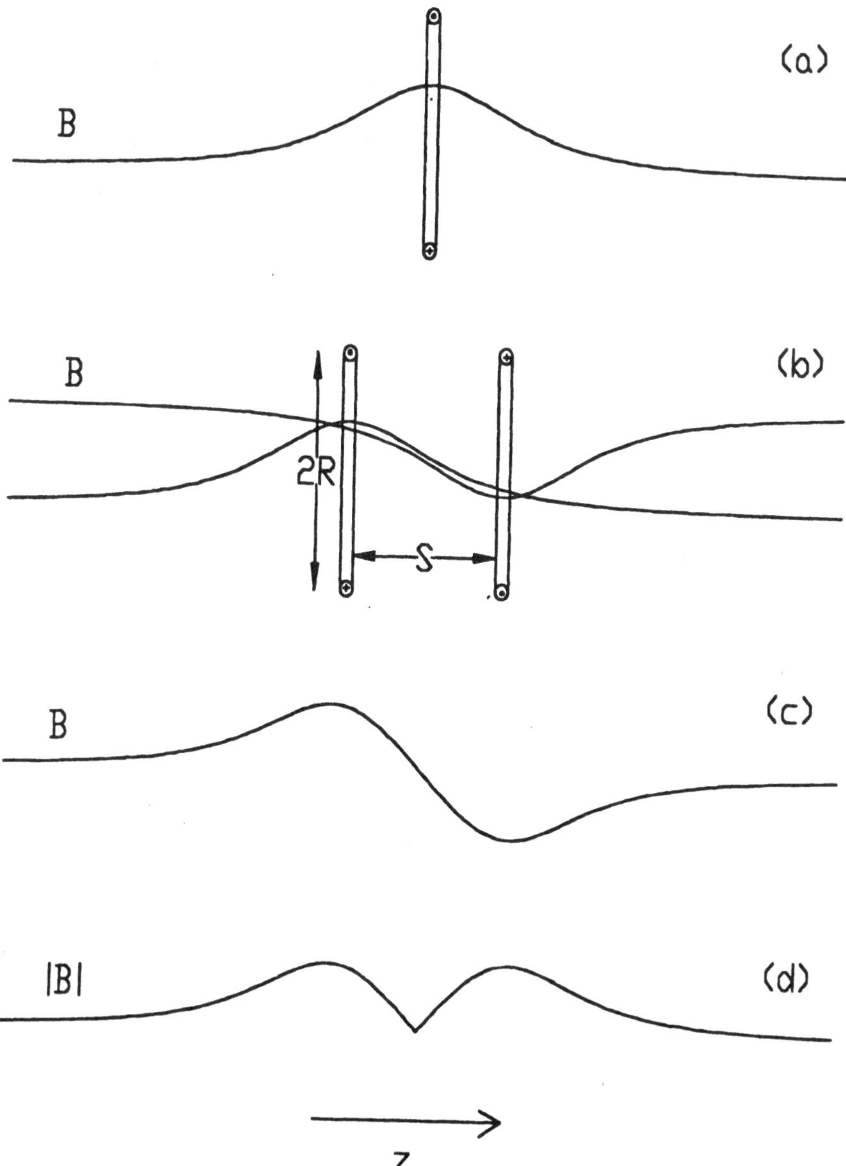

Figure 2. Schematic representation of quadrupole trapping produced by a two coil configuration (a) z component of the magnetic field, **B**, due to a single coil (b) z component of **B** due to two identical coils with spatial separation S = 1.25R and carrying opposite currents (c) z component of the quadrupole field along the axis (d) potential (which is proportional to |**B**|) as seen by an atomic dipole. Here R is the radius of coils and D is the separation between the coils.

304

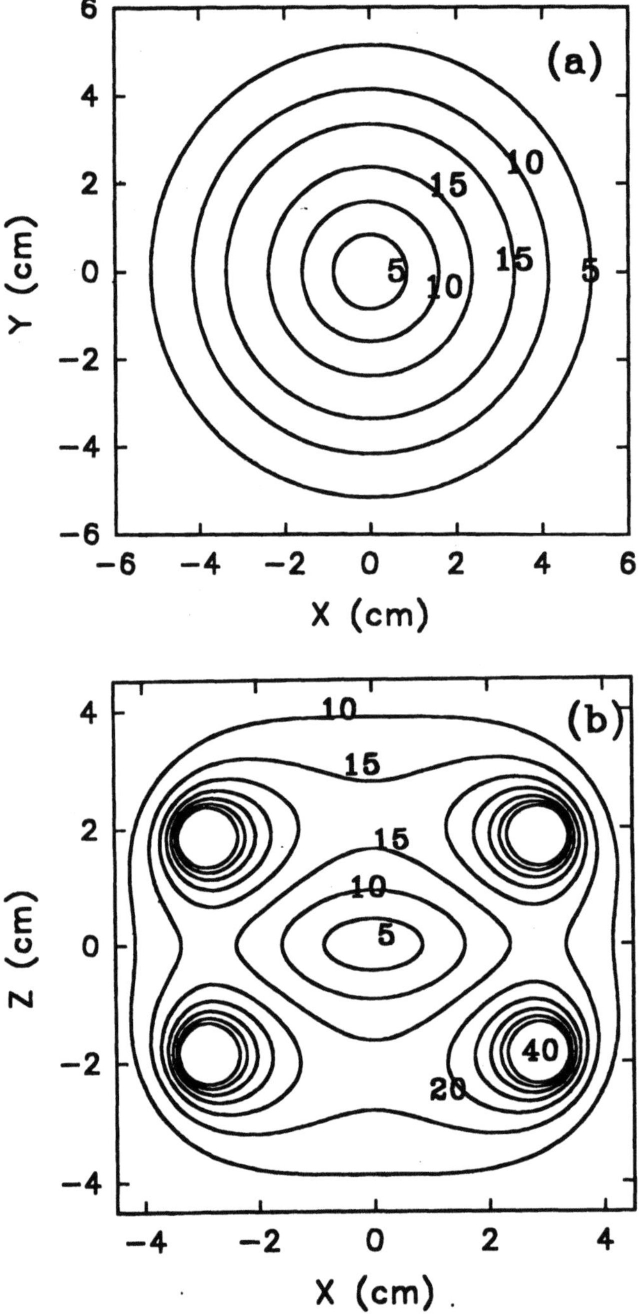

Figure 3. Contours of |**B**| simulated for a two coil SQMT in (a) x-y plane and in (b) x-z plane. In these calculations, R = 30 mm, S = 37.5 mm and current = 100 A. The contours are drawn at 5 G interval and the value of |**B**| is indicated on each contour.

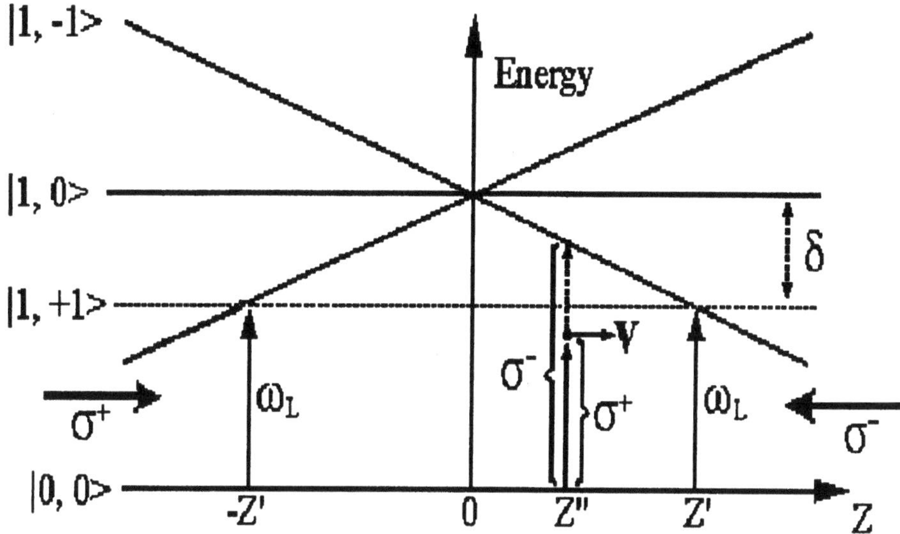

Figure 4. Principle of MOT in one dimension. Inhomogeneous magnetic field gives rise to space dependent Zeeman splitting. $|J, M\rangle$ denote the Zeeman states. Counter propagating laser beams of frequency ω_L and polarisation σ^+ and σ^- (shown by horizontal arrows), resonant with the atomic transitions at $z = \pm z'$, produce space dependent force on the atom which is directed towards the origin. At $z = z''$ an atom moving to the right with velocity V sees σ^+ beam red shifted and σ^- beam blue shifted. The resonance therefore occurs at $z'' < z'$.

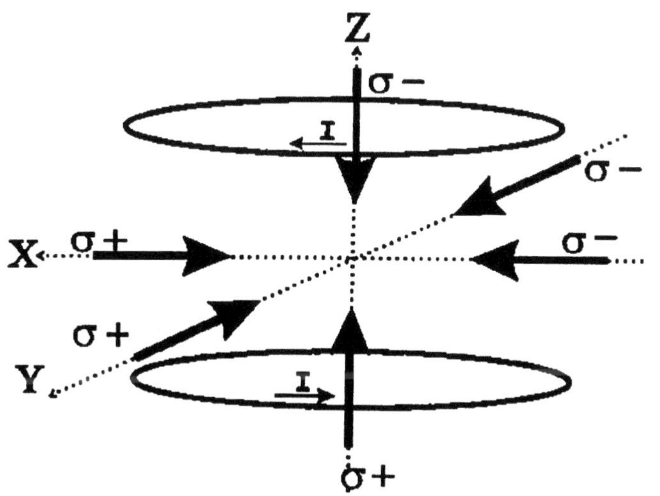

Figure 5. Schematics of a MOT in three dimensions. Spherical quadrupole field is produced by a pair of coils carrying current (I) in the opposite directions. Six laser beams in $\pm \hat{x}, \pm \hat{y}$ and $\pm \hat{z}$ directions with specified polarisations are used.

on and is filled for a few seconds. For additional cooling of the atomic sample, Sisyphus cooling mechanism may be invoked. To this end the magnetic field of MOT is switched off, the laser frequency is red shifted and laser intensity is reduced. After a few milliseconds, the lasers are quickly switched off in microsecond time scales (20 μs). A circularly polarised light pulse of millisecond duration is sent through the atomic sample to optically pump the atoms into the low field seeking state. At the end of the pulse, the magnetic field of the SQMT is switched on in microsecond time scale (20 μs).

Recently the technique of buffer gas loading of neutral atoms has been demonstrated to load a SQMT with europium and chromium atoms [23]. In this technique the trapping region is filled with helium buffer gas maintained at cryogenic temperatures. The atomic species of interest are introduced in the trap where they diffuse through the buffer gas and thermalise by elastic collisions. The atoms in the weak field seeking states are contained by the magnetic field. Atom densities of the trapped atoms (10^{12} atoms/cc) attained in these experiments is comparable to the highest density achieved in a MOT. This technique promises the extension of magnetic trapping to a wider variety of atomic and molecular species.

3. VARIANTS OF SQMT

The life time of the atoms in a SQMT is governed by two mechanisms which lead to the loss of atoms from the trap [7]. Firstly the collisions with the background gas can lead to the trap loss. Clearly a vacuum of the order of 10^{-8} torr or better is a must for the operation of a SQMT. Secondly the Majorana spin flip transitions near the trap centre limit the life time of the atoms in SQMT. This loss by far is of much concern in the experiments designed for realisation of Bose condensates.

The loss of atoms due to spin flip transitions may be understood as follows: An atom remains in the trap as long as its spin remains aligned with the local field. Very near the trap centre, the field direction changes sharply (see Fig. 2) and all the spin orientations become nearly degenerate. Thus an atom crossing the zero field region can find its spin flipped and as a consequence it is ejected out of the trap. This spin flip transition (Majorana transition) makes the SQMT leaky as if the trap has a hole at the zero field. Moreover this zero field hole affects the very cold atoms in the trap since these atoms are more likely to be near the trap centre.

It is clear then that for a successful trapping experiment for BEC, the zero field hole needs to be "plugged". Time orbiting potential (TOP) trap [1] is a variant of SQMT where the leaky hole is plugged by adding a small uniform transverse rotating field to the spherical quadrupole field. The rotation rate is selected in such a way that it is slow enough to align the atomic spin with the instantaneous magnetic field, but fast enough to govern the atomic motion through space by the time average of the potential. Such an arrangement is shown to result in an effective average potential which is axially symmetric, 3-D harmonic potential providing a tight and stable confinement. Another idea to plug the leaky hole is to use "optical plug" as demonstrated by Ketterle and his group [2]. In this variation of SQMT, the trap loss is suppressed by adding a repulsive optical potential at the centre of the SQMT, the trap loss is suppressed by adding a repulsive optical potential at the centre of the SQMT. This is achieved by tightly focussing an intense blue detuned laser which generates a repulsive optical dipole force.

4. MAGNETO-OPTICAL TRAP (MOT)

MOT is based on the scattering or spontaneous radiation pressure force in an inhomogeneous magnetic field [9-18]. An advantage of MOT is that it has a large trap depth which is relatively insensitive to the experimental imperfections. A MOT for Cs atoms having a trap depth of ~ 1 K can be easily realised with a few mW of laser power. A potential of this depth is capable of confining atoms moving with velocity 2×10^3 cm/s. Moreover the number density in a MOT can be as high as 10^{11} atoms/cc. These characteristics make MOT a standard and indispensable tool in any laser cooling and trapping experiments. As shown in Fig. 1, a MOT is used to obtain a cold and dense atomic sample which serves as a good starting point to cross the phase boundaries for BEC.

The basic principle of a MOT [9, 10] is illustrated in Fig. 4 where we consider a hypothetical two-level atom of transition frequency ω_a and having $J_g = 0$ ground level and $J_e = 1$ excited level moving in one dimension. The atom is placed in a weak inhomogeneous magnetic field $B_z(z) = Az$. The energy levels of the atom are Zeeman split by an amount $\Delta E = \mu_B g_J m_J B_z$ where μ_B is the Bohr magneton, m_J is the magnetic quantum number and g_J is the Landé g factor for the level whose angular momentum is J. The atom is illuminated by two counter propagating laser beams of frequency ω_L. The laser frequency is detuned below the zero field atomic resonance by $|\delta| > \gamma$ where $\delta = \omega_L - \omega_a$ and γ is the homogeneous linewidth of the transition. The polarisations of the beams are such that the beam propagating in $+\hat{z}(-\hat{z})$ direction has polarisation $\sigma^+(\sigma^-)$. The σ^+ and σ^- beams drive the transitions $|m_g = 0\rangle \rightarrow |m_e = 1\rangle$

308

and $|m_g = 0\rangle \to |m_e = -1\rangle$ respectively. Atomic resonance therefore occurs only near two points (see Fig. 4) $z = \pm z'$ where the laser frequency is in resonance with the Zeeman tuned atomic states. Consequently an atom at $z = +z'$ absorbs a photon preferentially from σ^- beam and experiences a scattering force in $-z'$ direction. On the other hand, an atom at $z = -z'$ absorbs a photon mostly from the σ^+ beam and is driven in the $+z$ direction. In effect the atom experiences a net time averaged force towards the origin. The turning points $\pm z'$ can be moved by either tuning the laser or changing the magnetic field gradient. Having the laser beams tuned to the red of the resonance transition also provides the damping force. For a moving atom the optical frequencies in the rest frame of the atom are changed by the Doppler effect. In general for a given velocity v of the atom, the Zeeman and Doppler shifts combine to produce resonance with $\sigma^\pm$ beams when $\delta = \pm \left(|kv| + \mu A z / \hbar \right)$. As a consequence, in the region bounded by $\pm z'$, resonance occurs over an extended velocity range as the atoms move in the inhomogeneous magnetic field.

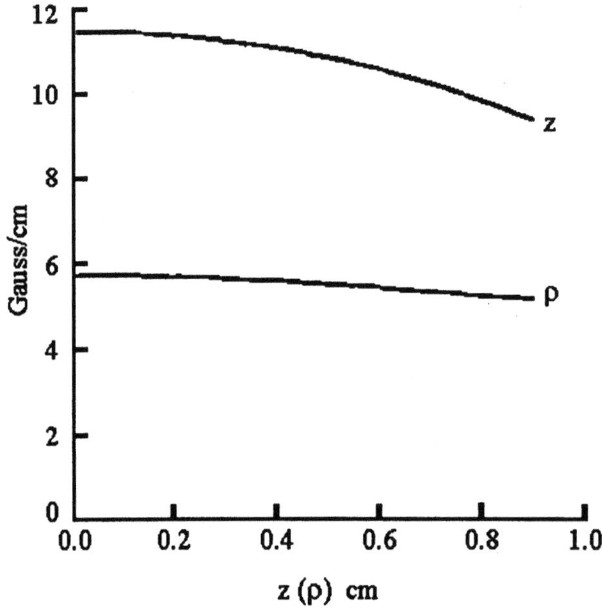

Figure 6. Magnetic field gradient along the axial (z) and radial (ρ) directions for a SQMT excited by 100 ampere. Identical coils are of R 30 mm and D = 37.5 mm.

The trapping scheme as discussed above can be readily generalised to three dimensions as shown in Fig. 5. This 3-D generalisation involves six laser beams propagating along the Cartesian axes $\pm\hat{x}$, $\pm\hat{y}$ and $\pm\hat{z}$ and a spherical quadrupole magnetic field. The polarisations of the laser beams are such that every counter propagating pair has beams of opposite helicity. The spherical quadrupole field is produced by a pair of coils in "anti-Helmholts" configuration as discussed in section 2. For example, for a pair of coils used in our MOT [18], the simulated magnetic field gradient is shown in Fig. 6. Typically one requires field gradient in the range of 5-10 G/cm. We may see from this figure that the field gradient is nearly constant over a radius of 5 mm. This region essentially forms the trapping region in the MOT [18]. The MOT installed in our laboratory [18] consists of an octagonal UHV chamber (see Fig. 7) made of SS and is provided with ten ports. Six ports of this chamber along the Cartesian axes are for passing the six trapping beams. One port is connected to the UHV pumping system which consists of an ion pump (24 l/s) and a turbo molecular pump (150 l/s). One port of this system is used for connecting a Cs/Rb reservoir and two optical ports are kept for observation of the cold cloud of atoms through CCD and fluorescence. The coils producing the spherical quadrupole magnetic field are mounted appropriately on this chamber.

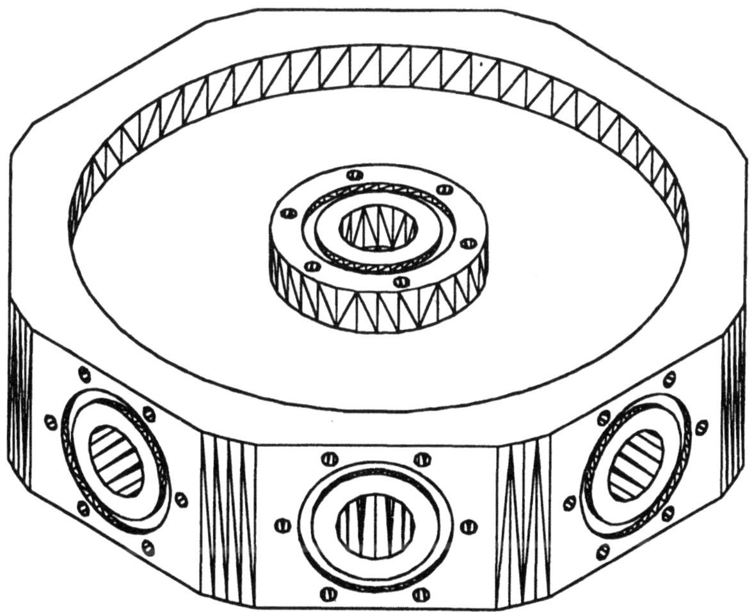

Figure 7. Schematic drawing of UHV chamber of MOT installed in BARC [18]. For details see text.

A MOT is loaded by introducing atomic vapour in the UVH chamber. MOT captures the atoms directly from the low velocity tail of the thermal distribution of the atomic vapour. Monroe et al [12] have demonstrated that using such an arrangement, they could trap about 1.8×10^7 atoms in a few seconds. Clearly the number of atoms trapped in a MOT is determined by a balance between the capture rate into the trap and the loss rate from the trap [12-16]. The loss rate is governed by a number of collision processes involving the ground and excited states of atoms in the trap and the background gas. The need for UHV ($< 10^{-8}$ torr) stems out of the considerations of the lifetime of the atoms in the trap which is governed mainly by the collisions with the background gas.

In case of the alkali atoms, the atomic levels are split by the hyperfine interaction $\left(\vec{F} = \vec{I} + \vec{J} \right)$. For example, in case of Cs (I = 7/2), the ground level $6s^2\, S_{1/2}$ splits into two hyperfine levels corresponding to F = 3 and F = 4. Similarly the excited level $6p^2\, P_{3/2}$ splits into four hyperfine levels F = 2, 3, 4 and 5. Of the various transitions possible involving this hyperfine manifold, the transition $^2S_{1/2}\, F = 4 \rightarrow {}^2P_{3/2}\, F = 5$ is the trapping transition. The trapping laser is locked to the red of this transition using saturated absorption spectroscopy. Usually a detuning of 15-30 MHz is provided depending on the magnetic field gradient. Due to non-resonant spontaneous Raman process $^2S_{1/2}\, F = 3$ level is usually populated during the trapping process. In order to prevent the atoms from accumulating in this state, a second laser corresponding to the transition $^2S_{1/2}\, F = 3 \rightarrow {}^2P_{3/2}\, F = 4$ is used. This hyperfine pumping laser is locked using saturated absorption spectroscopy and sent through one of the ports of the MOT to overlap with the trapping region formed by the trapping laser beams.

5.   DARK SPOT

Typical number density achievable in a MOT is about 10^{11} atoms/cc. The number density is limited by two processes occurring in the MOT. The first process corresponds to the two-body collisions between the ground and excited state atoms in the trap. In these collisions, the excitation energy is transferred to kinetic energy resulting in the trap loss. Experiments reveal that the rate constant for these collisions is $(1-5) \times 10^{-11}$ cm^3/s which sets up a limit of 10^{11} for the number density of the atoms in a MOT [11]. The second process corresponds to the re-absorption of fluorescence photons by the atoms which results in the repulsive forces between the atoms [17]. In other words at certain densities ($> 10^{11}$ atoms/cc) the outward radiation pressure of the fluorescence light balances the confining force of the trapping beams. Further increase in the number of atoms leads to larger atomic clouds but not to a higher number density. At high outward radiation pressures the sample breaks into a central cloud surrounded by an orbiting ring [13].

The outward radiation pressure due to fluorescence of the trapped atoms can be minimised using a variation of MOT, namely, the Dark Spontaneous force optical trap (Dark SPOT) [17]. Let us recall from the discussion in section 4, that for the trapping of alkali atoms, laser beams of two frequencies are used. Of these, one frequency is for optical trapping (i.e. $^2S_{1/2}$ F = 4 $\rightarrow$ $^2P_{3/2}$ F = 5 in Cs) and the other one is for hyperfine pumping ($^2S_{1/2}$ F = 3 $\rightarrow$ $^2P_{3/2}$ F = 4 in case of Cs). In Dark SPOT the laser beam corresponding to hyperfine pumping is tailored in such a way that it has zero intensity at the centre. As a consequence, the atoms trapped at the centre of the MOT remain in the hyperfine level ($^2S_{1/2}$ F = 3 in Cs) which is not optically connected to other relevant levels. These atoms remain shelved in the "dark" hyperfine level and do not undergo absorption emission cycles. As a result they do not experience an outward radiation pressure. These atoms can drift freely in the "dark spot" at low velocities and as soon as they emerge out of the dark region, they see both trapping and hyperfine pumping lasers and begin to absorb again. Thus outside the dark spot the atoms experience the trapping force and are driven back into the dark region of the trap. Such a trap can operate at densities very close to 10^{12} atoms/cc [17]. This is clearly the upper limit of number density in a MOT. The ultimate limit comes obviously from various collision processes involving atoms in the excited state and those in the ground state.

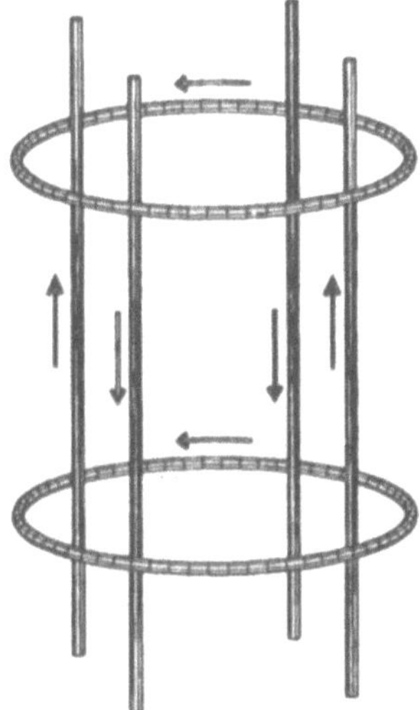

Figure 8. Schematic representation of Ioffe Pritchard Trap. It consists of two identical coils and four current rods. Typically, R = 1.5 cm, S = 4.5 cm and four parallel rods are placed symmetrically along a circle of 2 cm radius, concentric with the coils. Arrows indicate the current direction.

312

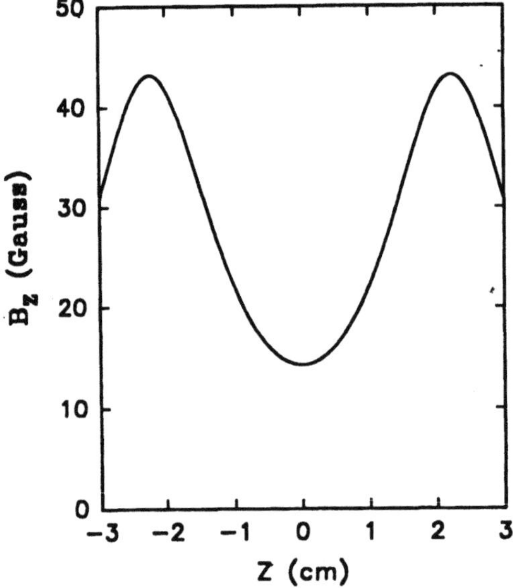

Figure 9. Magnetic field B_z along the z direction calculated for IPT excited by 100 A current. The design parameters are as given in figure 8. The bias field at z = 0 is 14.3 G.

6. IOFFE-PRITCHARD TRAP (IPT) AND CLOVER LEAF TRAP (CLT)

In section 2, we discussed extensively the properties of a SQMT. One of the drawbacks of the SQMT is that at its trap centre $B = 0$ which allows non-adiabatic spin flip transitions to take place resulting in a trap loss. In section 3, we discussed two solutions which "plug" this hole in the SQMT. Another solution to this problem is to use a magnetic trap which has a bias magnetic field at the trap centre. In the BEC experiments of Hulet [3], where the trap consists of permanent magnets, such a bias field exists naturally. A trap configuration consisting of permanent magnets, however, does not offer flexibility and does not allow easy measurement of the properties of the Bose condensates.

IPT is one such trap where the trap centre has a bias field. This trap was originally suggested by Ioffe for plasma confinement [24] and later used by Pritchard for confinement of cold atoms [25]. The configuration of IPT is shown in Fig. 8; it consists of two "pinch" coils with parallel currents and four parallel straight conductors with currents in alternating directions. In Fig. 9 we show the axial magnetic field for typical design parameters [6]. Note here that the trap has an axial bias field and the axial confinement from

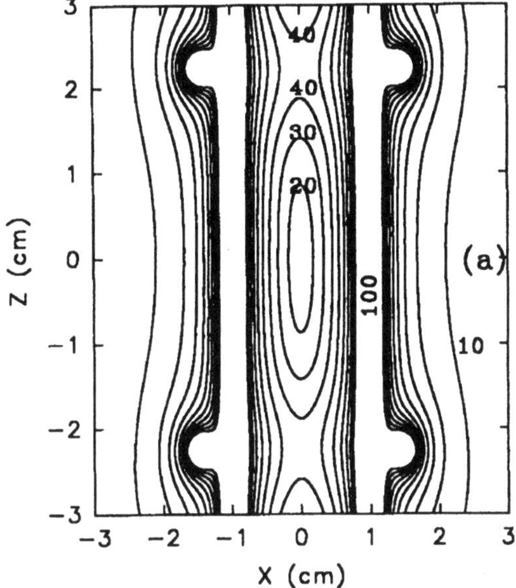

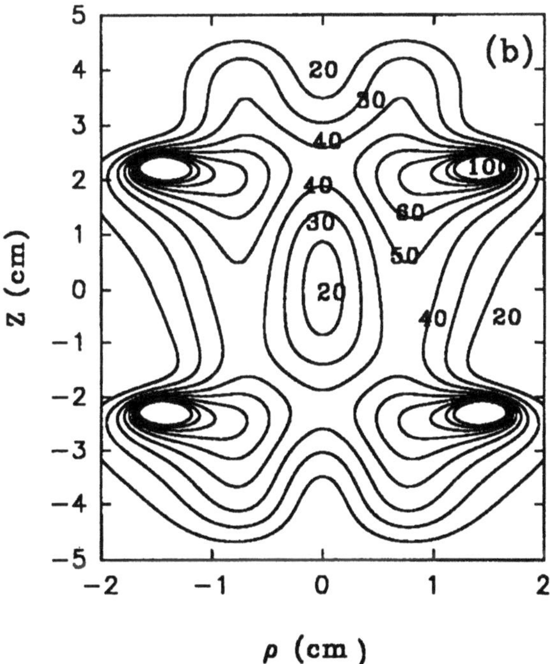

Figure 10. Contours of |**B**| calculated for IPT of figure 8 and excited by 100 A. (a) Contours in x-z plane (b) contours in the plane containing z-axis and x = y line. ($\rho = \sqrt{2}$ x = $\sqrt{2}$ y.) The contours are drawn at 10 G interval and the value of |B| is indicated on each contour.

a magnetic bottle field due to pinch coils. Transverse confinement is provided by four wires giving quadrupole focussing field. Because of the non-zero field at the centre the Larmor frequency of the atoms at the trap centre can be larger than the orbital frequency. As a consequence the spin flip Majorana transitions are avoided. In Fig. 10 a and b we show the iospotential lines of the IPT in x-z plane and x-ρ plane respectively. The B_z near the minimum of the trap varies as $B_z(z, \rho) \approx B_0 + (B_2/2)z^2$ where B_0 is the bias field and B_2 is the axial curvature component. Similarly close to the trap minimum the radial field behaves as $B_\rho(z, \rho) \approx B_1 \rho$ and vanishes at the trap centre. Here B_1 refers to the radial gradient of the field.

It then appears that an IPT may be used instead of a SQMT or its variant in BEC experiments (see Fig. 1). Although this is possible in principle, design constraints do not permit the use of this trap in a straightforward manner. Firstly the four vertical conductors of the trap do not allow an easy optical access. Secondly, the design constraints do not permit mounting of such a trap on the chamber of design shown in Fig. 7. These problems in IPT (Fig. 8) are overcome in "Clover leaf" trap (CLT); a design used by Ketterle and his group for BEC experiments [19]. In CLT, the vertical current bars of IPT (fig. 8) are replaced by coils, which contribute to the axial curvature and radial gradient. The resultant coil structure resembles a clover leaf. The trap is provided with bias coils to compensate for the axial bias field at the centre. Two such sets of coils, separated by a specified distance, forms a CLT. This trap thus allows independent variation of the gradient field, curvature field and the axial bias field. This facilitates the adjustment of confinement of the trapped atoms.

7.   CONCLUSIONS

BEC of an ideal gas is a paradigm of quantum statistical phase transitions. During past four years, several groups have demonstrated BEC in a weakly interacting dilute gas of alkali atoms at sub-microkelvin temperatures. These experiments combine ideas from two, hitherto, independent approaches of achieving BEC, namely the evaporative cooling of spin polarised hydrogen in magnetic traps and laser cooling of neutral atoms. This "hybrid" approach is essential to achieve the right sort of temperatures and phase space densities to create a new state of matter – the Bose-Einstein condensate. These observations have opened up many new possibilities for further exploration such as gaseous superfluidity, anomalous light scattering and experiments with the fermionic atoms like ^{6}Li etc [26]. These developments are a result of many pioneering ideas in magnetic and optical trapping of neutral atoms. In this article, we have discussed the ideas underlying these developments and also sketched some of our efforts in building these traps. These neutral atom traps allow one to have an ultimate control over atomic matter and study various other areas of physics such as atom optics, coherent matter waves, many body physics and precision experiments to test the fundamental symmetries [4]. They also find use in several applications such as atomic clocks and nano-technology.

ACKNOWLEDGMENTS

Authors thank Dr. A. P. Roy, Head, Spectroscopy Division and Dr. N. Vankataramani, Head, Laser and Plasma Technology Division for their keen interest in this work.

REFERENCES

1. M. H. Anderson, J. R. Ensher, M. R. Matthews, C. E. Wieman and E. A. Cornell, Science **269**, 198 (1995).

2. K. B. Davis, M. O. Mewes, M. R. Andrews, N. J. van Drutent, D. S. Durfee, D. M. Kurn and W. Ketterle, Phys. Rev. Lett. **75**, 3969 (1995).

3. C. C. Bradly, C. A. Sackett, J. J. Tollet and R. G. Hulet, Phys. Rev. Lett. **75**, 1687 (1995).

4. A. S. Parkins and D. F. Walls, Phys. Rep. **303**, 1 (1998).

5. K. Huang, *Statistical Mechanics*, Wiley, New York (1987).

6. T. Bergeman, G. Erez and H. J. Metcalf, Phys. Rev. **A35**, 1535 (1987).

7. H. Metcalf and P. van der Straten, Phys. Rep. **244**, 203 (1994).

8. M. Kasevich, K. Moler, E. Riis, E. Sundermann, D. Weiss and S. Chu, At. Phys. **12**, 47 (1991); D. N. Stacy, At. Phys. **13**, 46 (1993).

9. E. L. Raab, M. Prentiss, A. Cable, S. Chu and D. Pritchard, Phys. Rev. Lett. **59**, 2631 (1987).

10. H. Metcalf, J. Opt. Soc. Am. **B6**, 2206 (1989).

11. D. Sesko, T. Walker, C. Monroe, A. Gallaghar and C. Wieman, Phys. Rev. Lett. **63**, 961 (1989).

12. C. Monroe, W. Swann, H. Robinson and C. Wiemann, Phys. Rev. Lett. **65**, 1571 (1990).

13. D. Sesko, T. Walker and C. Wieman, J. Opt. Soc. Am. **B8**, 946 (1991).

14. A. M. Stean, M. Chowdhury and C. Foot, J. Opt. Soc. Am. **B9**, 2142 (1992).

316

15. K. E. Gibble, S. Kasapi and S. Chu, Opt. Lett. **17**, 526 (1992).

16. K. Lindquist, M. Stephens and C. Wieman, Phys. Rev. **A46**, 4082 (1992).

17. W. Kitterle, K. B. Davis, M. A. Joeffe, A. Martin and D. Pritchard, Phys. Rev. Lett. **70**, 2253 (1993).

18. a) A. P. Marathe, K. G. Manohar, B. N. Jagatap, S. G. Nakhate, S. A. Ahmad and R. C. Sethi, *Design and Development of Spherical Quadrupole Magnetostatic Trapping Fields*, Proc. Of DAE Nuclear Physics Symposium, **41B**, 394 (1998).

    b) B. N. Jagatap, K. G. Manohar, S. G. Nakhate, A. P. Marathe and S. A. Ahmad, Curr. Sci. **76**, 207 (1999).

19 M. O. Mewes, M. R. Andrews, N. J. van Druten, O. M. Kurn, D. S. Durfee and W. Ketterle, Phys. Rev. Lett. **77**, 416 (1996).

20. D. S. Hall, M. R. Matthews, J. R. Ensher, C. E. Wieman and E. A. Cornell, Phys. Rev. Lett. **81**, 1539 (1998).

21. D. M. Stampr-Kurn, M. R. Andrews, A. P. Chikkatur, S. Inonye, H. J. Miesner, J. Steinger and W. Ketterle, Phys. Rev. Lett. **80**, 2027 (1998).

22. B. N. Jagatap, S. A. Ahmad, U. K. Chatterjee and A. P. Roy, in Seminar on "Physics with Cooled and Trapped Atoms and Ions", BARC, March 5-6, 1998.

23. J. D. Weinstein, R. de Carvalho, J. Kim, D. Patterson, B. Friedrich and J. M. Doyle, Phys. Rev. **A57**, R3171 (1998).

24. Y. V. Gott, M. S. Ioffe and V. B. Tel'kovskii, Nucl. Fusion, 1962, Suppl. Pt. 3, 1045 (1962).

25. D. E. Pritchard, Phys. Rev. Lett. **51**, 1336 (1983).

26. See for example, O. Morice, Y. Castin and J. Dalibard, Phys. Rev. **A51**, 3896 (1995); B. V. Svistunov and G. V. Shlyapnikov, Sov. Phys. JETP **71**, 71 (1990).

Correlation and Photoionization: Retrospect and Prospect

S. N. Tiwary

University Department of Physics
B.R.A. Bihar University, Muzaffarpur, Bihar, India.

A recent theoretical as well as experimental study of a simultaneous ejection of two electrons by a single photon scattering by two-electron systems is reviewed, with particular emphasis on the electronic correlation in the vicinity of the threshold. The interaction of a photon with each electron is independent from the others so that double-photoionization (DPI/PDI) is forbidden process unless the electronic correlation is taken into account. If two electrons with small kinetic energies leave the residual positive ion, the motion is strongly influenced and controlled by their mutual repulsion due to Coulomb interaction ($1/r_{12}$). The interaction leads to the exchange of energy and angular momentum over long distances and therefore implies a correlation between outgoing electrons. The DPI can be described by the reaction as follows:

$$h\nu + X \rightarrow X^{++} + e^- + e^- \quad \text{or} \quad \gamma + X \rightarrow X^{++} + e^- + e^- \,, \quad (\gamma, 2e)$$

where X stands for atoms or molecules or ions. The initial state wave function can be obtained using configuration interaction (CI) method. The final state consists of a positive ion and two continuum electrons can not be described by Hylleraas, CI, MCHF and MCDF type wave functions. Accurate double-continuum wave functions (DCWF) have been a long standing and challenging problem for theorists. DCWF can be obtained solving the Schrödinger wave equation for two electrons without imposing any constraint but unfortunately this is not possible. For this reason, several existing possible DCWF which are valid in different physical situations will be described.

The validity of existing DCWF is analysed and the importance of electronic correlation in both the initial as well as final states wave functions involved in the transition amplitude for DPI process is

Trends in Atomic and Molecular Physics,
Edited by Sud and Upadhyaya. Kluwer Academic/Plenum Publishers, New York, 2000.

demonstrated. At present, we do not have comprehensive and practical DCWF which account the full correlation of two-electron in the continuum. Basic difficulties in making accurate theoretical calculations of DPI by a single high-energy photon especially in the vicinity of the threshold, where the correlation plays an important role, are discussed. Illuminating, illustrative and representative examples are presented in order to show the present status and the progress in this field. Future challenges and directions, in high precision double-photoionization cross sections, have been discussed and suggested.

1. INTRODUCTION

The subject of various forms of double-continuum wave functions is of growing interest to theorists because it is needed to solve a very broad range of problems, for example, double-photoionization, photo double detachment of negative ions, inner-shell photoionization followed by Auger process, electron impact single ionization, inner-shell excitation by electron impact which leads to autoionization, threshold law, etc. To obtain full solution of Schrödinger equation for two-electron system in any co-ordinate system is practically not feasible until now. Asymptotic solution is feasible. Knowledge of accurate asymptotic double-continuum wave functions is indispensable in order to perform a reliable theoretical calculation employing elaborate method. For example, R-matrix [1] method requires accurate functions in the outer region i.e. asymptotic region. Asymptotic wave functions are also used in deriving threshold law. Wannier hypothesis, which is built also in the Wannier-Rau-Peterkop (WRP) [2-4] theory, is that the probability for the double escape of two electrons is only determined by the long range interaction i.e. the Coulomb interaction in the zone II and III as shown below.

I	II	III
Reaction Zone	Coulomb Zone	Outer Zone

Due to this assumption only the asymptotic part of the wave function should be considered. Rosenberg [5], Rudge and Seaton [6], Rudge [7], Peterkop [8] and Burke et al [9] have investigated the asymptotic double-continuum wave functions. In the threshold law, the exponent of the excess energy depends only on the final state wave function i.e. asymptotic part of the double-continuum wave function.

Double photoionization consists in the absorption of a single photon by an atom or a molecule followed by simultaneous ejection of two electrons. The interaction of a photon with each electron is independent from the others so that double photoionization is a forbidden process unless the electronic correlation is taken into account. If two electrons with small kinetic energies leave the residual positive ion, the motion is strongly influenced and controlled by their mutual repulsion due to the Coulomb interaction ($1/r_{12}$).

The interaction leads to the exchange of energy and angular momentum over long distances and therefore implies a correlation between outgoing electrons. The final state consists of an ion and two continuum electrons i.e.

$$\gamma + X = X^{++} + e^- + e^-$$

For He atomic system, extensive investigation [10-26] of double photoionization process has been made. For H$^-$ system, Donahue et al. [27] have studied in details. For H_2 system, Dujardin et al [28] and Le Rouzo [29] have studied the double-photoionization cross section. Threshold law for the double photoionization has also been discussed [30-51]. In the case of complex atoms and molecules, the double photoionization (DPI) process can be divided into two classes (a) the normal Auger process via core ionization and (b) the resonant double Auger process via resonant core excitation. From theoretical point of view, multi-electron atoms, molecules and ions are extremely difficult because of the core.

The main reasons for choosing double-continuum wave functions and double-photoionization of He, H$^-$ and H$_2$ in this review are:

1. Various forms of asymptotic double-continuum wave functions are available but in the close vicinity of threshold of double *photoionization* [52-73], where full correlation plays an important role, accurate double continuum wave function is not obtained until now. Accurate double-continuum wave functions have been long standing and challenging problem for theorists. Several possibilities are explored in this review.

2. The double-photoionization has the advantage that this is dominated by electric dipole transition and the resulting final state is pure and well defined. In the case of He, H$^-$ and H$_2$, the final state consists of nucleus and two outgoing electrons and hence there are no complications due to core and offers the best opportunity to test the validity of double-continuum wave function [56-78].

3. A number of experimental observations and theoretical predictions of double photoionization cross sections [79-91] are available for two-electron systems but there is considerable discrepancy between experiment and theory which indicates that the probability for the DPI process in atoms, molecules and ions is significant because of electronic correlations. Electronic correlations have been extensively investigated by Tiwary and his co-workers [92-118].

2. ALL EXISTING ANALYTICAL DOUBLE-CONTINUUM WAVE FUNCTIONS (ADCWF)

2.1   Rudge-Seaton wave function
2.2   Product of two Plane wave functions
2.3   Product of Plane wave and Spherical wave function
2.4   Product of Plane wave and Coulomb wave function
2.5   Product of two unscreened Coulomb wave functions
2.6   Product of two screened Coulomb wave functions
2.7   Rau wave function
2.8   Redmond wave function
2.9   Peterkop wave function
2.10 Merkuriev wave function
2.11 Altick monopole wave function
2.12 Altick dipole wave function
2.13 Garibotti and Miraglia wave function
2.14 Burke et al wave function
2.15 Brauner, Briggs and Klar (BBK) wave function
2.16 Crothers wave function
2.17 Alt and Mukhamedzhanov wave function
2.18 Maulbetsch and Briggs wave function
2.19 Berakdar and Briggs wave function
2.20 Berakdar ψ
2.21 Mukhamedzhanov and Lieber wave function
2.22 Kunikeev and Senashenko wave function
2.23 Engelns, Klar and Malcherek wave function
2.24 Gasaneo, Colavecchi, Garibotti, Miraglia and Macri ψ
2.25 Kunikeev ψ
2.26 Schillaci and Lieber ψ
2.27 Whelan et al ψ

details about ADCWF are given elsewhere [116-118].

3. THRESHOLD LAWS

There has been interest in the extensive theoretical and experimental investigations of threshold laws for escape processes, for example, double photoionization, electron impact ionization, etc. in atoms, molecules and ions because these laws provide answers of many fundamental questions in physics. A number of theoretical and experimental studies of these laws have been made for different escape processes. As we have mentioned earlier in this review, our main emphasis will be to study double photoionization threshold law in atoms, ions and molecules i.e.

$$h\nu + X \rightarrow X^{++} + e^- + e^-, \quad h\nu + X \rightarrow X^{++} + e^- + e^- + h\nu'$$

where X stands for atom or molecule or ion.

This requires a solution of three charged particles with Coulomb forces acting between them. The final state of reaction consists of $X^{++} + e^- + e^-$ where there are two attractive Coulomb forces, each electron being attracted by residual ion X^{++} and one Coulomb repulsive force, electron-electron interaction $1/r_{12}$. If we ignore the repulsive force, the quadratic dependence on E would be reduced to a linear dependence on E. If the double-continuum electrons are represented by a product of two Coulomb wave functions, one can expect a linear threshold law because only two attractive Coulomb forces have been taken into account. Complication arises when three charged particles with Coulomb interaction are considered because of the electron correlation. To represent the final two-electron into the continuum as a product of single particle functions is only an approximation and even then there are different choices. For example, one could argue that in the neighbourhood of the threshold, as two electrons escape, there is some discrepancy in how the energy is partitioned between them, so that the slower one sees the full Coulomb field of the residual ion but the faster sees a completely screened and therefore neutral field. In this case, the final state is a product of a Coulomb wave function and a plane wave function and then there is only one $1/k$ factor, so that

$$\sigma^{2+} \propto \varepsilon^2$$

Different assumptions on the relative screening in the escape process lead to different threshold laws. A successful threshold theory while provides information on the mutual dynamic screening and mutual Coulomb repulsive interaction.

The double photoionization threshold law is

$$\sigma^{2+} \propto \varepsilon^\beta$$

where $\varepsilon = \varepsilon_1 + \varepsilon_2$ = excess energy available to two-continuum electrons.

$\varepsilon = (E_\gamma - I^{2+})$ if the residual ion is in the ground state,

E_γ = incident photon energy,

I^{2+} = double ionization threshold,

and β = exponent = 1.056, and we can express σ^{2+} as

$$\sigma^{2+} = \sigma_0 \varepsilon^{\beta}.$$

where σ_0 is the constant of proportionality and $\sigma_0 = \sigma^{2+}$ at $\varepsilon = 1$ eV.
The exponent β depends only on the final state wave function i.e. the double-continuum wave function. It reflects that accuracy of double-continuum wave function can be tested calculating the exponent i.e. β offers the best opportunity to test the accuracy of the final state wave function. The question is how the accuracy of the initial state wave function plays an important role in obtaining reliable double photoionization cross section. One can argue that σ_0 depends on the accuracy of the ground state wave

function. Another question, immediately arises that does σ_0 depend on only the initial state or both initial and final states wave functions? There are many questions, which one can ask. Byron and Joachain [10] have calculated the double photoionization cross sections using the correlated wave function for the ground state and uncorrelated wave function for the final state. Their results are not in agreement with the recent experiment, which indicates that the correlation in the final state is necessary. Le Rouzo [29] has performed similar calculation for the DPICS of hydrogen molecule. His result is in good agreement with the experimental data of Dujardin et al. [28]. He has obtained the linear threshold law and is valid up to about 10 eV above the threshold. Tiwary [20] has performed calculation of DPICS of He using correlated initial state wave function and partially correlated final state wave function and obtained in good agreement with experimental data in the intermediate and high energy range and in the vicinity of the threshold the situation is unsatisfactory. This may be due to the lack of full correlation in the final state wave function because the correlation is extremely important in the neighbourhood of the threshold. Carter and Kelly [22] have calculated the DPICS of He using the many-body perturbation theory (MBPT). They have obtained good agreement with the experiment. Since their approach is non-wave function approach it is difficult to draw any definite conclusion.

3.1   Experimental Test of WRP Threshold Law for Double-Photoionization of He

Very recently, Kossmann et al [44] have performed an extensive experimental investigation of the threshold law for the cross section of double photoionization of He. Figure 1a represents experimental results of Kossmann et al [44] for the threshold cross section of double photoionization of He from threshold to 83 eV photon energy. Figure 1b exhibits the same data but smaller and enlarged energy scale. The solid line in both figures represents in a limited energy range a least square fit of the experimental data. Because of the small difference between a linear threshold law ($\beta = 1$) and the expected $\beta = 1.056$ non-linear threshold law, it appears from the Figure 1 that linear threshold law is valid for the double photoionization of He. Quantitative analysis of results clearly show that WRP threshold law is valid because theoretical $\beta = 1.056$ and experimental $\beta = 1.05 + 0.002$ clearly demonstrates that it is indispensable to include full correlations in both initial as well as final states involved in the transition in order to obtain reliable results.

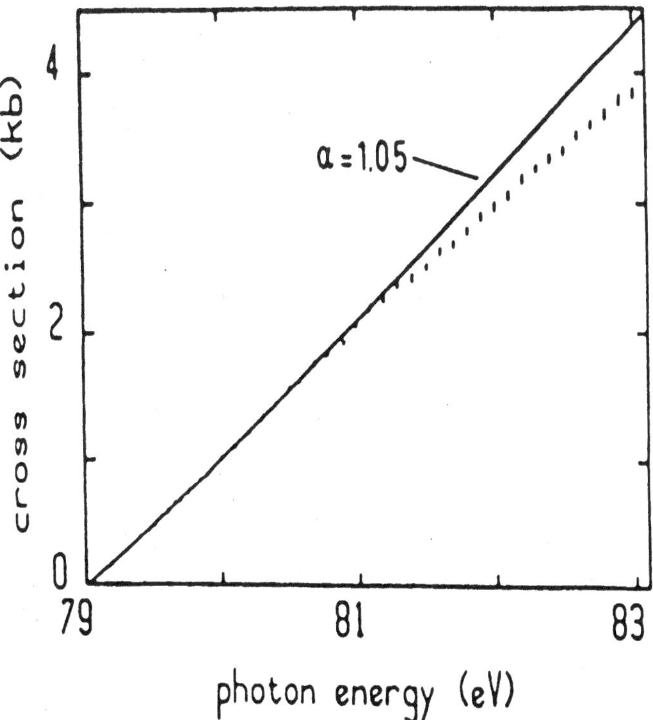

Figure 1a. Double-photoionization cross sections of He from threshold to 83 eV photon energy. .... Experimental data of Kossmann et al [44], — Least square fit.

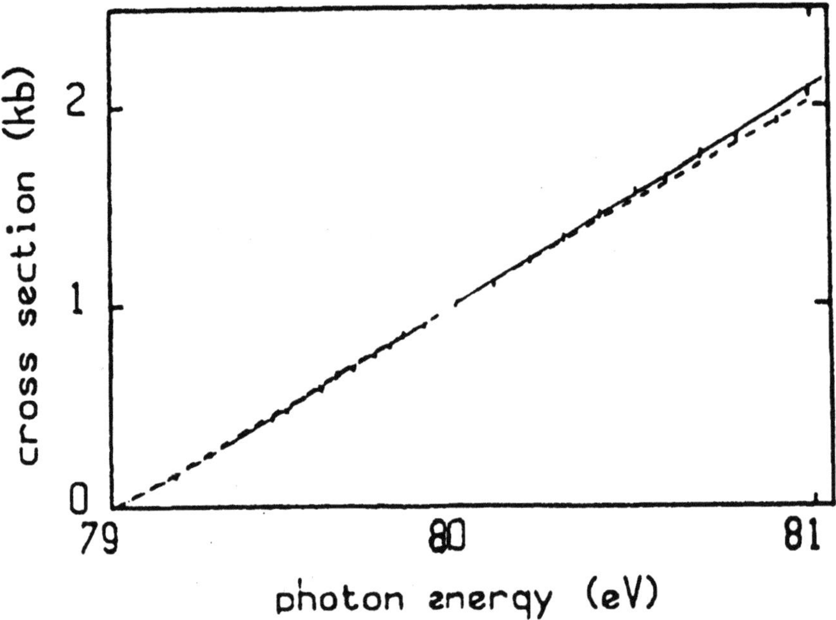

Figure 1b. Same as Figure 1a with smaller and enlarged energy scale.

3.2   Double-photoionization of He (1S^e)

Double photoionization of noble gas atoms has been of great interest to both experimentalists as well as theorists because double-electron photoionization in noble gases gives fundamental information on the electronic correlation. Helium, which is the simplest noble gas atom is more interesting because there is no complication due to core in the double photoionization process. A number of experiments and calculations have been carried out for the DPI of He. Figure 2 displays all available experimental as well as theoretical ratio of double to single photoionization cross sections. Figure 3 shows yield of very low energy electrons following photon impact on helium. For the first time, Byron and Joachain [10] performed calculation for the DPI of He using uncorrelated wave functions for both ground as well as the final states and correlated wave function for the ground state and uncorrelated product of two Coulomb wave functions with effective charge 2 for both outgoing electrons for the final state (reaction is shown below).

$$hv + He \rightarrow He^{++} + e^- + e^-$$

Results of DPI of He with uncorrelated wave functions for both initial and final states involved in the transition amplitude are extremely small which indicate that the probability of DPI process is very poor without correlations. Results with almost fully correlated wave function for the ground state and uncorrelated wave function for the final state are in excellent agreement with the first experiment of Carlson [11] (see Figure 2). Agreement suggests that the correlation is important in the initial state, not in the final state. The recent experimental observations of Holland et al [12] disagree considerably throughout the energy range of consideration with the experimental data of Carlson [11] and theoretical prediction of Byron and Joachain [10]. This experimental result suggests that correlation, probably, is equally important in both initial and final states involved in the transition. Results of Holland et al [12] are in accord with the experimental points of Schmidt et al [13] at low energies and tend to lie lower than the curve of Wight and Van der Wiel [14].

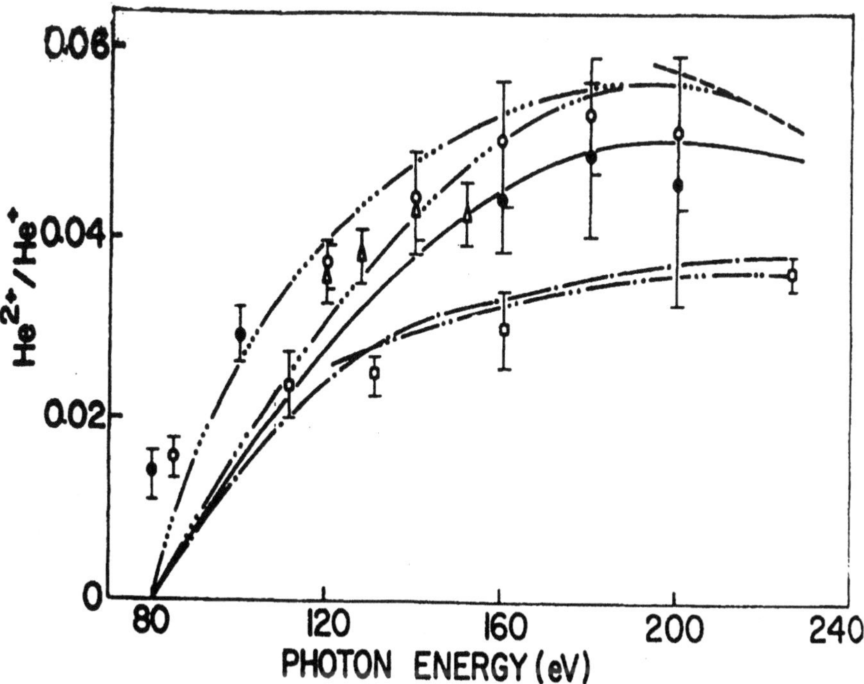

Figure 2. The ratio of cross sections He^{++} / He^{+}. Experimental curves are of Holland et al (●), Wight and Van der Wiel (Δ), Schmidt et al (O) and Carlson (□). The theoretical curves are those of Tiwary obtained from momentum matrix elements (——), Tiwary obtained from position matrix elements (— ··· —), Byron and Joachain with almost full correlation in the ground state (— · —), Byron and Joachain with no correlation in both initial and final states (------), Brown (— ·· —), Amusia et al (- - -) and Carter and Kelly (— ···· —).

326

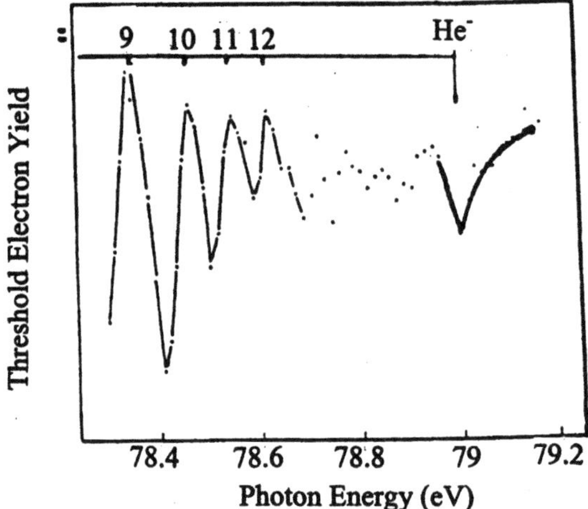

Figure 3. Yield of very low energy electrons following photon impact on He.

Brown [15] has re-evaluated the DPI cross section of He using a Hylleraas type wave function without decomposition into partial waves and Coulomb function for the final state. His results favour the oldest experimental data and the theoretical results of Byron and Joachain [10]. Amusia et al [16] have also investigated this problem in the limit of high, non-relativistic photon energies. Their method leads to a greatly overestimated cross section in the energy range of recent measurements. Yurev [18] and Varnavshikh and Labzovskii [19] have performed the calculations for the DPI cross section of He in the threshold energy region using perturbation and variational methods respectively. Their results are in qualitative agreement with each other, but are limited to the low-energy range (not shown in the figure). Tiwary [20] has performed calculations for the DPI cross section of He using the position and momentum dipole matrix elements. Tiwary has employed almost fully correlated wave function for the ground state and partially correlated wave function of Altick for the final state to evaluate dipole matrix elements. The values obtained from the length formulation tend to lie higher than those obtained using the velocity formulation. Both differ considerably from the first experimental observations and first theoretical predictions and tend to lie close to the recent reliable experimental curve of Holland et al [12] especially in the high energy range. There is considerable discrepancy between theoretical results of Tiwary and experimental data of Holland et al [12] in the vicinity of the threshold. This clearly indicates that (1) inclusion of correlation in the final state is important, (2) partial correlation is not adequate to obtain accurate results in the vicinity of the threshold i.e. two outgoing electrons are very slow. The effect of correlation decreases with increase of incident photon energy. It seems to be plausible because when escaping electrons are slow, they have enough time to develop correlations. Carter and Kelly [22] have performed calculation for the DPI cross section of He using the many-body perturbation theory (MBPT) incorporating full correlation in both initial and final states. Their results are in good agreement in the entire energy range with the most recent and reliable experimental data of Holland et al [12].

3.3   Double-photoionization of H⁻ (^{1}S)

Double photoionization of H⁻ negative ion by single photon impact has been measured by Donahue et al [27] using a crossed-relativistic beam technique with sufficient energy resolution and close enough to threshold to yield an exponent of excess energy in the threshold law. Intercepting them with laser photons Doppler shifts the photon energy in the frame of the ion to energies greater than 14.35 eV required for double detachment.

$$hv + H^-(^1S) \rightarrow H^+ + e^- + e^-$$

The threshold cross section data are shown in the Figure 4.

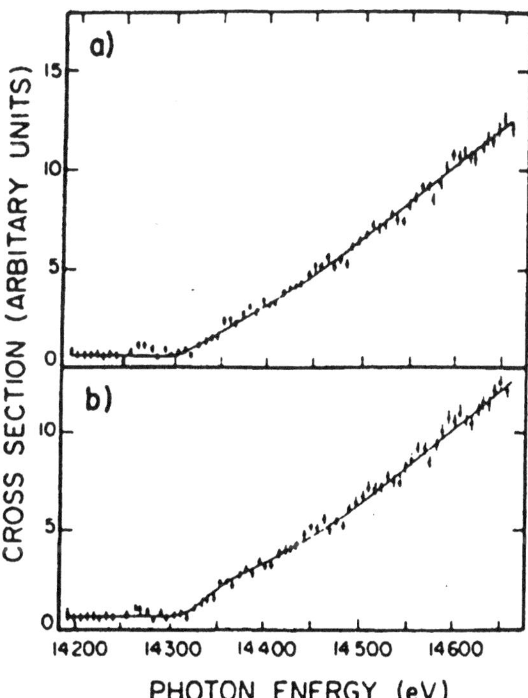

Figure 4. Double detachment cross sections for H⁻. Experiment points are of Donahue et al [27] and the solid line is the least square fit.

328

The data can be fitted quite accurately to the form

$$\sigma(E) = A(E - E_t)^m + B$$

The fit results are: $A = 38.5 + 1.5$ and $B = 0.68 + 0.05$.
Although Donahue et al [27] noted that they can also fit to an alternative result such as

$$\sigma(E) = A(E - E_t)(A + D\,Sin(C\ln(E - E_t) + F)) + B$$

The results of the fit are given details by Donahue et al [27].

The experimental arrangement suffers, unfortunately, from a spurious two-electron signal which sets in at energies slightly lower than 14.35 eV. This uncertainty in the threshold position, coupled with undetermined parameters prevents an unambiguous discrimination between the two results. Laboratory experiments on double detachment of negative ions can overcome some of these problems but results of sufficient accuracy in the close vicinity of the threshold are still unavailable.

3.4 Double-photoionization of $H_2\left({}^1\Sigma_g^+\right)$

It is interesting to both experimentalists as well as theorists to extend the studies of the double photoionization of helium atomic system to another system with only two electrons with molecular symmetry, i.e., hydrogen molecule. Although molecular hydrogen will be more complicated because of more degrees of freedom i.e. vibrations and rotations. For the first time, Dujardin et al [28] have measured the cross section of double photoionization of hydrogen molecule by single photon impact

$$hv + H_2 \rightarrow H^+ + e^- + e^-$$

using the photoion-photoion coincidence (PIPICO) method in the energy range 47.5 eV to 140 eV. Le Rouzo [29] treated this problem exactly in the same way as Byron and Joachain [10] treated the double-photoionization (DPI) of helium atomic system. Le Rouzo [29] performed calculation for the DPI of hydrogen molecule in the length and velocity forms using almost fully correlated wave function for the ground state and product of two unscreened Coulomb wave functions for the final state. Figure 5 displays the experimental observations of Dujardin et al [28] and theoretical prediction of Le Rouzo [29]. It is seen from the figure that there is an excellent agreement between experiment and theory. Le Rouzo has also

calculated cross sections in the close vicinity of the threshold as shown in
Figure 6.

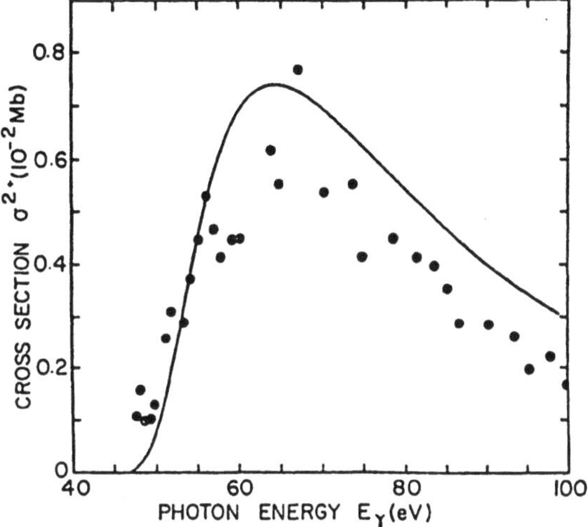

Figure 5. Double photoionization of molecular hydrogen.   The experimental points are of
Dujardin et al [28] and the solid line represents the R-averaged theoretical results of
Le Rouzo [29].

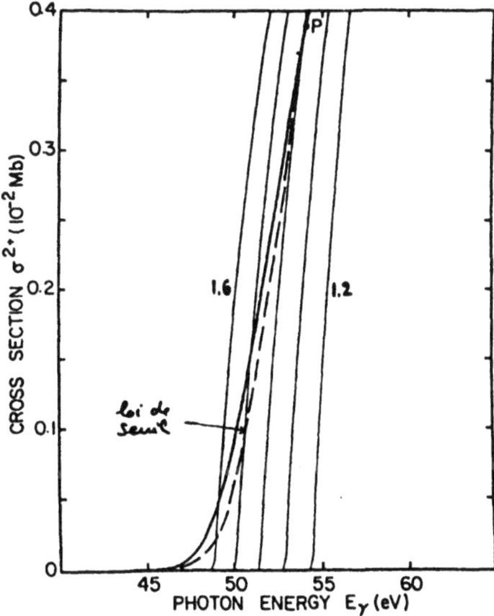

Figure 6. Double photoionization of molecular hydrogen in the threshold region. The dashed
curve represents molecular threshold law matched onto the exact theoretical result
at point P.

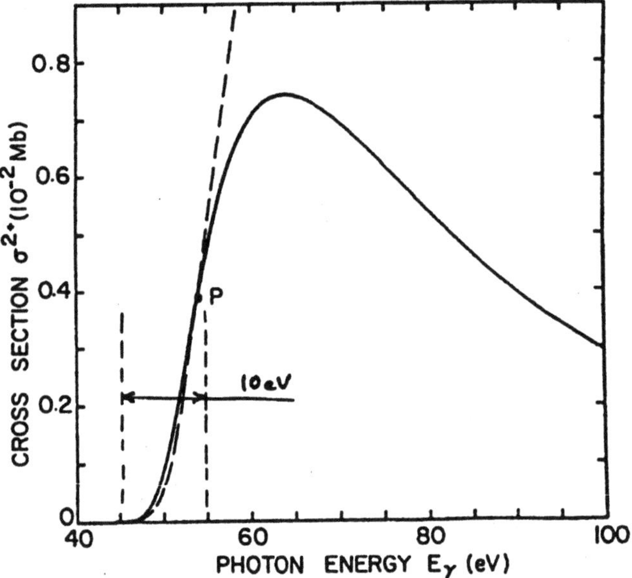

Figure 7. Comparison of the cross sections derived from the molecular threshold law (dashed curve) with the R-averaged one (solid curve) over the whole photon energy range.

This suggests that it would be interesting to perform more refined experiments and calculations with full correlations in both the initial as well as final states involved in the transition. It would also be interesting to see the effect of two-centre wave function for the final state in the case of hydrogen molecule. It is well known that the exponent of excess energy depends only on the asymptotic part of the final state wave function and molecular threshold law modified due to vibrations. The product of two unscreened Coulomb wave functions gives linear threshold law and hence the result of Le Rouzo obeys the linear threshold law which disagrees with experiment as shown in Figure 7.

3.5 Threshold Triple Differential Cross Section of Double-photoionization of Two-electron Systems

The threshold triple differential cross section (TDCS) for the double-photoionization (DPI) of two-electron systems by impact of light is very sensitive to the electron-electron correlations. Most recently, Pont et al [67] have applied two different methods to the calculation of TDCS for double-photoionization of helium. In one method, the 3C method, the final state is described by a product of 3 Coulomb continuum wave functions, while in the other method, the 2SC method, the final state is described by a product of 2 screened Coulomb wave functions employing effective charge.

Figures 8 and 9 show the theoretical results of Pont et al [67] along with the data of Lablanquie et al [49], for both (a) equal and (b) and (c) unequal energy sharing. The different plots have been rescaled so that the TDCS has the same value at its maximum for all sets of data in a given case. The agreement between the (rescaled) results of the 2SC calculation and the length-gauge version of the 3C calculation is good in all cases apart from the two unequal energy sharing cases at higher excess energy. Finally, it is seen that the qualitative agreement between the results of the 2SC calculation and experimental data is rather poor; the 2SC results lie, for the most part, well outside the error bars of the experiment.

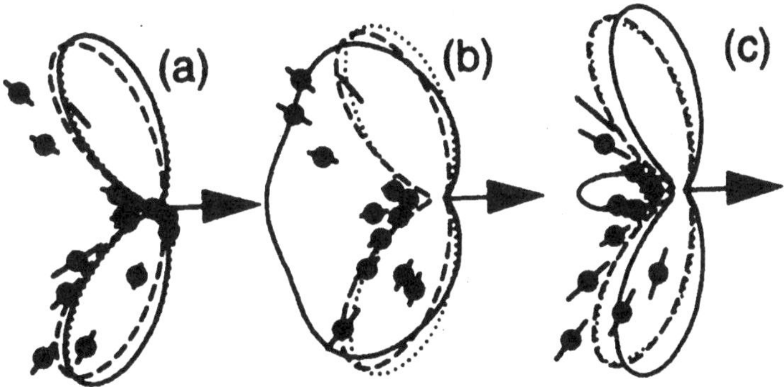

Figure 8. Polar plots of coplanar TDCS. Electron 1 emerges along the polarization axis (see arrow), and E = 4.0eV, with (a) E1 = E2 = 2.0eV; (b) E1 = 3.3 eV, E2 = 0.7 eV; and (c) E1 = 0.7 eV, E2 = 3.3 eV. Experimental data are from Ref. [51]. Solid and dashed lines are from velocity and length gauge version of 3c theory, respectively, and the plotted line is from 2SC theory. Plots have been rescaled so that TDCS has same value at its maximum.

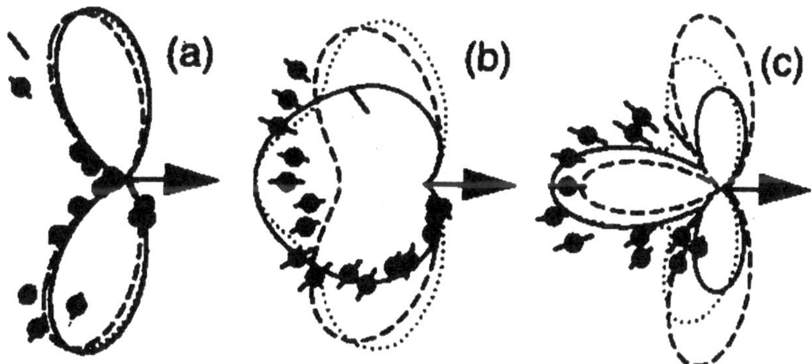

Figure 9. Same as figure 8 but with E = 18.6 eV and (a) E1 = E2 = 9.3 eV; (b) E1 = 15.6 eV, E2 = 3.0 eV; and (c) e1 = 3.0 eV, E2 = 15.6 eV.

4. CONCLUSIONS AND FUTURE DIRECTIONS

Great experimental and theoretical advancement has been made in the case of double photoionization (DPI) of two-electron systems but there is considerable discrepancy in the close vicinity of the threshold where correlations play an extremely important role. To obtain accurate double-continuum wave function in the vicinity of threshold is still challenging problem for the theorists. Kossmann et al [44] have measured the slope (σ_1) for the DPI of He but there is no theory to evaluate the slope directly. Dynamic screening i.e. energy dependent screening is crucial especially when two slow electrons are escaping the positive ion but there is no method available to include the dynamic screening in the Coulomb wave function, Altick wave function, distorted wave function or any other wave functions. Unscreened Coulomb wave function gives linear threshold law which disagrees with experiment. This suggests that a method should be developed to incorporate dynamic screening. In the case of even the simplest H_2, the DPI process has not been extensively investigated either experimentally or theoretically. There is one experimental result and one theoretical calculation with Coulomb wave function. Agreement is excellent between experiment and theory but this may be fictitious. The agreement reflects that the DPI of H_2 should be reinvestigated. In the case of He many body perturbation theory, which includes full correlations in both initial and final states, yields results which are in good agreement with the recent experiment. It clearly indicates that it is indispensable to incorporate full correlation of equal amount in both states involved in the transition in order to obtain reliable results. Angular distribution and energy sharing of two escaping electrons have not been extensively studied but these can offer the opportunity to test the validity of a theoretical model as the exponent of excess energy does in the threshold law.

It is clear that there are numerous difficulties in obtaining the accurate double-continuum wave functions which play an extremely important role in reliable double-photoionization cross sections. However, we would like to make some constructive and fruitful suggestions which may be the future directions:

- new reaction $(\gamma, \gamma'\, 2e)$
- new reaction (γ, ne), $(\gamma,\, \gamma'\, ne)$
- n-particle continuum wave function
- 3n-fold DCS $\dfrac{d^{3n}\sigma^{n+}}{dk_1 dk_2 dk_3 ......dk_n}$
- spin polarization of projectile, target, scattered and ejected particles, spin-polarized collision partners and measurements of orientation, alignment, 3n-fold differential cross section, spin-polarization

    parameters, asymmetry parameter, helicity and chirality, new photoionization parameters
- more complex systems and processes
- dichroic effect
- coalescence features
- unified theory, simultaneous and sequential emission of particles
- relativistic and non-relativistic photon operator, exact multipole operator
- gauge discrepancies in calculations
- marry the exact asymptotic wave function with close-coupling
- R-matrix with pseudostates (RMPS)
- a new R-matrix developed by Burke et al [9] may be very useful for the double-continuum wave function (DCWF),
- the standard R-matrix [1] may be combined with the Altick asymptotic DCWF, hyperspherical DCWF, BBK DCWF and GDCWF
- solving the Schrödinger equation including the higher terms of the Neumann series,
- developing some new ideas and techniques which provide to include the dynamic screening in the Coulomb, Altick and distorted wave functions
- developing some sophisticated numerical procedure to describe two continuum electrons in the entire configuration space.
- Exact ADCWF is not possible but exact NDCWF is possible using quantum computer.

In short, our knowledge of high-precision double-continuum wave function as well as double-photoionization cross section of atoms, molecules, clusters and ions, particularly for heavy atoms, molecules and ions, is by no means complete and future holds many challenges for both experiment and theory.

REFERENCES

1. K. A. Berrington, P. G. Burke, M. Le Dourneuf, W. D. Robb, K. T. Taylor and Vo Ky Lan, Comput. Phys. Commun. **14**, 367 (1978).
2. G. H. Wannier, Phys. Rev. **90**, 817 (1953).
3. A. R. P. Rau, Phys. Rev. A **4**, 207 (1971).
4. R. Peterkop, J. Phys. B **4**, 513 (1971).
5. L. Rosenberg, Phys. Rev. D **8**, 1833 (1973).
6. M. R. H. Rudge and M. J. Seaton, Proc. Roy. Soc. **283**, 262 (1965).

7. M. R. H. Rudge, Rev. Mod. Phys. **40**, 564 (1968).

8. R. Peterkop, Phys. Lett. A **62**, 81 (1977).

9. Burke, C. J. Noble and P. Scott, Proc. Roy. Soc. London, A **410**, 289 (1987).

10. W. Byron and C. J. Joachain, Phys. Rev. **164**, 1 (1967).

11. T. A. Carlson, Phys. Rev. **156**, 142 (1967).

12. D. M. P. Holland, K. Codling, J. B. West and G. V. Marr, Phys. B: **12**, 2465 (1979).

13. V. Schmidt, N. Sander, H. Kuntzmueller, P. Dhez, F. Wulleumier and E. Kaellne, Phys. Rev. A **13**, 1748 (1976).

14. G. R. Wight and M. J. Van der Wiel, J. Phys. B:**9**, 1319 (1976).

15. R. L. Brown, Phys. Rev. A **1**, 586 (1970).

16. M. Ya Amusia, E. G. Drukarev, V. G. Gorshkov and M. P. Kazachkov, Phys. B: **8**, 1248 (1975).

17. Drukarev, Phys. Rev. A **51**, R2684 (1995).

18. M. S. Yurev, Opt. Spectrosc. **38**, 4 (1975).

19. S. M. Varnavshik and L. N. Labzovskii, OPt. Spectrosc. **47**, 24 (1979).

20. S. N. Tiwary, J. Phys. B: **15**, L323 (1982).

21. S. N. Tiwary and G. Dujardin, ICPEAC, New York, USA, 1989.

22. S. L. Carter and H. P. Kelly, Phys. Rev. A **24**, 170 (1981).

23. C. Pan and H. P. Kelly, J. Phys. B: **28**, 5001 (1995).

24. G. C. King, M. Zabek, P. M. Rutter, F. H. Read, A. A. McDowell, B. West and D. M. P. Holland, J. Phys. B: **21**, L403 (1988).

25. G. Dawber, L. Avaldi, A. G. McConkey, H. Rojas, M. A. McDonald and G. C. King, J. Phys. B: **28**, L271 (1995).

26. J. M. Bizau and F. J. Wuilleumier, J. Electron Spectrosc. Relat. Phenom. 205 (1995).

27. J. B. Donahue, P. A. M. Gram, M. V. Hynes, R. W. Hamm, C. A. Fost, C. Brynt, K. B. Butterfield, D. A. Clark, and W. W. Smith, Phys. Rev. Lett. **48**, 1538 (1982).

28. G. Dujardin, M. J. Besnard, L. Hellner and Y. Malinovitch, Phys. Rev. A **35**, 5012 (1987).

29. H. Le Rouzo, Phys. Rev. A **37**, 1512 (1988).

30. A. R. P. Rau, J. Phys. B: **9**, L283 (1976).

31. A. R. P. Rau, Opt. Soc. Am. B **4**, 784 (1987).

32. C. H. Green and A. R. P. Rau, J. Phys. B: **16**, 99 (1983).

33. H. Klar and W. Schlect, J. Phys. B: **9**, 1699 (1976).

34. H. Klar, J. Phys. B: **14**, 3255 (1981).

35. J. M. Feagin, J. Phys. B: **17**, 2433 (1984).

36. A. Temkin, *Electronic and Atomic Collisions*, edited by J. Eichler, V. Hertel and N. Stolterhoft, Elsevier, Amsterdam, 1984.

37. A. Temkin, Phys. Rev. Lett. **49**, 365 (1982).

38. A. D. Stauffer, Phys. Lett. A **91**, 114 (1982).

39. D. S. F. Crothers, J. Phys. B: **19**, 463 (1986).

40. S. Cjevanovic and F. H. Read, J. Phys. B: **7**, 1841 (1974).

41. F. H. Read, J. Phys. B: **17**, 3965 (1984).

42. F. H. Read, *Electron Impact Ionization*, edited by T. D. Mark and H. Dunn, Springer, Berlin, 1985.

43. P. Lablanquie, J. H. D. Eland, I. Nenner, P. Morin, J. Delwiche and M. J. Hubin-Franskin, Phys. Rev. Lett. **58**, 992 (1987).

44. Kossmann, V. Schmidt and T. Anderson, Phys. Rev. Lett. **60**, 1266 (1988).

45. P. Hammond, F. H. Read, S. Cvejanovic and G. G. King, Phys. B: **18**, L141 (1985).

46. S. Spence, Phys. Rev. A **11**, 1539 (1975).

47. P. Selles, A. Huetz and J. Mazeau, J. Phys. B: **20**, 5195 (1987).

48. A. Huetz, P. Selles, D. Waymel and J. Mazeau, J. Phys. B: **24**, 1917 (1991).

49. P. Lablanquie, J. Mazeau, L. Andric, P. Selles and A. Huetz, Phys. Rev. Lett. **74**, 2192 (1995).

50. V. Schmidt, Rep. Prog. Phys. **55**, 1483 (1992).
    Proceedings of the Workshop on Double Photoionization of He, Publication No. ANL/PHY-94/1 (Argonne National Laboratory, Argonne, IL, 1994).

51. O. Schwarzkopf and V. Schmidt, J. Phys. B: **28**, 2847 (1995).

52. M. Biagini, Phys. Rev. A **46**, 656 (1992).

53. O. Schwarzkopf, B. Kraessig, J. Elmiger and V. Schmidt, Phys. Rev. Lett. **70**, 3008 (1993).

54. P. Lablanquie, Ph. D. Thesis, LURE, Orsay, France, 1989.

55. I. Nenner, P. Morrin and P. Lablanquie, Comm. At. Mol. Phys. **22**, 51 (1988).

56. J. A. R. Samson, in *Corpuscles and Radiation in Matter*, edited by Mehlhorn, Handbuch der Physik, Vol. **32**, p. 23, Springer, Berlin, 1982.

336

57. A. F. Starace, in *Corpuscles and Radiation in Matter*, edited by Mehlhorn, Handbuch der Physik, Vol. **31**, p. 1, Springer, Berlin, 1982.

58. B. Crasemann, Comm. At. Mol. Phys. **22**, 163 (1989).

59. C. O. Almbladh and L. Hedin, in *Handbook on Synchrotron Radiation*, Vol. **1b**, ed. E. E. Koch, 1983.

60. G. Dawber, R. I. Hall, A. G. McConkey, M. A. MacDonald and G. C. King, J. Phys. B: **27**, L341 (1994).

61. A. R. P. Rau, Comm. At. Mol. Phys. **14**, 285 (1984).

62. Z. Teng and R. Shakeshaft, Phys. Rev. A **47**, R3487 (1993).

63. D. Proulx and R. Shakeshaft, Phys. Rev. A **48**, R875 (1993).

64. M. Pont and R. Shakeshaft, Phys. Rev. A **51**, R2676 (1995)

65. M. Pont and R. Shakeshaft, J. Phys. B: **28**, L571 (1995).

66. F. Maulbetsch, J. S. Briggs, M. Pont and R. Shakeshaft, Phys. B: **28**, L341 (1995)

67. M. Pont, R. Shakeshaft, F. Maulbetsch and J. S. Briggs, Phys. Rev. A **53**, 3671 (1996).

68. J. A. R. Samson, R. J. Bartlett and Z. X. He, Phys. Rev. A **46**, 7277 (1992).

69. J. A. R. Samson, in *Many-body Theory of Atomic Structure and Photoionization*, ed. by T. N. Chang, World Scientific, New York, 1992.

70. J. A. R. Samson, C. H. Greene and R. J. Bartlett, Phys. Rev. A **71**, 201 (1993).

71. R. J. Bartlett, P. J. Walsh, Z. X. He, Y. Chung, E. M. Lee and A. R. Samson, Phys. Rev. A **46**, 5574 (1992)

72. J. A. R. Samson, Phys. Rev. Lett. **65**, 2861 (1990).
Berkowitz, Photoabsorption, Photoionization and Photoelectron Spectroscopy, Academic, New York, 1979.

73. P. L. Altick, Phys. Rev. A **21**, 1381 (1980).

74. P. L. Altick, Phys. Rev. A **25**, 128 (1982).

75. P. L. Altick, J. Phys. B: **16**, 3543 (1983).

76. C. Garibotti and J. E. Miraglia, Phys. Rev. A **21**, 572 (1980).

77. M. Brauner, J. S. Briggs and H. Klar, J. Phys. B: **22**, 2265 (1989).

78. V. Schmidt, Appl. Opt. **19**, 4080 (1980).

79. A. F. Starace, Appl. Opt. **19**, 4051 (1980).

80. U. Fano, Rep. Prog. Phys. **46**, 97 (1983).

81. A. Dalgarno and H. R. Sadeghpour, Phys. Rev. A **46**, R3591 (1992).

82. T. T. Scholz, H. R. J. Walters, P. G. Burke and M. P. Scott, Mon. Not. R. Astr. Soc. **242**, 692 (1990).

83. F. Maulbetsch and J. S. Briggs, Phys. Rev. Lett. **68**, 2004 (1992).

84. U. Becker and D. A. Shirley, Phys. Scr. T **31**, 56 (1990).

85. R. Wehlitz, F. Heiser, O. Hemmers, B. Langer, A. Menzel and U. Becker, Phys. Rev. Lett. **67**, 3764 (1991)

86. J. C. Levin, D. W. Lindle, N. Keller, R. D. Miller, Y. Azuma, Mansour, H. G. Berry and I. A. Sellin, Phys. Rev. Lett. **67**, 968 (1991).

87. J. C. Levin, D. W. Lindle, N. Keller, R. D. Miller, Y. Azuma, B. Mansour, H. G. Berry and I. A. Sellin, Phys. Rev. A **47**, 16 (1993).

88. P. Lablanquie, K. Ito, P. Morrin, I. Nenner and J. D. Eland, Phys. D **16**, 77 (1990).

89. G. Wendin, Comm. At. Mol. Phys. **17**, 115 (1986).

90. G. Wendin, Photoionization of Atoms and Molecules, Proceeding of the Daresbury one-day Meeting on Photoionization of Atoms and Molecules, Report DL/SCI/R11, ed. by B. D. Buckley, 1978.

91. S. N. Tiwary, Chem. Phys. Lett. **93**, 47 (1982).

92. A. Hibbert, A. E. Kingston and S. N. Tiwary, J. Phys. B: **15**, L643 (1982).

93. S. N. Tiwary, Chem. Phys. Lett. **96**, 333 (1983).

94. S. N. Tiwary, A. E. Kingston and A. Hibbert, J. Phys. B: **16**, 2457 (1983).

95. S. N. Tiwary, Astrophys. Journal **269**, 803 (1983).

96. S. N. Tiwary, Astrophys. Journal **272**, 781 (1983).

97. S. N. Tiwary, P. G. Burke and A. E. Kingston, ICPEAC, Berlin, West Germany, 983.

98. S. N. Tiwary, Proc. Ind. Acad. Sci. **93**, 1345 (1984).

99. S. N. Tiwary, Invited Talk, Book-World Book Publisher, New York, 1987.

100. A. E. Kingston, A. Hibbert and S. N. Tiwary, J. Phys. B: **20**, 3907 (1987).

101. S. N. Tiwary,A. P. Singh, D. D. Singh and R. J. Sharma, Can. J. Phys. **66**, 405(1988).

102. S. N. Tiwary and G. Dujardin, Symposium on Auger Process, Paris, 1989.

103. S. N. Tiwary, Praman-J. Phys. **35**, 89 (1990).

104. S. N. Tiwary, Fizika, **22**, 577 (1990).

338

105. S. N. Tiwary and P. Kumar, Acta Phys. Pol. **80**, 23 (1991).

106. S. N. Tiwary, Int. J. Theor. Phys. **30**, 825 (1991).

107. S. N. Tiwary, Fizika **23**, 27 (1991).

108. S. N. Tiwary, Nuovo Cimento D, **13**, 1073 (1991).

109. S. N. Tiwary, Fizika, 1, 181 (1992).

110. S. N. Tiwary and D. D. Singh, Nuovo Cimento D **14**, 739 (1992).

111. S. N. Tiwary, M. Kumar, D. D. Singh and P. Kumar, Nuovo Cimento D **15**, 77 (1993).

112. S. N. Tiwary, P. Kandpal and A. Kumar, Nuovo Cimento D **15**, 1181 (1993).

113. S. N. Tiwary and P. Kandpal, Nuovo Cimento D **16**, 339 (1994).

114. S. N. Tiwary and M. Kumar, Ind. J. Phys. **69**, 133 (1995)

115. S. N. Tiwary and B. Kumar, Nuovo Cimento D **17**, 393 (1995).

116. S. N. Tiwary, Nuovo Cimento, **18**, 1 (1995).

117. A. Engelns, H. Klar and A. W. Malcherek, J. Phys. B: **30**, L811 (1997), N. Chandra, Phys. Rev. A **56**, 1879 (1997).

118. S. N. Tiwary, Ind. J. Phys. B: **72**, 427 (1998).

Low Energy Photoionization in the Ar Isoelectronic Sequence: Complex Effects of Z

H. S. Chakraborty, P. C. Deshmukh and S. T. Manson[*]

Department of Physics,
Indian Institute of Technology - Madras, Chennai 600 036, India

[*]Department of Physics and Astronomy,
Georgia State University, Atlanta, GA 30303-3083, USA

Valence $3p$ photoionization of neutral Ar and several ions isoelectronic to it is investigated in the low photon-energy regime using the relativistic-random-phase approximation. With increasing Z, a steady shift of "Cooper minimum" towards the ionization threshold is seen in the cross section such that at sufficiently high Z it crosses the threshold to enter the discrete of $3p$ channel. Further, a remarkable threshold feature is observed for some ions that essentially originates from strong Z-dependent interaction with inner $3s$ photoexcitation series. Results strengthen our previous caution against any indiscriminate interpolation (or extrapolation) of ionic data at low energies.

1.   INTRODUCTION

Study of the interaction of electromagnetic wave with matter in its all forms is difficult owing to the complex nature of electron correlation. To understand the effects of this correlation, however, atomic photoionization stands out as an excellent device, mainly for two reasons. First, the weak coupling [1] between the radiation and target allows unambiguous determination of target properties; second, because the target is relatively simple, the forces are well-known, and incident photon vanishes in the final channel, the process is readily amenable to detailed *ab initio* theoretical treatment. Over the years, myriad theoretical [2, 3] and experimental [4-7]

Trends in Atomic and Molecular Physics,
Edited by Sud and Upadhyaya. Kluwer Academic/Plenum Publishers, New York, 2000.   339

studies have aimed at various many-body aspects of the atomic photoionization process.

Unfortunately, this has not been the case with atomic ions. On the experimental side, this is broadly due to the difficulties in sustaining target sample in a definite state of ionization for long enough period. On the theoretical side, while existing techniques for neutrals can, in principle, be extendible to address photoionization of ions, the absence of sufficient data for comparison deterred the urge. Also, the prevailing mind-set that ionic properties are easily obtainable by interpolation / extrapolation of existing results has dampened progress. On the other hand, in the context of applications involving various astrophysical as well as laboratory plasmas, the importance of ionic studies is growing by leaps and bounds in recent times due to high accumulation of quality data [8]. What then very much in order is to generate database of accurate and extensive experimental / theoretical information. Although laboratory techniques have of late gone through improvements in sophistication and precision, there still exist practical problems in preparing and sustaining higher state of target ionization. Emphasis, therefore, is on theoretical programs to provide quality data, in particular, for cases unachievable experimentally; an ongoing program aiming at this goal is the so-called opacity project [9].

With the above viewpoint in mind, our current efforts have mostly focussed on investigating the photoabsorption properties of ions by considering isoelectronic sequences; wherever possible, parallel experimental studies are also being carried out to affirm the findings. Many new results have already emerged from this effort. In a recent study [10, 11], the relativistic resonances between the $2p_{\frac{3}{2}}$ and $2p_{\frac{1}{2}}$ thresholds of the first three members of the Ne isoelectronic sequence have been shown to evolve in a complicated manner along the sequence. For the remainder of the sequence, however, a steady Z-dependent evolution is seen [11]. In another theoretical-plus-experimental study [12] on the same isoelectronic sequence, a seemingly anomalous threshold behaviour of $2p$ photoionization is detected for a number of ions; in Si^{4+} this is most dramatic. The generality of all these phenomena over the entire Periodic Table has also been established, thus putting across a note of caution against any simplistic attempt, such as, interpolation (or extrapolation) of existing data to obtain new information.

Current article re-stresses the point by illustrating new results in the Ar isoelectronic sequence. In this work, valence $3p$ photoionization is investigated using the relativistic-random-phase approximation (RRPA) methodology [13, 14]; single electron dipole transitions are considered. RRPA is a reasonably sound many-body theory that treats ground state correlation to a good extent and also provides an appropriate framework for interchannel coupling. In the present calculations, we employ two levels of interchannel coupling scheme: (i) all seven channels arising from *3p* and *3s*, and (ii) five channels from *3p* only. The explicit designations of the relativistic channels are:

- For 3*p* subshell:

$$3p_{\frac{3}{2}} \quad \rightarrow \quad \varepsilon d_{\frac{5}{2}}, \ \varepsilon d_{\frac{3}{2}}, \ \varepsilon s_{\frac{1}{2}}$$

$$3p_{\frac{1}{2}} \quad \rightarrow \quad \varepsilon d_{\frac{3}{2}}, \ \varepsilon s_{\frac{1}{2}}$$

- For 3*s* subshell:

$$3s_{\frac{1}{2}} \quad \rightarrow \quad \varepsilon p_{\frac{3}{2}}, \ \varepsilon p_{\frac{1}{2}}$$

2.   RESULTS AND DISCUSSION

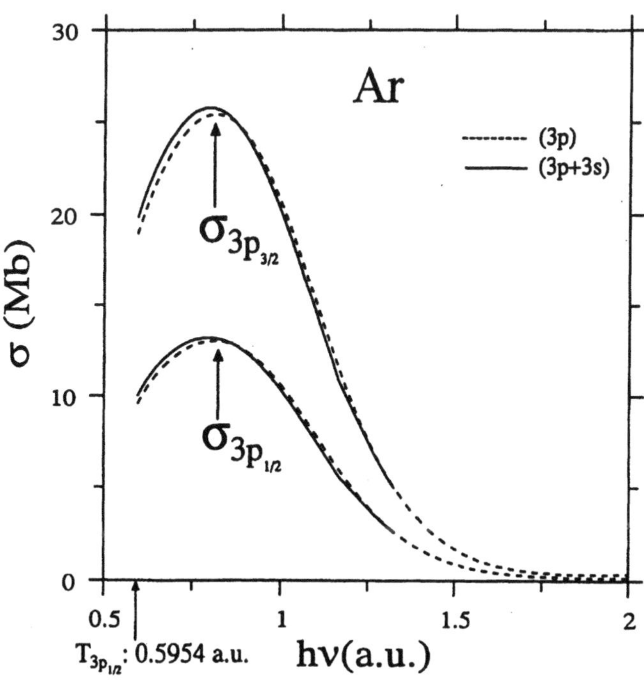

Figure 1. Ar $3p_{\frac{3}{2}}$ and $3p_{\frac{1}{2}}$ cross sections calculated with (solid line) and without (dashed line) 3*s* coupling and shown from $3p_{\frac{1}{2}}$ threshold (T).

$3p_{\frac{3}{2}}$ and $3p_{\frac{1}{2}}$ differential cross sections starting from $3p_{\frac{1}{2}}$ Dirac-Fock threshold (0.5954 a.u.) are presented in Fig. 1 for neutral argon. Calculations with both $(3p+3s)$ and *only-3p* coupling show similar qualitative profile. The smooth delayed-maximum at photon-energy ~ 0.7 a.u. actually results as the final d-wave circumvent the characteristic centrifugal barrier with increasing energy [9]. However, this feature does not exist for any other ion in the sequence. This is simply because even a unit increment of the nuclear charge pulls the d-wave sufficiently *inward* so as to enable it overcome the barrier at zero photoelectron kinetic energy itself.

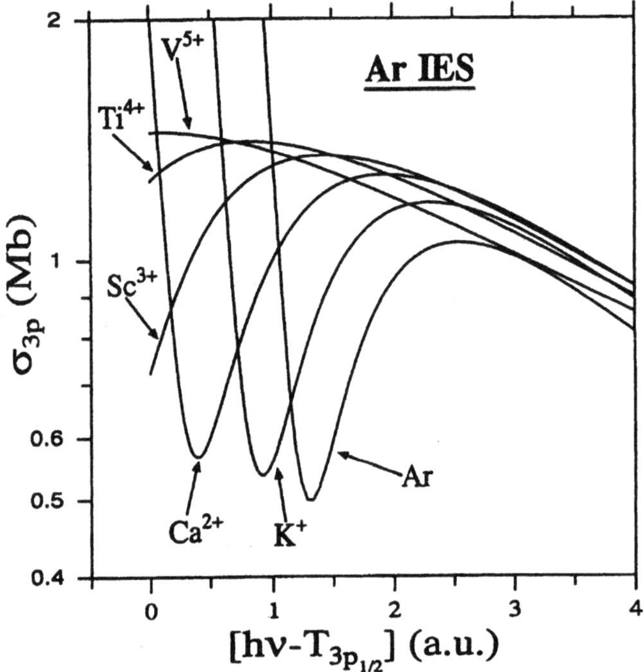

Figure 2. 3p cross section for the first few members of the Ar isoelectronic sequence plotted as a function of $3p_{\frac{1}{2}}$ photoelectron kinetic energy.

Another important low energy feature attributable to the properties of initial and final state wave functions is the "Cooper minimum" [15]. The radial *3p* wave function has one node. Therefore, this wave and d radial wave of departing electron will have opposite signs in their overlap region to affect some degree of cancellation. For Ar, this overlap leads to zero oscillator strength (complete cancellation) at an energy in the ionization continuum giving rise to the "Cooper minimum". At higher photon-energies, this overlap matrix element changes its sign. For ions, on the other hand, increased nuclear attraction makes the wave functions compact. This indicates that with increasing Z, such cancellation effect in the overlap integral, and hence the "Cooper minimum", will occur more and more close to the threshold to eventually migrate to the discrete part of the channel spectrum. Fig. 2, total *3p* cross sections (calculated in *only-3p* coupling) for Ar through V^{5+} as a function of photoelectron kinetic energy, illustrates this point. For Ar, K^+ and Ca^{2+}, the "Cooper minimum" appears in the continuum with its position gradually shifting towards threshold. Coming to Sc^{3+}, the minimum moves below the threshold and continues going deeper as Z increases further. It may be noted that the value of cross section at its minimum is very small but not exactly zero, and this is because different relativistic matrix elements corresponding to *d*-final state channels do not hit their respective zeros at the same energy.

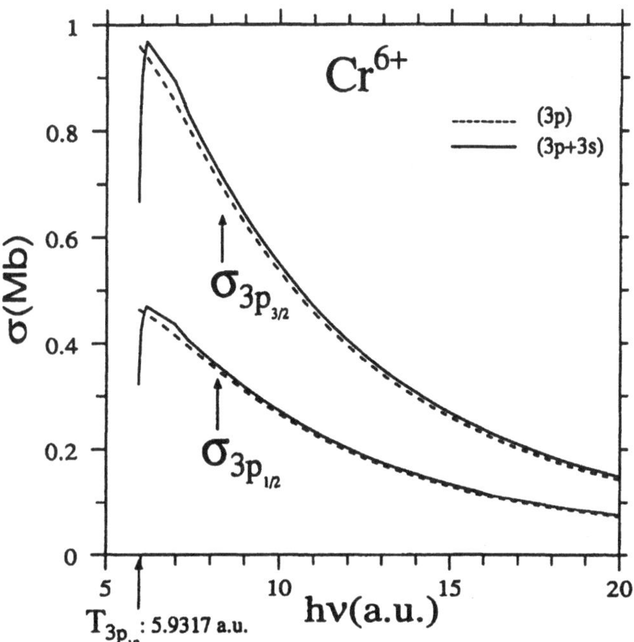

Figure 3. Cross sections as in Fig. 1 but for Cr^{6+} ion that shows strong threshold effect.

Finally, in the calculations with *(3p+3s)* coupling, a sudden and remarkable behaviour of threshold cross section is detected that appears first in the sequence at the level of Cr^{6+} (see Fig. 3), while the *only-3p* result is smooth. Also, none of the previous V^{5+} and latter Mn^{7+} ions (not shown) exhibit any "threshold-anomaly" of this kind. This phenomenon is a combined effect of Z-dependent change in the separation of the *3p* and *3s* thresholds plus the evolution of the Rydberg series emanating from *3s*

344

threshold as a function of Z. What actually happens is that the autoionizing resonances, whose energies are well above the $3p$ thresholds in neutral Ar, move closer to the $3p$ thresholds with increasing Z along the sequence. This is because these Rydberg states move lower below the $3s$ ionization threshold as the square of the effective charge (roughly $z+1$, where z is the ionic charge), while the $3p$ thresholds increase their separation with the $3s$ threshold to move deeper only as the first power of $z+1$. Thus eventually each $3s \rightarrow np$ resonance moves below the $3p$ thresholds one after another with Z increasing and becomes a true bound state. In the process, at Cr^{6+}, the $3s \rightarrow 4p$ resonance becomes very nearly degenerate with the threshold, leading to the feature seen in Fig. 3. Similar feature returns at Co^{9+} when the next, $3s \rightarrow 5p$, resonance occurs nearly at the threshold, see Fig. 4. The effect, thus, is re-enacted for a number of ions along the sequence when subsequent resonances move past the threshold. These degeneracies are indeed accidental, as they occur nearly at integral values of a generalized Z. It can be mentioned here that similar feature in ionic photoionization has been seen previously, both theoretically as well as experimentally, in the Ne isoelectronic sequence [12]. Present finding supports our predictions made earlier on the generality of this phenomenology.

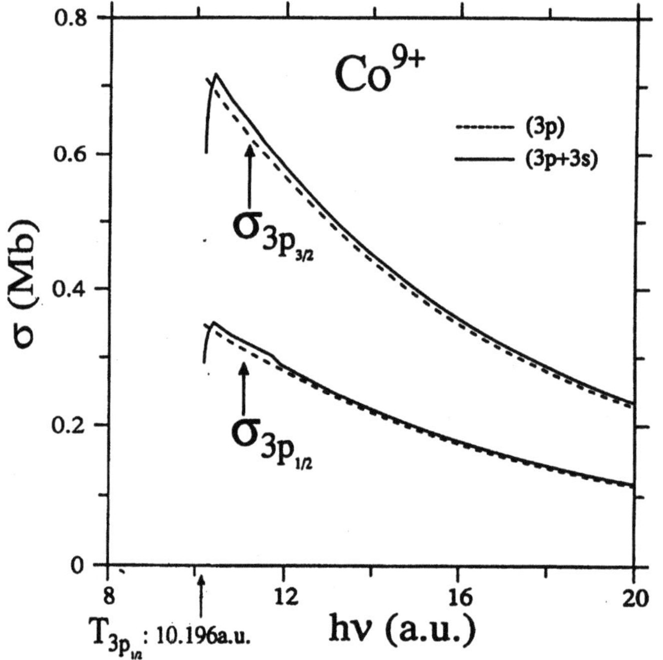

Figure 4. Same as Fig. 3 but for Co^{9+}.

The most important revelation of these dynamical features described above is that it is simply impossible to explain photoionization properties at low energies by any numerical fitting method characterized by smooth Z-scaling law in the Ar isoelectronic sequence; at higher photon energies, however, the change is monotonic, and hence simple.

3.   CONCLUSIONS

We have presented in this article a set of new results of our ongoing photoionization studies in the Ar isoelectronic sequence. The effect of centrifugal barrier in the form of a hump in Ar near-threshold cross section is found to be fully overcome from the next ion itself. A steady shift in the position of "Cooper minimum" towards the ionization threshold and its eventual entry into discrete is displayed over the first few members of the sequence. As a crucial generality, this feature is expected for any nl (with $l < n - 1$) subshell ionization of any isoelectronic sequence across the Periodic Table that has at least one "Cooper minimum" in the continuum for the neutral member. We have also seen a very strong threshold effect for a number of ions along the sequence which essentially originates from an accidental near-degeneracy of $3s \to np$ resonances with $3p$ ionization thresholds. This result too satisfies our contention that such anomalies must occur for any non-s subshell and perhaps, in some cases, even for inner s subshell photoionization. Further detailed investigation has been considered and will be published elsewhere. As a final note, like our early result on the Ne isoelectronic sequence, this result is also completely at odds with the results of opacity project [9]. This is most likely because of non-relativistic (and hence in-exact) placement of resonances in those calculations. This fact clearly points at a need for relativistic techniques to be brought into to ensure quality of standard in applications.

ACKNOWLEDGMENTS

The work was supported by National Science Foundation of USA.

REFERENCES

1.   H. A. Bethe and E. E. Salpeter, *Quantum Mechanics of One- and Two-Electron Atoms*, Springer-Verlag, Berlin, p. 248ff (1958).

2.   R. F. Reilman and S. T. Manson, Astrophys. J. Supp. **40**, 815 (1979).

346

3. C. Mendoza, Phys. Scripta **T65**, 198 (1996).

4. V. Schmidt, Rep. Prog. Phys. **55**, 1483 (1992); *Electron Spectrometry of Atoms Using Synchrotron Radiation*, Cambridge University Press, New York (1997).

5. B. Sonntag and P. Zimmermann, Rep. Prog. Phys. **55**, 911 (1992).

6. M. Ya Amusia, *Atomic Photoeffect*, Plenum Press, New York (1990).

7. E. B. Saloman, J. H. Hubbell, and J. H. Scofield, At. Data Nucl. Data Tables **38**, 1 (1988).

8. J. C. Raymond, *Proceedings of NASA Laboratory Space Science Workshop*, Harvard-Smithsonian Center for Astrophysics, April 1-3, p. 1 (1998).

9. A. K. Pradhan, Phys. Scripta **35**, 840 (1987).

10. H. S. Chakraborty, P. C. Deshmukh, E. W. B. Dias, and S. T. Manson, Submitted to J. Phys. B.

11. P. C. Deshmukh, H. S. Chakraborty, E. W. B. Dias, and S. T. Manson, published in this proceeding.

12. H. S. Chakraborty, P. C. Deshmukh, and S. T. Manson, Pramana-J. Phys. **6**, 607 (1998); S. T. Manson, Z. Altun, H. S. Chakraborty, E. W. B. Dias, P. C. Deshmukh, and C. S. Turner, *Atomic Processes in Plasmas*, Ed. E. Oks and M. S. Pindzola, AIP, New York, p. 19 (1998); H. S. Chakraborty, P. C. Deshmukh, S. T. Manson, A. Gray, and E. T. Kennedy, to be submitted to Phys. Rev. Lett.

13. W. R. Johnson and C. D. Lin, Phys. Rev. A **20**, 964 (1979).

14. W. R. Johnson, C. D. Lin, K. T. Cheng, and C. M. Lee, Phys. Scripta **21**, 409 (1980).

15. J. Berkowitz, *Photoabsorption, Photoionization, and Photoelectron Spectroscopy*, Academic Press, New York, p. 45 (1979).

Z-Depencence of Photoabsorption Properties in Isoelectronic Sequences

P. C. Deshmukh, H. S. Chakraborty, E. W. B. Dias[*] and S. T. Manson[*]

Department of Physics,
Indian Institute of Technology - Madras, Chennai - 600036, India

[*]Department of Physics and Astronomy,
Georgia State University, Atlanta, GA 30303-3083, USA

Isoelectronic sequences are often studied and used to extrapolate and/or interpolate atomic/ionic properties. The present work shows that while such extrapolation/interpolation procedure may be relied upon for high-Z members of an isoelectronic sequence, it cannot be used for the lower members which show striking quantitative and even qualitative differences in atomic properties when Z changes even as much as mere unity.

1.   INTRODUCTION

Atomic photoabsorption studies provide substantial information about some of the most fundamental interactions that take place in the universe. Experimental [1-3] and theoretical [3-5] studies on ions is of importance to understand astrophysical processes, but experiments are difficult to be performed on ions and hence relatively less data are available on ionic systems. It is thus natural to study isoelectronic sequences with the hope that atomic properties change smoothly across members of such sequences and then interpolate / extrapolate the values of ionic physical parameters. It is demonstrated in this paper that for lower members of the isoelectronic sequences, atomic properties do not, however, change smoothly. For higher members, the changes are generally smooth allowing at least a qualitative interpolation / extrapolation.

In continuation of our studies on various atoms and ions [6-8], we have made use of the relativistic random phase approximation (RRPA) [9] to include both relativistic effects and many-body correlations. This method takes into account major electron correlations both in the initial state as well as the final state of a photoabsorption process. The electron correlations give rise to configuration interactions between bound-bound transitions and bound-continuum transitions causing autoionization resonances. These are included in the formalism *a-priori* and are studied efficiently using the relativistic multichannel quantum defect theory (RMQDT) [10] using the quantum defect parameters obtained *ab-initio* from the RRPA.

2.   RESULTS AND DISCUSSION

Essentially, in the present work, we have made use of the RRPA + RMQDT formalism to study the atomic excited states in the neon isoelectronic sequence. BOTH low-Z and intermediate-Z members of the sequence have been studied. A primary objective of this study was to determine how the excited states are characterized, a question that can be well answered fairly rigorously using the present formalism. It is superfluous to add that the autoionization resonances are difficult to be designated, since electron correlations prevent the normally used single-particle labels from being correct, and it then becomes difficult to determine the correct labels for the resonances. In a full-RRPA calculation, all relativistically permitted dipole channels would be coupled; however, we have done a *truncated*-RRPA calculation in which the dipole channels originating from the 1s shell have been left out. The channels that were coupled are:

$$2p_{\frac{3}{2}} \quad \rightarrow \quad kd_{\frac{5}{2}}, \; kd_{\frac{3}{2}}, \; ks_{\frac{1}{2}},$$

$$2p_{\frac{1}{2}} \quad \rightarrow \quad kd_{\frac{3}{2}}, \; ks_{\frac{1}{2}},$$

$$2s_{\frac{1}{2}} \quad \rightarrow \quad kp_{\frac{3}{2}}, \; kp_{\frac{1}{2}},$$

where k = n for discrete (autoionizing) and ε for continuum states. At photon energies above the $2p_{\frac{3}{2}}$ threshold and below the $2p_{\frac{1}{2}}$ threshold, the $2p_{\frac{3}{2}}$ photoionization cross-section is strongly influenced by the $2p_{\frac{1}{2}} \rightarrow kd_{\frac{3}{2}}, ks_{\frac{1}{2}}$ autoionization resonances. These resonances alter the photoionization cross-section very dramatically and any study of these atoms/ions, even in the low-Z range, must include relativistic effects without which these major features simply cannot be addressed.

Of first and foremost concern is the designation and ordering on the energy scale of the $2p_{\frac{1}{2}} \to kd_{\frac{3}{2}}$ and $2p_{\frac{1}{2}} \to ks_{\frac{1}{2}}$ resonances. The continuum kd states have significantly less quantum defect compared to the continuum ks states since the former do not penetrate deep enough into the atomic/ionic core due to the centrifugal barrier potential. Furthermore, the quantum defects change not merely with the orbital angular quantum number, but also with Z, and most importantly, the Z-dependence of these quantum defects is different for the kd states compared to that for the ks states. This results in a spectral profile which is *both* quantitatively and qualitatively different in the cases studied.

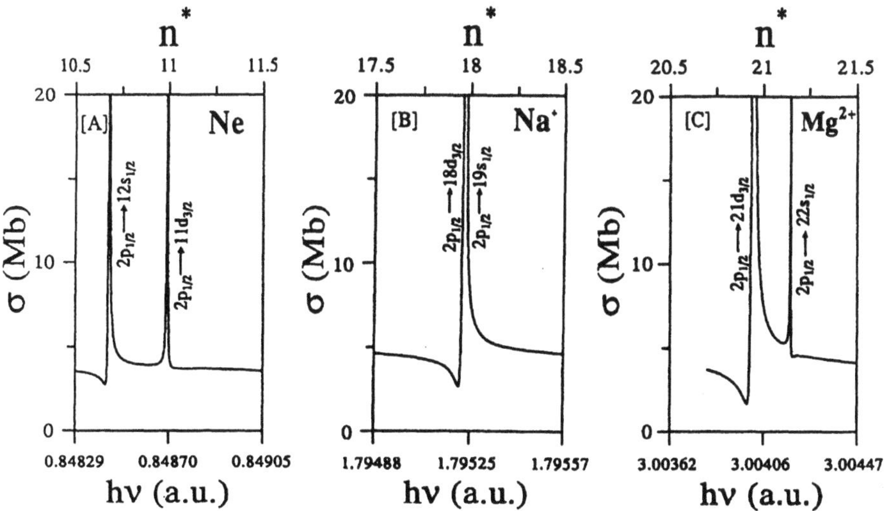

Figure 1. Threshold photoionization cross section showing one resonance doublet for Ne, Na^{1+}, and Mg^{2+} as a function of photon energy hv and effective quantum number n^*.

The $2p_{\frac{1}{2}} \to kd_{\frac{3}{2}}, ks_{\frac{1}{2}}$ resonances in Ne, Na^{1+} and Mg^{2+} are shown respectively in Fig. 1A, 1B and 1C. These are the three lowest members of the isoelectronic sequence. The corresponding set of resonances for somewhat higher-Z members, namely Si^{4+}, Ar^{8+} and Ni^{18+} are shown in Fig. 2A, 2B and 2C respectively.

It is immediately seen that whereas the resonances shown in Fig. 2 are atleast qualitatively similar to each other, those shown in Fig. 1 are very

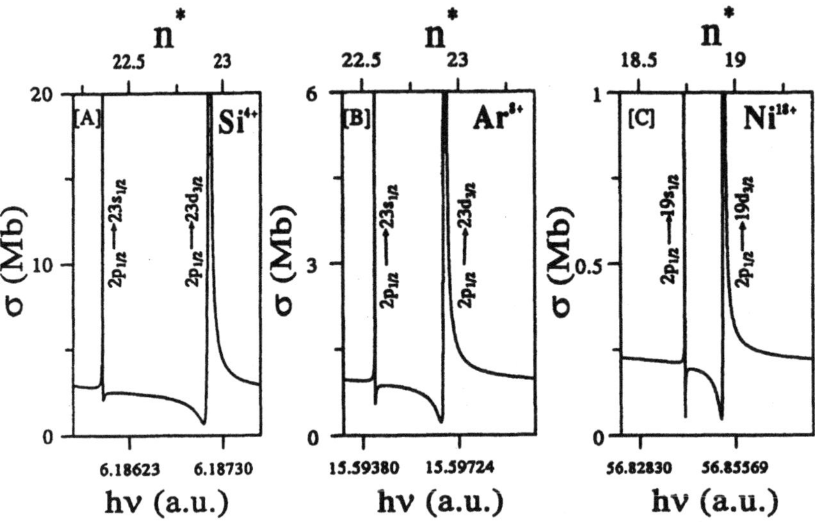

Figure 2. Threshold photoionization cross section showing one resonance doublet for Si^{4+}, Ar^{8+}, and Ni^{18+} as a function of photon energy $h\nu$ and effective quantum number n^*.

different from each other. In particular, whereas $2p_{\frac{1}{2}} \rightarrow (n+1)s_{\frac{1}{2}}$ resonances appear on the lower energy side of the $2p_{\frac{1}{2}} \rightarrow nd_{\frac{3}{2}}$ resonances for neutral neon, their ordering is reversed for Mg^{2+} and for the intermediate ion, Na^{1+}, these two resonances are practically degenerate. Naturally, for the Na^{1+} case, one cannot label these resonances by the LSl designation. Across this intermediate ion, since the ordering of the excited kd and ks states reverses, any simple extrapolation of spectral patterns is not possible. On the other hand, at intermediate-Z and/or high-Z, most atomic properties change rather smoothly, as suggested by the results presented in Fig. 2. The quantum defects for the kd states and the ks states for the ions studied in this work are presented in Table 1.

Table 1. Quantum defects for s and d resonances for the systems considered.

Systems	s	d
Ne	1.322	0.012
Na^{1+}	1.039	0.041
Mg^{2+}	0.853	0.060
Si^{4+}	0.642	0.075
Ar^{8+}	0.439	0.073
Ti^{12+}	0.265	0.061

3. CONCLUSION

The degeneracy of $(n+1)s$ resonance with nd resonance in the Na^{1+} case is due to the fact that the quantum defect of the $(n+1)s$ states differs with that of the nd states by unity in this case. Thus, the Na^{1+} resonance spectrum shows strong $s - d$ mixing. Similar strong mixing is then to be expected whenever the d quantum defect differs by *any* integer from the s quantum defect. This is a peculiar situation and suggests a generalization: *Whenever the quantum defect of the d states differs from that of the s states by an integer at some member of an isoelectronic sequence, the spectral ordering of the corresponding resonances would reverse across that member, and furthermore, for that particular member, the corresponding resonances would be degenerate.* This generalization will have important consequences in (a) spectral patterns and (b) characterization of the resonances, as seen in the present case.

ACKNOWLEDGMENTS

This work was supported by the National Science Foundation, USA.

REFERENCES

1. V. Schmidt, Rep. Prog. Phys. **55**, 1483 (1992).

2. V. Schmidt, *Electron spectrometry of atoms using synchrotron radiation*, Cambridge University Press, Cambridge (1997).

3. M. Ya Amusia, *Atomic Photoeffect*, Plenum Press, New York (1990).

4. U. Becker and D. A. Shirley (Eds.) *VUV and soft x-ray photoionization*, Plenum Press, New York (1996).

5. R. F. Reilman and S. T. Manson, Astrophys J. Supp. **40**, 815 (1979).

6. G. N. Haque, Ph. D. Thesis, Georgia State University (1991), unpublished; E. W. B. Dias, Ph. D. Thesis, Indian Institute of Technology -Madras (1998), unpublished.

7. H. S. Chakraborty, P. C. Deshmukh, and S. T. Manson, Pramana-J. Phys. **6**, 607 (1998); S. T. Manson, Z. Altun, H. S. Chakraborty, E. W. B. Dias, P. C. Deshmukh, and C. S. Turner, *Atomic Processes in Plasmas*, Eds. E. Oks and M. S. Pindzola, AIP, New York, p. 19 (1998).

8. H. S. Chakraborty, P. C. Deshmukh, and S. T. Manson, published in this proceeding.

9.   W. R. Johnson and C. D. Lin, Phys. Rev. A **20**, 964 (1979).

10.   C. M. Lee and W. R. Johnson, Phys. Rev. A **22**, 979 (1980).

Multiphoton and Multistep Laser Ionization Spectroscopy of Atoms

S. A. Ahmad and S. G. Nakhate

Spectroscopy Division, Bhabha Atomic Research Centre
Mumbai 400 085, India

1.   INTRODUCTION

Most of our knowledge about the structure of atoms and molecules has been provided by spectroscopic investigations carried out on these systems. Spectroscopy has, thus, played an outstanding role in the development of knowledge about atomic and molecular physics. The development of tunable lasers has brought about phenomenal growth in the field of laser spectroscopy, during the last three decades. The field of photoionization spectroscopy has received perhaps the greatest attention with a wide range of application in many branches of physics.

During the last few years we have initiated setting up experimental facilities for carrying out photoionization spectroscopy of atoms using tunable pulsed dye lasers. This article gives a brief introduction to the photoionization processes in atoms, describes some of the experimental facilities set up by us and presents some of the interesting features of the multiphoton ionization (MPI) studies and the multistep ionization studies carried out respectively on Barium and Europium atoms using the tunable dye lasers pumped by the excimer laser.

2.   MULTIPHOTON IONIZATION OF ATOMS

It has been known for nearly a century that an atom can be photoionized by absorbing a single photon from electromagnetic radiation whose frequency (ω) is such that the photon energy ($\hbar\omega$) is higher than or equal to

Trends in Atomic and Molecular Physics,
Edited by Sud and Upadhyaya. Kluwer Academic/Plenum Publishers, New York, 2000.   353

354

the ionization energy (E_i). Single–photon absorption has a linear dependence on the weak intensity of radiation. Multiphoton ionization (MPI) results from the simultaneous absorption of several photons whose energy is less than the ionization potential of atoms. That is, $\hbar\omega < E_i$. Since multiphoton process has a non-linear response to the intensity of radiation, the MPI process requires a source of intense radiation, and could be observed only after the advent of lasers. MPI was first observed by Hall, Robinson and Branscomb in 1965 [1a]. Since then many interesting aspects of MPI has been studied and these could be found in the some of the excellent reviews [1 b-f].

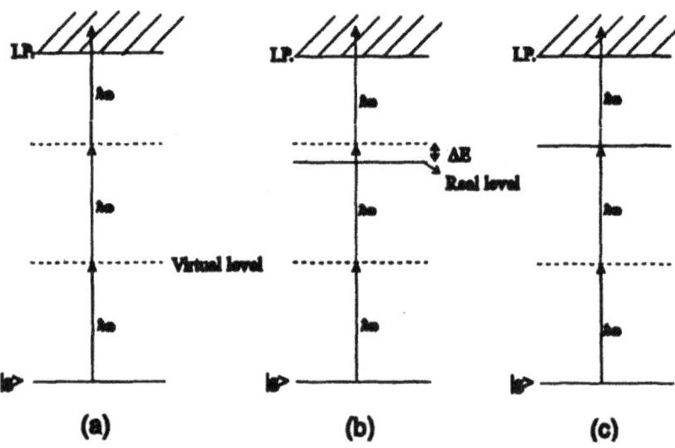

Figure 1. Single-color multiphoton ionization (MPI) of an atom, (a) Nonresonant MPI, (b) Quasi-resonant MPI, (c) Resonant (MPI).

An atom can be photoionized using different pathways (Fig. 1) which require laser intensity of different magnitudes. One such path (Fig. 1) referred as nonresonant multiphoton ionization of an atom, requires high laser intensity ($> 10^6$ watts cm^{-2}) and can be performed with a single laser. Multiphoton absorption from the ground state to the continuum takes place through laser-induced virtual states, which are not eigenstates of the atom. In principle such multiphoton ionization process does not require any intermediate atomic state. The lifetime of the virtual state is very short and is of the order of one optical cycle i.e.~ 10^{-15}s for visible photon excitation. Consequently, the absorption of photons through laser-induced virtual states must occur within time scale $<10^{-15}$s. Therefore the photon flux has to be high enough so that next photon gets absorbed within 10^{-15} s.

According to the perturbation theory of non resonant multiphoton ionization, N-photon ionization rate, W_N, is given by,

$$W_N = \sigma_N I^N \qquad (1)$$

Where σ_N is the generalized N-photon ionization cross-section expressed in units $cm^{2N} S^{N-1}$, I is the laser intensity expressed in units photon / cm^2 s and N is called the 'order of nonlinearity'. From the above equation one can see that the log-log plot of number of ions generated (n_i) versus laser intensity (I) is a straight line with a slope giving the number of photons absorbed in the multiphoton process.

The cross-section for nonresonant N-photon ionization process can be estimated as follows. Let us consider the simplest case of two photon ionization of an atom with a laser frequency ω and intensity I. The 1-photon transition will take place at the rate $W = \sigma_1.I$, where σ_1 is the 1-photon absorption cross-section and is typically of the order 10^{-17} cm^2. Second photon can be absorbed if it is incident within the time τ which is of the order of ω^{-1} i.e. 10^{-15} s. Again the rate of the second event is $\sigma_1.I$ so that the over all rate for the 2-photon ionization will be $W_2 = \sigma_1.I.\omega^{-1}.\sigma_1.I$. Thus $\sigma_2 = W_2/I^2 = \sigma_1.\omega^{-1}.\sigma_1 = 10^{-49}$ cm^4 s. This estimated value is in good agreement with the reported experimental data.

The above discussed facts are valid only for nonresonant ionization. If a real atomic state falls within small proximity of laser induced virtual state, the aforementioned time can be determined by $1/\Delta E$, where ΔE is the energy defect as shown in Fig. 1b; for example $\tau = 3 \times 10^{-11}$ s for the $\Delta E = 1$ cm^{-1}. Because of such quasi-resonant process the absorption cross-section increases and multiphoton ionization takes place at a lower laser intensity. A general expression for σ_N is given by,

$$\sigma_N = C\omega \left| \sum_{a_{k-1}} \cdots\cdots \sum_{a_1} \frac{\langle f|r|a_{k-1}\rangle\cdots\cdots\cdots\langle a_2|r|a_1\rangle\langle a_1|r|g\rangle}{(\Delta E_{k-1})\cdots\cdots\cdots(\Delta E_2)\,(\Delta E_1)} \right|^2 \quad (2)$$

where C is a constant, ω is the angular frequency of laser , $\langle b|r|a\rangle$ are dipole matrix elements between atomic states while ΔE are the energy differences, the first one being $(E_{a1} - E_g - \hbar\omega)$ etc.; with E_g being the energy of the initial state $|g\rangle$. The multiple summations extend over complete sets of atomic states $|a_i\rangle$.

If $\Delta E = 0$ (Fig. 1c), the resonant multiphoton ionization of an atom leads to many interesting results. The atomic level shifts as well as broadens because of higher laser field. The atomic level shift (Δ) is linear with respect to the laser intensity I, that is, $\Delta = \alpha I$. This atomic level shift brings a dynamics resonance detunning $\delta = \Delta E + \alpha I$, which can tune the resonance closer or away depending on whether ΔE is positive or negative. Consequently the order of nonlinearity $N = \delta \log n_I / \delta \log I$ mentioned above, does not reflect the number of photons absorbed in the multiphoton process.

For a long time, multiphoton ionization of atoms has been described successfully using perturbation theory, the laser electric field representing the perturbation with respect to the Coulomb field. With the development of increasingly powerful lasers which are now available, the electric field produced by the lasers is of the order of the Coulomb field of an atom in its ground state. The electric field E produced at the laser spot is given as

$$E\,(V\,/\,cm) = 27.4\sqrt{I} \qquad\qquad (3)$$

Where I is laser intensity in W/cm^2. With I = 3.5 × 10^{16} W/cm^2 one gets E = 1.79 × 10^{10} V/cm which is the Coulomb field experienced by 1s electron in the hydrogen atom. This new situation no longer allows the use of perturbative treatment and motivated large efforts in the development of non-perturbative theoretical approaches.

2.1 High Intensity Laser Atom Interaction

There has been a substantial shift in the research activity in the field of multiphoton ionization in last few years. Most of the recent efforts at relatively high laser intensities (10^{10} – 10^{15} W/cm^2) have been directed in three main areas: (i) Above–Thershold ionization (ATI), (ii) Production of multiply-charged ions and (iii) High order harmonic generation. Multiphoton ionization of atom starts at laser intensities of about 10^6 W/cm^2. If the intensity is further increased to tens of gigawatt/cm^2 (10^{10} W/cm^2) the atom starts absorbing photons in ion continuum and one observes a number of electron energy peaks separated by energy $\hbar\omega$. This process is known as Above-Threshold Ionization (ATI). At terawatt laser intensity (10^{12} W/cm^2) one observes multiphoton multiple ionization of atom. At 10^{13} - 10^{15} W/cm^2 laser intensity, high-order harmonic generation of laser light starts appearing. We discuss briefly only ATI and production of multiply charged ions as these are relevant to studies carried out by us.

2.1.1 Above–threshold ionization (ATI) of atoms

Above–threshold ionization results from the absorption of a number of photons larger than the minimum number N required to ionize the atom. It was believed that the kinetic energy of the emitted electron would be given by N$\hbar\omega$ - E$_i$, where $\hbar\omega$ is the photon energy and E$_i$ is the ionization energy of the atom. This is a simple extrapolation of the Einstein picture of the photoelectric effect. In 1979, Agostini et al [2] observed that the energy spectrum of electrons generated in 6-photon ionization of xenon atom, using

Nd-YAG laser of 1064 nm at 10^{-12} W/cm^2 intensity, consisted of two peaks corresponding to the absorption of 6 and 7 photons. The presence of this second peak means that the atom has absorbed more than the minimum number N = 6 photons. Later MPI studies carried out in a number of laboratories showed that the electron energy spectrum consisted of series of evenly spaced peaks, the spacing between these peaks corresponding to the energy of the photon used for MPI (see Fig. 2).

One of the key points of ionization in an intense laser field lies in the fact that the free electron immediately starts to oscillate at the frequency ω of the electromagnetic field of the laser radiation. The electron thus acquires an oscillation energy, E_{osc}, which is generally referred as 'quiver energy'

$$E_{OSC} = \left\langle \frac{e^2 \varepsilon^2}{4 m \omega^2} \right\rangle \tag{4}$$

$$= 9.3.\,10^4 \, I \, \lambda^2 \text{ (where } E_{osc} \text{ is in eV, I is in W/cm}^2 \text{ and } \lambda \text{ is in } \mu m)$$

For Nd-YAG laser ($\lambda = 1.064$ μm) the values of E_{osc}, for different values of I are:

I (W/cm^2)	10^{11}	10^{13}	10^{15}
E_{osc} (eV)	0.01	1.0	100

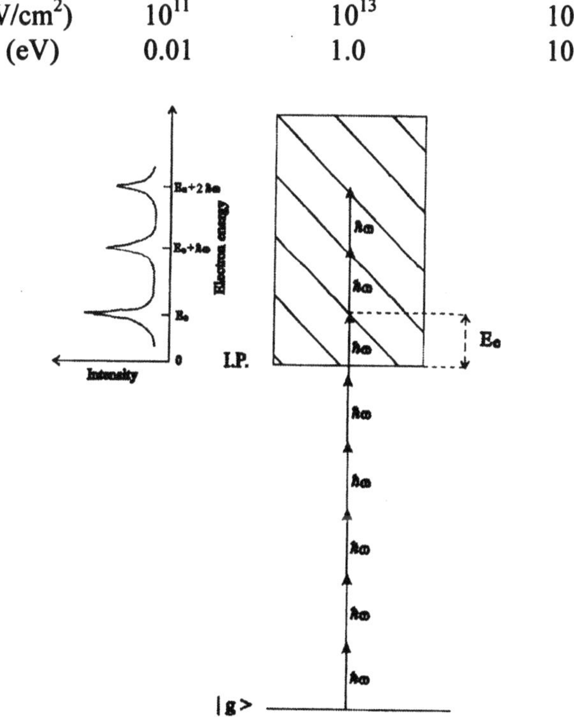

Figure 2. Above-threshold ionization of an atom.

An electron created with an initial kinetic energy $E_o < E_{osc}$ cannot escape from the Coulomb potential well of the parent ion. It is as if the ionization energy of the atom in a weak field i.e. E_i, now becomes equal to $E_i + E_{osc}$ (Fig. 3). This corresponds to a new state half-free and half-bound in the sense that the electron cannot escape to infinity. The electron, which remains in the immediate neighborhood of its parent ion can now easily absorb more photons from the laser field and thus gain energy. The electron can escape from the potential well as soon as it has absorbed 's' additional photons such that the energy $E_o + s\hbar\omega > E_{osc}$.

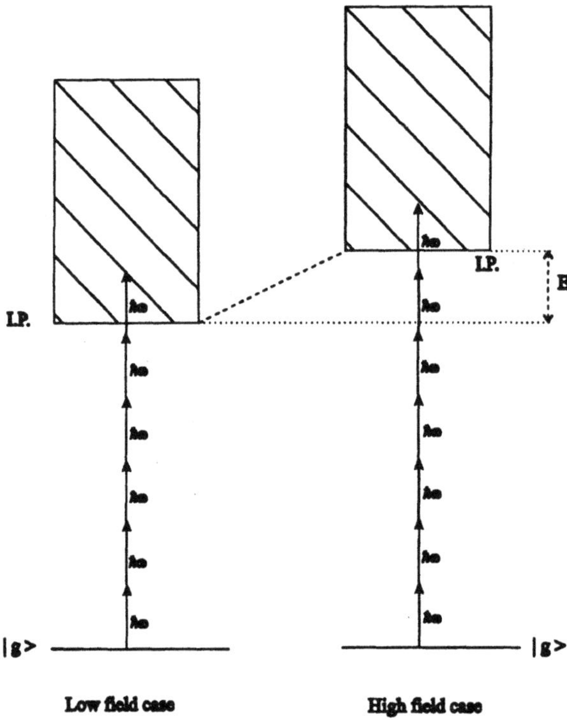

Figure 3. Multiphoton ionization of an atom in low and high laser field. The ionization limit is lifted up by an amount E_{osc} in the presence of high laser field.

2.1.2 Multiphoton multiple ionization of atoms

When an atom is photoionized with an intense laser, several electrons may be removed. The formation of multiply-charged ions by multiphoton absorption had been observed for the first time for atoms with two outer-shell electrons, namely alkaline earths, rare earths and lead.

In the formation of doubly-charged ions two types of paths may be involved, either a (i) direct process or (ii) stepwise process:

$$(i) \quad A + n_1 \, . \, \hbar\omega \rightarrow A^{++} + 2e^-$$

$$(ii) \quad A + n_2 \, . \, \hbar\omega \rightarrow A^+ + e^-$$

followed by

$$A^+ + n_3 \, . \, \hbar\omega \rightarrow A^{++} + e^-$$

It was observed that certain species could be doubly-ionized at surprisingly low laser intensity and the ratio A^+/A^{++} was found to be very low. Also, the effective order of nonlinearity was found to be smaller than number of photons required to doubly-ionize the species either by direct or stepwise process. This could be due to presence of resonances in the multiphoton ionization process. In this perspective, resonances either above the first ionization limit i.e. autoionizing states or resonances in the spectrum of singly-charged ions could be responsible for lowering the slope observed for A^{++} yield. Measurement on electron energy revealed the pathways of ionization. These two mechanisms i.e. stepwise and direct double ionization co-exist and which one is predominant depends on the atom and the laser parameters.

The formation of highly charged ions has been observed in MPI of rare gases [3]. The nonlinear coupling of an intense laser field with closed-shell atom is so efficient that a large amount of energy (as high as 1 keV) can be transferred from the laser to the atom to remove most of the electrons of the outer-shell, for example, stripping of all 5p subshell electrons of Xe at 10^{13} W/cm^2 using Nd-YAG frequency doubled laser at 532 nm and 50 ps pulse width. Multiply–charged ions can be left in either their ground or excited states. Since the emission of the excited multiply-charged ions lie in the VUV or even the soft X-ray region, it could thus be possible to obtain an intense emission in the XUV region. This prospect has motivated the interest of several laboratories in the field and lead to the realization of tabletop X-ray laser [4].

2.2   Stepwise Resonance Ionization Spectroscopy

The development of tunable laser has immensely increased the range of their applications in many branches of spectroscopy. Perhaps the greatest expansion has occurred in the field of photoionization spectroscopy. Before

360

the advent of lasers photoionization spectroscopy had limited applications because it relied on the use of electrical discharge and arc lamps. These conventional light sources were of limited utility because one needed short wavelengths and large spectral intensities required for inducing atomic or molecular photoionization. With the development of tunable dye lasers the efficient photoionization of atoms / molecules through resonant stepwise multiphoton laser excitation was recognized. This method is also widely referred as Resonance Ionization Spectroscopy (RIS).

In RIS the atoms are excited by multiple frequency laser radiation through stimulated quantum transitions into a high lying energy state in one or more stages, via intermediate states. These excited atoms are subsequently ionized by any one of the following methods: (a) photoionization, (b) electric field ionization, (c) collisional ionization (d) by absorbing microwave or infrared radiation etc. These various schemes for resonance excitation and ionization of atom are shown pictorially in Fig. 4. The process of laser excitation of an atom to high lying state and its subsequent ionization has been discussed briefly in the following subsections.

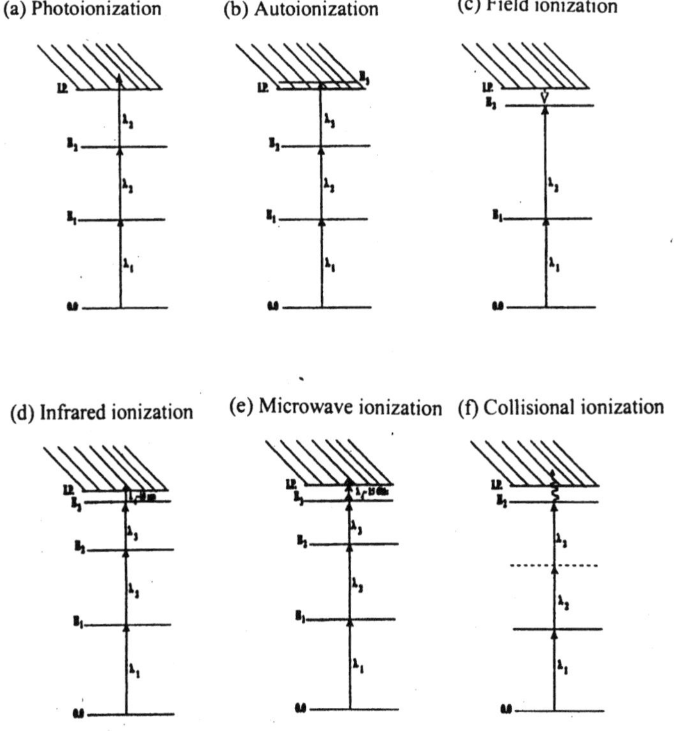

Figure 4. Different pathways of stepwise resonance ionization.

2.2.1   Multistep Resonant Excitation of Multi–level System

Resonant excitation of an atom to a high lying state forms the basis for stepwise laser multiphoton ionization spectroscopy. For an atom with large number of quantum levels, the most simple way to do this is to use multiple frequency resonant laser radiation ω_{kn} to successfully excite the multi-level quantum systems up the ladder of the discrete energy levels to some state near the ionization limit as shown in Fig. 5. The multistep excitation scheme employed must provide the maximum probability of exciting to this final high lying state f, i.e. it should saturate the transition. To achieve the saturation, the intensity of the exciting laser must satisfy the following condition:

$$I(\omega_{kn}) > I_{sat}^{kn} = \frac{\hbar\omega_{kn}}{2\sigma_{kn}\tau_k} \tag{5}$$

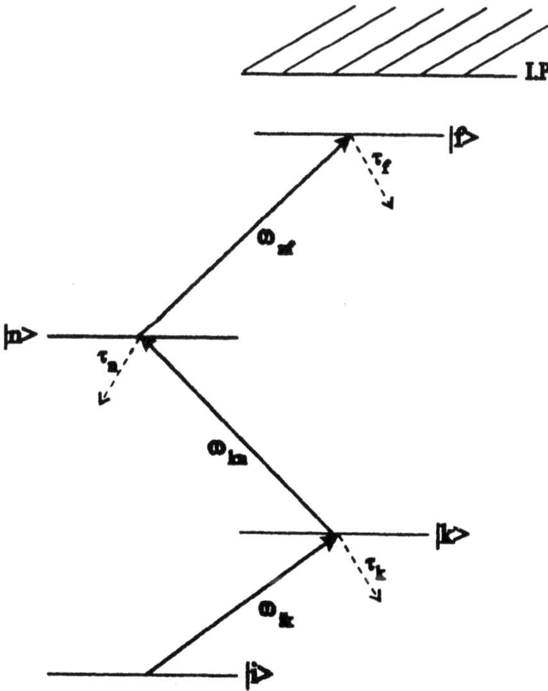

Figure 5. Multistep resonance excitation of an atom close to the ionization limit.

where τ_k can be the minimum of the relaxation time of the level k or the transition time of the atom through laser beam or the laser pulse width. σ_{kn} is the cross-section of the radiative transitive k → n. In case of pulsed laser excitation, when the duration of the laser pulse at each excitation stage is less than the decay time of initial and final levels of transition, the saturation energy density is given by

$$\varepsilon_{sat}^{kn} = \frac{\hbar \omega_{kn}}{2\sigma_{kn}} \qquad (6)$$

The value of ε_{sat}^{kn} lies in the range 10^{-8} - 10^{-5} J/cm^2 for allowed quantum transitions, which can be easily obtained with the presently available dye lasers.

2.2.2 Ionization of the Excited Atom

Qualitatively, the ionization of an atom from the excited state may be represented as shown in Fig. 4. The excited states which may be far from the ionization limit (Fig. 4a, b) can be ionized effectively only by laser radiation and the method of ionization is known as laser photoionization. Following two possibilities exist in this case: direct non-resonant photoionization by transition to the continuum (Fig. 4a) whose cross-section is of the order of 10^{-17} - 10^{-19} cm^2 and the resonant photoionization by transition to an autoionizing state (Fig. 4b) which has a better cross-section and of the order of 10^{-16} cm^2. Autoionizing states are bound atomic states whose energy lies above the ionization limit of the outer valence electron. Due to the interaction of such states with the continuum state, they decay by ejecting one electron. The highly excited states lying closer to the ionization limit (i.e. Rydberg state) are comparatively easy to ionize with a high efficiency by one of the following methods: electric field ionization (Fig. 4c), infrared radiation ionization (Fig. 4d) microwave radiation ionization (Fig. 4e) and also collisional ionization with other particles or with wall (Fig. 4f). For the field ionization (Fig. 4c), the critical field E_{cr} and the Rydberg state having effective principal quantum number n* bears the relation

$$E_{cr} = \frac{1}{16 n^{*4}} \qquad \text{(in atomic units)}$$

$$= \frac{3.21 \times 10^8}{n^{*4}} \qquad \text{(in V/cm)} \qquad (7)$$

If the electric field exceeds the critical value E_{cr}, for a given Rydberg state, the ionization efficiency can be brought closer to unity. Thus ionization cross-section of an atom excited to an intermediate state is governed by the cross-section of its resonant excitation to a Rydberg state. This cross-section is of the order of 10^{-13} cm^2, which is four to five orders of magnitude more than that of nonresonant ionization to the continuum.

For microwave ionization (Fig. 4e) the cross section is given by relation

$$\sigma_{ni} = \left(\frac{64\pi}{3\sqrt{3}}\right)\alpha\left(\frac{E_i}{\hbar\omega}\right)^3\left(\frac{a_0^3}{n^5}\right) \tag{8}$$

where α is fine structure constant and a_0 is the Bohr radius.

Increase in effective principle quantum number n^*, increases the atomic collision cross-section because the atomic size grows in proportion to n^{*2} ($\langle r \rangle = 3/2 \times n^2 a_0$) and thus its cross-sectional area $\sim n^{*4}$. Taking into account the weak bond of an excited electron, one can expect large values for cross-section of collisions with atoms which result in ionization (Fig. 4f). Moreover, lifetime of the Rydberg atoms increases with the increase of 'n' (approximately as n^{*3}) which promotes its ionization. Because of these factors collisional ionization cross-sections for Rydberg are very high. Collisional ionization efficiency close to one for Rydberg state having $n \geq 20$ in thermionic diode has been reported.

2.2.3 Unique features of multistep multiphoton Resonance Ionization (RIS) Spectroscopy

Resonance ionization spectroscopy, based on the multistep resonant excitation and subsequent ionization of atoms or molecules, is a unique method with which all the ultimate characteristics can be realized. The sensitivity of the method is high enough to permit detection of single atom or molecule. If all the resonant transitions are simultaneously saturated by several laser pulses and the final excited state of a multistep process decays into a narrow autoionizing state, it is possible to deplete completely the ground state through excitation and thus achieve 100% ionization yield. These ions can be detected with almost 100% efficiency and thus even the detection of a single atom is possible using RIS.

An important feature of the resonant multistep excitation and ionization method is its exceptional high selectivity. Selectivity gets multiplied at each excitation stage and the total selectivity is given by

$$S = S_1 \times S_2 \times S_3 \ldots\ldots S_n \tag{9}$$

364

Where $S_1.S_2.S_3.........S_n$ is the selectivity achieved in first, second, third and n^{th} excitation step.

Use of narrow band lasers for excitation along with Doppler-free techniques makes it possible to increase the selectivity to isotopic species. To enhance the selectivity further, the photoions are mass resolved, generally by employing Time–of–Flight Mass Spectrometer. This technique is then termed as Resonance Ionization Mass Spectroscopy (RIMS). All these factors enable to achieve a total selectivity of the order of 10^{17} - 10^{20}. This means a single atom can be detected in the background of 10^{-3} - 1 gm of matter.

3.   EXPERIMENTAL TECHNIQUE

We have set up various experimental facilities for carrying out photoionization studies of atoms and molecules. We have developed the heat-pipe oven and thermionic diode systems for RIS of atoms and molecules [5a]. We have also designed and developed atomic beam source with time-of-flight mass spectrometer for RIMS studies [5b]. Presently we briefly describe the heat-pipe oven and thermionic diode systems developed by us and used for carrying out MPI and RIS studies respectively on Ba and Eu atoms.

3.1   Heat Pipe Oven and Thermionic Diode

Our studies have been carried out on the vapors of Eu, Ba, Mg and K and it is well known that it is not easy to handle these highly reactive vapors. In order to handle the corrosive vapor of these elements we have used Heat-pipe Ovens (HPO). The HPO provides a constant atom density over a well defined zone for a long period of time. It also avoids the problem of the deposition of the vapor on optical windows of the cell. The design and construction of the heat-pipe oven is discussed in some detail in the following section. Photoions produced in HOP were detected by thermionic diode which is best suited for detecting ions produced in vapor cell operating at low pressures. We have used thermionic diode of different types suitable for specific purpose in conjunction with the heat-pipe oven. The principle of thermionic diode and the design and the construction of various types of thermionic diodes used in the present work is also discussed.

3.1.1　Design, Construction and Working of Heat-Pipe Oven

The heat–pipe oven is based on the original idea of Grover et al. [6] and later it was implemented for spectroscopic application by Vidal and Cooper [7]. It uses the well–known fact that large amount of heat with a small temperature drop can be transferred if one evaporates a liquid, transports the vapor through a duct and condenses it again. In order to operate such a device continuously it is necessary to return the condensate back to the evaporator. This is achieved in a very elegant way by using the effect of surface tension with wire mesh.

A heat-pipe oven (Fig. 6) usually consists of a closed tubing, the inner wall of which is covered by a capillary structure which acts as a wick. This wick is saturated with a wetting liquid of metal. During the operation the metal evaporates at the center of the tube due to heat provided by an external source. By providing a heat sink at the ends, the vapor is driven at the ends of the tube and condenses again. The condensate is then returned by the capillary action through the wick to the evaporator. In this manner the same material gets circulated and heat-pipe oven works for long period continuously without the depletion of the sample. The design of a heat-pipe oven built by us shown in Fig. 6.

It consists of crossed stainless steel (SS 304) tubes of diameter 36 mm and length 500 mm. Crossed heat-pipe oven is built to make available more number of ports needed for various purposes. In our case, the opposite ports are used for sending two counter-propagating laser beams and the fluorescence is observed through the third port orthogonal to laser beams. The fourth port is used for inserting the thermocouple for temperature measurement or for placing an ion detector inside the oven. Double layer of fine stainless steel mesh role is inserted inside the pipe extending up to cooler zone of the heat-pipe which acts like a wick. The ends of the heat-pipe are vacuum sealed by O-rings and demountable flanges. All the ends of heat-pipe are cooled by flowing water through jackets in order to provide heat sink to condense the metal vapor and also to protect O-rings. To evaporate the metal loaded in the center of the heat-pipe, the tube is heated by a resistive heating element. D. C. power supply of 1.5 kW power was used for this heater which is made up of Kanthal wire (1 mm diameter, 2.82 Ω/meter) wounded over the heat-pipe. Two separate Kanthal wire elements of length $\sim$ 6.5 meter each and resistance $\sim$ 18 ohm are wound over four arms of the heat-pipe oven and were connected in series to give total resistance of 36 Ω. Electrical insulation is provided by a layer of alumina powder cement (Whyte heat cement, Type: A; ACC make). Heater wire is sandwiched between the two layers of alumina cement. Outer cement layer is finally covered with alumina blanket to provide thermal insulation. With this heating arrangement we could operate the heat-pipe oven easily

366

upto1000 °C. A small stainless steel tube, provided at one end of the heat-pipe oven, is connected to the vacuum system and buffer gas filling system. A diaphragm gauge which measures pressure in the range 0-50 torr is attached to this side tube to measure buffer gas pressure. Argon has been normally used as a buffer gas at about 5 torr pressure.

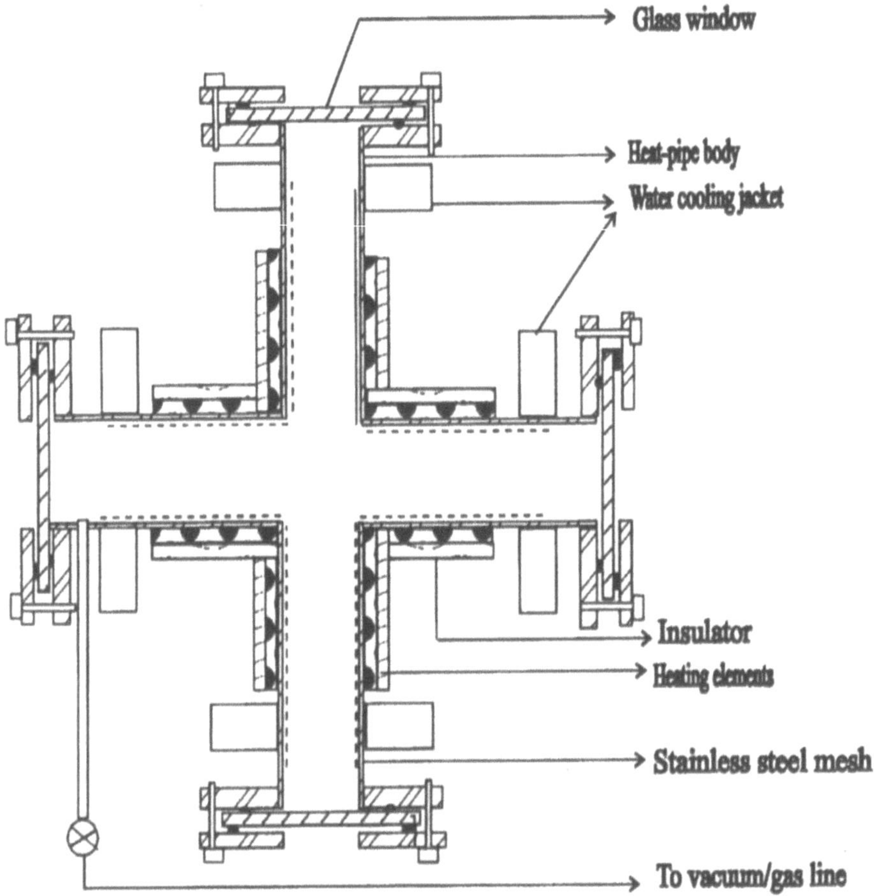

Figure 6. Schematic cross-section of the heat-pipe oven.

The advantages of a heat-pipe oven are: (1) Heat-pipe oven can be operated continuously under extremely well-defined conditions i.e. well defined temperature profile and density distribution, (2) Buffer gas between the window and the vapor zone removes the problem of vapor deposition on the window. Thus no heat shields or baffles are required and (3) The metal vapor is very clean because the continuous evaporation and condensation purifies the sample.

3.1.2 Thermionic Diode Detector

The thermionic diode, as an effective ion detector, was discovered independently as early as in 1923 by Kingdon [8] and Hertz [9]. The applicability of the thermionic diode in spectroscopy was indicated by Foote and Mohler [10] and has now become a highly sensitive and simple technique for detecting highly excited atoms especially Rydberg atoms [11 and references therein]. In some of the earlier papers it has been described under the name of "space charge diode" or "Kingdon cage". In the usual fluorescence detection method the sensitivity is limited because of (i) line branching, (ii) limited solid angle of observation, (iii) loss of fluorescence intensity because of several optical surfaces before fluorescence light reaching detector. The thermionic diode for detection of photoions on the other hand, is more sensitive because of: (i) 4π solid angle detection efficiency and (ii) amplification of the order of 10^4 to 10^6.

Thermionic detector is essentially a diode operating in the space charge limited mode. In its simplest form it is made up of cylindrical anode with an axially mounted cathode filament usually made of tungsten wire. If the cathode is hot there is thermionic emission of electrons from the surface. With activated cathode (e.g. an alkali or alkaline earth element layer on the filament) or by externally heating the filament, the thermionic emission of electrons is so large that an electron cloud is formed around the cathode which limits the current flowing through the diode. In this condition the diode is said to be working in the "space-charge limited" mode. Under such conditions the current density Je is related to saturated Richardson-Dushman current $\mathbf{J_s}$ as follows,

$$J_e = J_s \times \exp\left(\frac{-e\phi_{el}}{k_b T}\right) \qquad (10)$$

Where ϕ_{el} is the space charge potential, T is the filament temperature and k_b is Boltzman constant. If the diode is filed with a gas or metal vapor at low pressure and if the ions are produced inside the diode e.g. by laser photoionization of atoms, these ions get attracted towards the filament and get trapped in the negative space charge potential well for a considerable long time. As a result the space charge potential decreases by $\Delta\Phi_{el}$. This decrease in the potential barrier results in the release of thermionic electrons in burst and thus increase in the electron current ΔJ given by

$$\Delta J = J_s \times \exp\left(\frac{-e\,\Delta\phi_{el}}{k_b T}\right) \qquad (11)$$

368

The gain of the diode is directly proportional to the trapping time of the ions in the space charge potential well. Since an individual ion can remain un-neutralized for a time interval many orders of magnitude longer than that of an electron, it is possible to set free about 10^4 to 10^6 electrons resulting in a gain of the same order of magnitude.

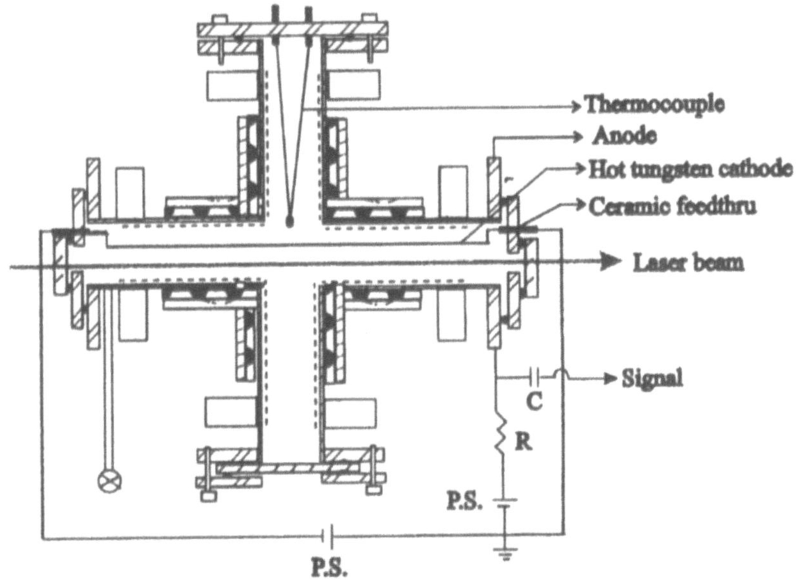

Figure 7. Heat-pipe oven-Thermionic diode system with externally heated cathode filament.

High thermionic emission of the electrons from the cathode surface is a crucial factor for the built up of space charge and subsequently for the amplification of the signal. For diode filled with alkali or alkaline earth metals, the work function for emission of electrons of the cathode material which is usually tungsten wire, are lowered by the adsorbed alkali or alkaline earth element layer on the cathode surface and thus one gets the required thermionic emission at operating temperature necessary to generate metal vapor. However while working with elements other than alkali or alkaline earth metals one has to heat the cathode wire externally to generate the required thermionic electrons (Fig. 7). In our experiments with barium the tungsten cathode was not heated and heat-pipe oven temperature of 750 °C was enough to get amplification in the diode signal (Fig 8a). However, in the photoionization experiments with europium, tungsten cathode was heated resistively to a temperature of 1100 °C to generate thermionic emission and thus amplification in the photoionization signal (Fig. 8b).

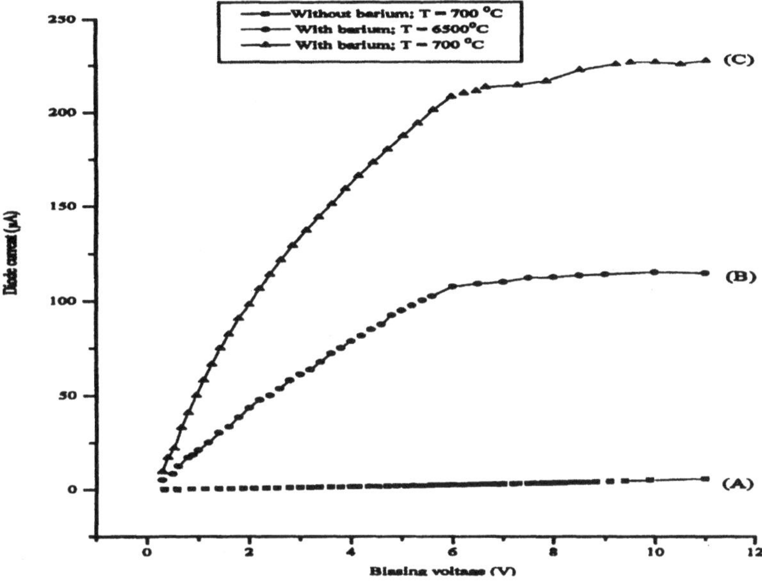

Figure 8a. Typical current-voltage characteristic curves for thermionic diode filled with barium, at different heat-pipe oven temperatures.

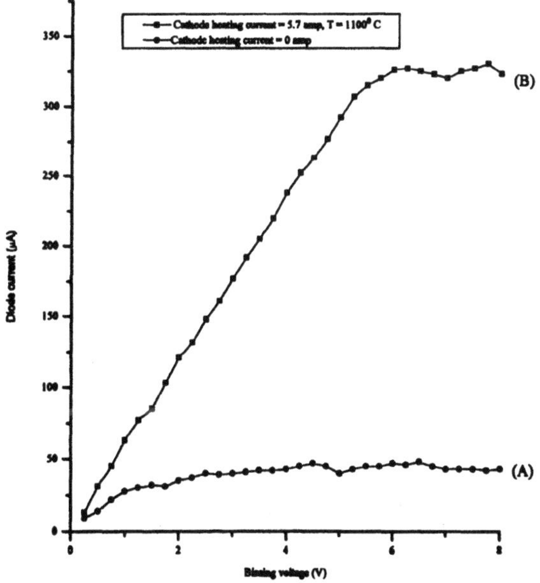

Figure 8b. Typical current-voltage characteristic curves for thermionic diode filled with europium, at different cathode filament currents.

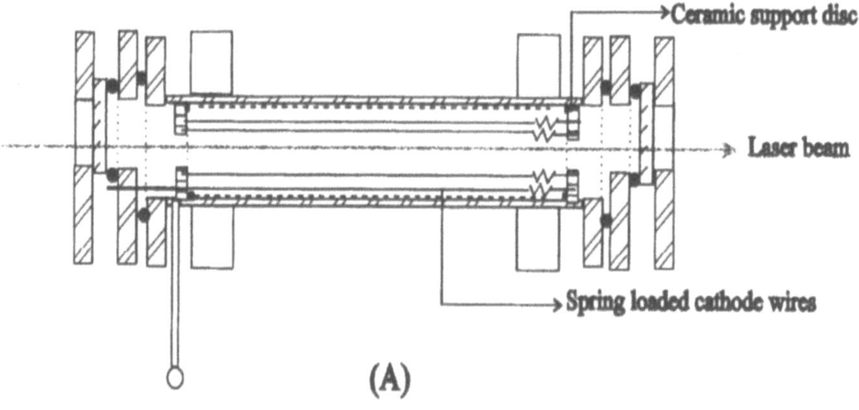

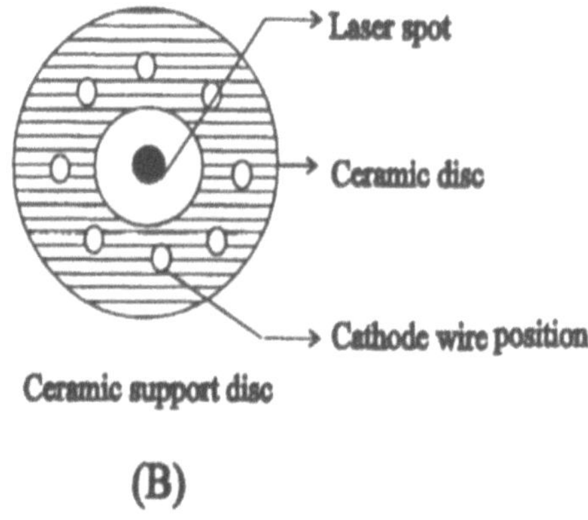

Figure 9. Schematic diagram of a thermionic ring diode. The positions of eight cathode wires.are shown in (B).

As mentioned earlier, thermionic diode detector is best suited for detecting highly excited atoms especially Rydberg atoms. However, since polarizabilities of highly excited states are large, a small electric field ($\sim$ 1 volt/cm) used for biasing the electrode, or even the electric field due to thermionic electrons is sufficient to broaden and shift the resonances because of the Stark effect. As a consequence it is impossible to observe Rydberg states having principle quantum number n greater than 50 in such type of thermionic diodes. Moreover additional Stark induced transitions make the spectrum more complicated. In order to have laser-atom interaction in an electric field free zone we have made a thermionic 'ring diode' as described by Beigang et al. [12]. The design of the thermionic ring diode is given in the Fig. 9. Our thermionic ring diode basically consists of eight stainless steel wires (0.25 mm diameter) arranged longitudinally on a radius of about 1 cm around the axis of the heat-pipe oven of 33 mm inner diameter (see Fig. 9A). The wires are supported on the machined ceramic discs shown in Fig. 9B. The wires are spring loaded to ensure a sufficient tension also at high temperature. All the wires are electrically connected in parallel to each other and the ion current is measured between the heat-pipe wall and the wires. The wires are heated indirectly by the oven which heats the whole pipe. The excitation by the laser beam takes place exactly at the center of the pipe. Although there are space charges and biasing voltage on the wires, the resulting field at the center of the pipe is almost zero due to the radial symmetry, as found by us by carrying out 2-photon excitation of Rydberg states in potassium [5a].

4. MULTIPHOTON IONIZATION OF BARIUM ATOM

The study of multiphoton double ionization of alkaline-earth elements by a laser pulse was initiated by a Russian group [13, 14] which revealed that such species could be doubly ionized at surprisingly low field strengths. The important question then arose whether the double ionization is a direct process of simultaneous excitation of two electrons through two-electron spectrum upto the double ionization limit or a two-step process, namely creation of singly charged ions and subsequent ionization of the ions by the same laser pulse. The situation is illustrated in Fig. 10 for the specific case of barium atom. Double ionization of Ba, Sr and Eu has been reported with the general conclusion that the double ionization is a direct process [15]. A similar conclusion has been drawn in multiphoton double ionization of Ca with picosecond pulses [16]. Double ionization of Ba have been reported by number of authors, suggesting the two-step process [17-20]. Of course these two mechanisms must co-exist and which one is predominant must depend on the atom and laser parameters. Direct double ionization could be made dominant by forcing the resonant transitions on two-electron states. Barium is an attractive candidate because of its richness in its doubly excited states below the first ionization limit and thus the expected large density of the autoionizing states through which direct double ionization can be favorable channel.

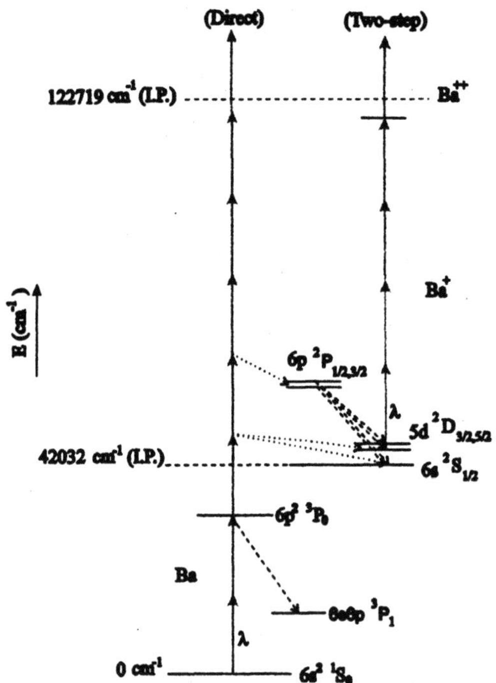

Figure 10. Energy level diagram for Ba I and Ba II. The two mechanisms, direct and two step, for double ionization of the neutral atom are illustrated.

In our studies the ionization pathways for the production of Ba⁺ have been studied by monitoring the fluorescence from excited ions formed by the multiphoton process in Ba. In order to facilitate the fluorescence measurements, Ba vapor was generated in a heat–pipe operate in a non-heat-pipe regime.

The MPI studies of Ba atoms were undertaken with the following objectives: (1) to probe the multiphoton ionization (MPI) pathways of Ba atom at laser intensity of $\sim 10^{10}$ W/cm² (10 GW/cm²) in 570-608 nm wavelength range; (2) to obtain fine-sturcture sublevel resolution of the final ionic states of Ba⁺ (usually not possible with electron spectroscopy technique due to its resolution limit) to probe the population distribution in the sublevels of Ba⁺ after photoionization.

4.1 Experimental Technique

The experimental set-up is shown in Fig. 11. The dye laser (Lambda Physik, FL 3002) is pumped by an XeCl excimer laser (Lambda Physik, EMG 201 MSC) at a repetition rate of 10 Hz. Rhodamine–6G dye is used to get tunability in the wavelength range 570 to 608 nm. The dye laser provided pulses of 20 ns duration and 0.2 cm^{-1} bandwidth. The beam is focused by a planoconvex lens of 22.6 cm focal length at the center of a crossed heat-pipe oven where Ba undergoes continuous evaporation and condensation. The maximum laser intensity at the focus is around 10 GW/cm^2, which is equivalent to an electric field of 2.7×10^6 V/cm, and the photon fluence is 2.8×10^{28} photons/cm^2 s at 580 nm. The typical operating temperature of heat-pipe oven is 675 °C at which at the Ba vapour pressure is 0.9×10^{-3} torr which corresponds to a number density of $\sim 10^{13}$ atoms/cm^3. Argon is used as a buffer gas at 5 torr pressure. Photoions generated in the multiphoton ionization process are detected by using heat-pipe oven as a thermionic diode described in detail in section 3.1.2. Barium heat-pipe oven operating temperature is sufficient to generate good thermionic emission from tungsten cathode (Fig. 8a). A negative 5 V d.c. biasing voltage is applied to collect the photoions. Using a proper RC circuit, the transient photoion signal is averaged using a box-car averager. Fluorescence light is collected orthogonally to the laser beam direction and analyzed with a 0.5 m monochromator (Pacific) and detected by a photomultiplier tube. Calibration of the fluorescence spectrum of barium was carried out by simultaneous recording of the emission spectrum of commercial Cu/Ne hollow cathode [21].

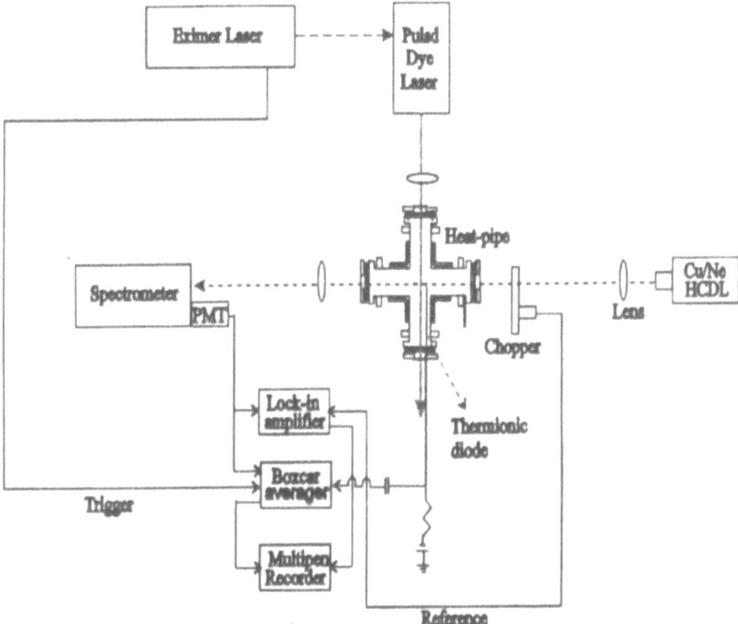

Figure 11. Schematic experimental layout showing simultaneous recording of laser induced fluorescence spectrum and calibration spectrum. Thermionic diode for collecting photoionization signal is also shown.

4.2   Results and Discussion

4.2.1   Photoionization Spectrum

The photoionization spectrum of Ba, as a function of laser wavelength, was recorded by collecting the ions produced using the heat-pipe as a thermionic diode. The photoionization spectrum is shown in Fig. 12. The resonances observed in the laser wavelength region 570-608 nm are listed along with their assignments in Table 1. The resonances observed at laser wavelengths 573.74, 575.75 nm do not have any coincidence with Ba I energy levels below the first ionization potential [22]. So these resonances may be due to direct three- or more-photon transition to the autoionizing states of Ba atom (Ba I).

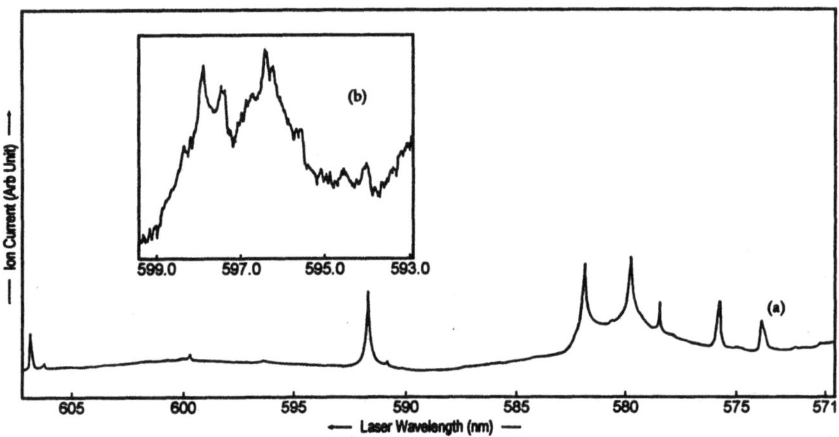

Figure 12. (a) Photoionization spectrum of Ba at laser focal intensity ~ 10 GW/cm^2 by collecting ions using thermionic diode; (b) part of the same spectrum at higher detection sensitivity.

Some additional photoion peaks, which are broad and weak (Fig. 12b; inset) could be tentatively assigned to the three- or four-photon resonances in Ba$^+$ (Ba II). A minimum of three photons are required to ionize Ba atom in the ground state which decay into to 5d ^{2}D$_{3/2,\,5/2}$ ionic state, whereas to reach 6p ^{2}P$_{1/2,\,3/2}$ state of Ba+, four-photons are required (see Fig. 10). Thus three- or four-photon ionization of Ba atom populates a variety of final ionic states (such as 5d ^{2}D$_{3/2,\,5/2}$, 6P ^{2}P$_{1/2,\,3/2}$ and ground state of barium ion). If the laser frequency is in coincidence with the three- or four–photon transition to the upper state of Ba II, it may result in the production of Ba^{2+} ions (Fig. 10) by stepwise process.   Our experimental set-up cannot completely distinguish between singly and doubly-charged ions but it is reasonable to consider that the doubly charged ions will neutralize more space charge than singly-charged ions to give an enhancement in the signal above non resonantly produced ions. We have tentatively assigned such resonances arising from $5\,{}^2D$ and $6\,{}^2P$ state and they are listed in Table 1.

Table 1. Observed resonances in the laser wavelength (λ_L) region $570 - 608$ nm: (a) in the photoionization spectrum obtained by collecting ions and (b) by monitoring the intensity of the D_1 line of Ba⁺.

Observed resonances			Resonant transitions[†]		
Ion detection λ_L (air) (nm)	Intensity of D_1 line λ_L (air) (nm)	E (cm⁻¹) (vac.)	Number of photons	E_{theo} (cm⁻¹) (vac.)	Assignment
573.74[a]	573.74	17424.665	4	-	$(0.0 - 69698.660\ \mathrm{cm}^{-1})$ Ba I
575.75[a]	575.75	17363.834	3	-	$(0.0 - 52091.502\ \mathrm{cm}^{-1})$ Ba I
-	577.24	17319.014	3	17318.939	$5d\,{}^2D_{5/2} - 5f\,{}^2F_{7/2}$ Ba II
579.66	579.66	17246.710	2	17246.949	$6s^2\,{}^1S_0 - 6p^2\,{}^3P_0$ Ba I
580.60[b]	-	17218.788	1	17219.742	$6s\,5d\,{}^3D_3 - 5d\,6p^1F_3$ Ba I
581.73	581.73	17185.341	2	17185.390	$6s^2\,{}^1S_0 - 6p^2\,{}^1S_0$ Ba I
-	585.40	17077.604	1	17078.572	$6d\,{}^2D_{3/2} - 6p\,{}^2P_{3/2}$ Ba II
590.75[b]	-	16922.945	1	16922.564	$6s\,5d\,{}^3D_1 - 5d\,6p\,{}^3P_2$ Ba I
591.64	591.64	16897.488	2	16897.920	$6s^2\,{}^1S_0 - 5d\,7s\,{}^1D_2$ Ba I
595.67	-	16783.170	3	16786.281	$6p\,{}^2P_{1/2} - 9d\,{}^2D_{3/2}$ Ba II
569.42	-	16762.065	4	16757.625	$5d\,{}^2D_{5/2} - 11s\,{}^2S_{1/2}$ Ba II
567.88[c]	-	16721.133	4	16702.133	$5d\,{}^2D_{3/2} - 7g\,{}^2G_{7/2,\,9/2}$ Ba II
599.71[b]	-	16670.109	1	16670.155	$6s\,5d\,{}^3D_1 - 5d\,6p\,{}^3P_1$ Ba I
606.33[b]	-	16488.104	1	16488.622	$6s\,5d\,{}^3D_2 - 5d\,6p\,{}^3P_1$ Ba I
606.93	606.93	16471.804	2	16471.800	$6s^2\,{}^1S_0 - 5d\,7s\,{}^3D_2$ Ba I

[†] Using the energy level listing of Moore [22].
[a] Newly observed autoionizing resonances (see section 4.2.4).
[b] Transition from metastable states of Ba I
[c] Observed and assigned by Bondar et al. [19] also.

4.2.2   Excitation Spectrum of Ba

Fluorescence spectrum at laser excitation wavelength at which photon resonances are observed (e.g. $\lambda_L = 581.73$ nm) provided information about the population of 6P, ${}^2P_{3/2,\,1/2}$ levels via four-photon ionization. We observe

376

strong fluorescence at 4554.0 Å and 4934.0 Å corresponding to the emission from 6p, $^2P_{3/2,\,1/2}$ levels to 6s, $^2S_{1/2}$ level (D$_1$ and D$_2$ lines). Barium atom number density dependent fluorescence intensity measurement of D$_1$ and D$_2$ lines of Ba$^+$ was carried out to confirm that the population of these excited ionic levels are due to multiphoton ionization processes.

4.2.3   Four-Photon Ionization of the Ba Atom: Preferential population of the $^2P_{3/2}$ level of the Ba$^+$ ion

Production of Ba+ ions due to four–photon absorption from the ground state of the Ba atom is monitored indirectly by monitoring the variations in intensity of fluorescence lines from 6p $^2P_{3/2,\,1/2}$ levels at 4554 and 4934 Å. Fig. 13 shows the change in fluorescence intensity for the D$_1$ (6p $^2P_{3/2} \rightarrow$ 6s $^2S_{1/2}$) and D$_2$ (6p $^2P_{1/2} \rightarrow$ 6s $^2S_{1/2}$) lines as a function of laser wavelength. We get the spectrum identical to the photoionization spectrum recorded by collecting ions (see Fig. 12 and 13). The observed resonances are tabulated in Table 1 itself. The extent of nonresonant four-photon ionization signal is also seen from Fig. 13, as the intensity of D$_1$ and D$_2$ lines as a function of laser wavelength is always above the baseline. Thus we can say that the 5d ^2D and 6p ^2P states of Ba$^+$ are playing an important role in two-step double ionization of Ba atom.

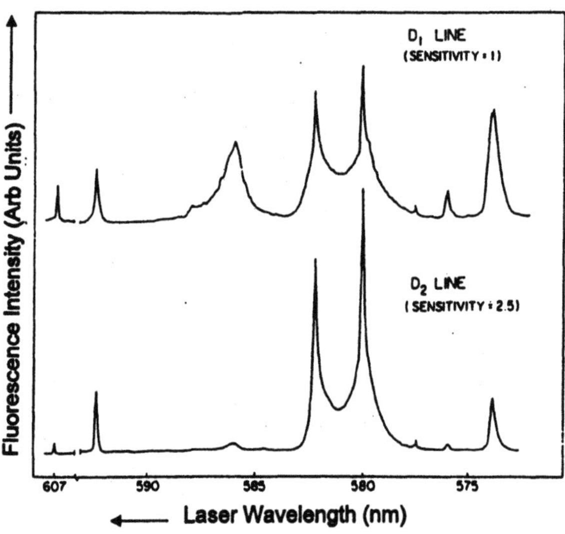

Figure 13. Fluorescence intensity against dye laser wavelength for D$_1$ (6p, $^2P_{3/2} \rightarrow$ 6s, $^2S_{1/2}$) and D$_2$ (6p, $^2P_{1/2} \rightarrow$ 7s, $^2S_{1/2}$) fluorescence lines of Ba$^+$ which gave fine-structure sublevel resolution of the final ionic state (6p ^2P) reached after four-photon ionization of Ba I. Spectra are taken at a laser focal intensity of $\sim$ 10 GW cm^{-2}.

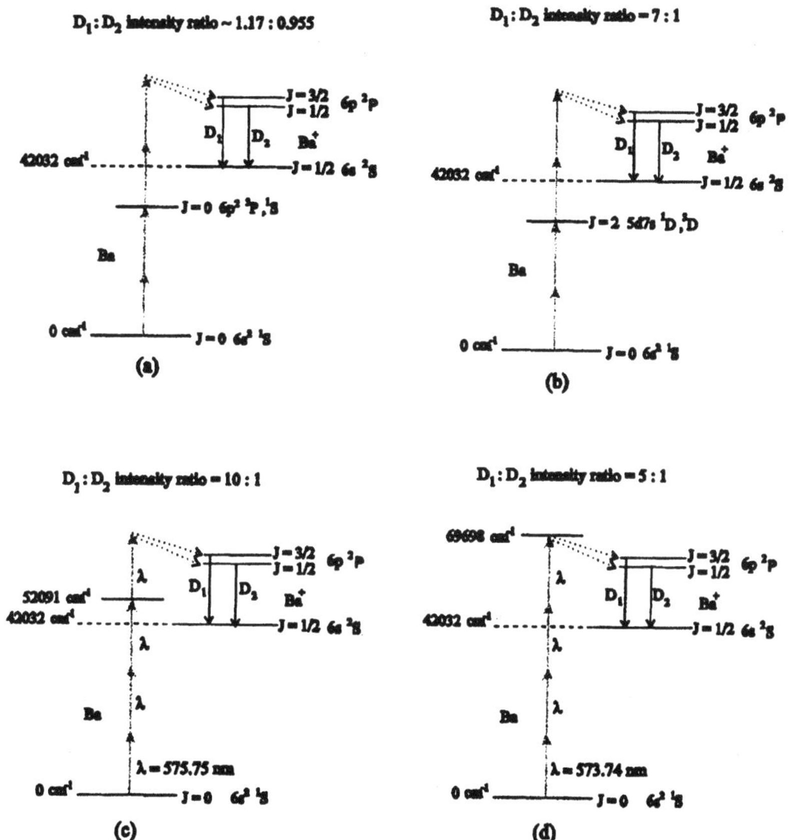

Figure 14. Relative population of $^2P_{3/2,\ 1/2}$ levels of Ba^+ (as reflected in the $D_1 : D_2$ intensity ratio) after four-photon ionization using different excitation paths; (a) $J = 0 \xrightarrow{2\hbar\omega} J = 0$ (579.66 and 581.73 nm), (b) $J = 0 \xrightarrow{2\hbar\omega} J = 2$ (591.64 and 606.93 nm), (c) $0.0\,cm^{-1} \xrightarrow{3\hbar\omega} 52091\ cm^{-1}$(newly discovered AI state, $J = ?$), (d) $0.0\ cm^{-1} \xrightarrow{4\hbar\omega} 69698\ cm^{-1}$(newly discovered AI state, $J = ?$). With normal statistical distribution the intensity ratio D_1 to D_2 line is expected to be 1.17:0.955. All the above excitation paths, except for (a), showed non-statistical distribution of population in $^2P_{3/2}$ and $^2P_{1/2}$ Ba^+ levels after multiphoton ionization. Preferential population in $^2P_{3/2}$ level was observed.

J-sublevel resolution, which was not possible in pervious studies by electron spectroscopy is achieved in our work very easily as the fluorescence

378

from $^2P_{3/2}$ and $^2P_{1/2}$ is quite separable. Thus the relative intensity of emission at 4554 and 4934 Å is a good indication of the relative populations of the $J = 3/2$ and 1/2 sublevels. With normal statistical distribution the intensity ratio of the D_1 and D_2 lines is expected to be 1.17:0.955. Some of our observations regarding the population distribution in these sublevels after four-photon ionization are as follows:

(i)   In general the relative population of $J = 3/2$ and $J = 1/2$ levels is non-statistical and we found preferential population of $J = 3/2$ level.

(ii)   We found that the intensity ratio of D_1 and D_2 lines is close to the normal statistical distribution (1.17:0.955) when the laser is tuned to $J = 0 \rightarrow J = 0$ ($^1S_0 \rightarrow {}^3P_0$ and $^1S_0 \rightarrow {}^1S_0$) two photon resonances in Ba atom (see Fig. 14a).

(iii)   When the laser is tuned to $J = 0 \rightarrow J = 2$ ($^1S_0 \rightarrow {}^3P_2$ and $^1S_0 \rightarrow {}^1D_2$) two–photon resonances in Ba I, the ratio of intensities of D_1 and D_2 lines found to be 7:1 showing preferential decay to the $^2P_{3/2}$ level after four-photon ionization (see Fig. 14b).

(iv)   Similar preferential decay to the $^2P_{3/2}$ level is found when laser is tuned to the newly discovered autoionizing (AI) resonances at 573.74 and 575.75 nm (see section 4.2.4 for further details about newly discovered AI states). The intensity ratio for D_1 to D_2 lines is found to be 10:1 and 5:1 at $\lambda_L = 575.75$ and 573.74 nm respectively (see Fig. 14c, d). Presently it is difficult to comment further on these observations.

4.2.4   Ionic Fluorescence Intensity (I_F) Dependence on Laser Pulse Energy (E_L)

The dependence of the intensity of ionic fluorescence I_F (D_1 line of Ba^+) on laser pulse energy E_L (laser intensity I_L is directly proportional to laser pulse energy) has been studied in six observed MPI resonances at 573.74, 575.75, 579.66, 577.27, 585.40 and 574.50 nm. The first three resonances are resonant in the neutral barium atom, the next two are in Ba^+ ion and the last one is the nonresonant MPI of Ba atom. For these six resonances the plots of $\log I_F - \log E_L$ are presented in Fig. 15a-f. The least–square fit of the data points was carried out and slope K_{exp} (also known as the order of nonlinearity of MPI process, discussed in section 2) of the curves (details mentioned in Fig. 15 a-f) were obtained for the linear parts of the plots.

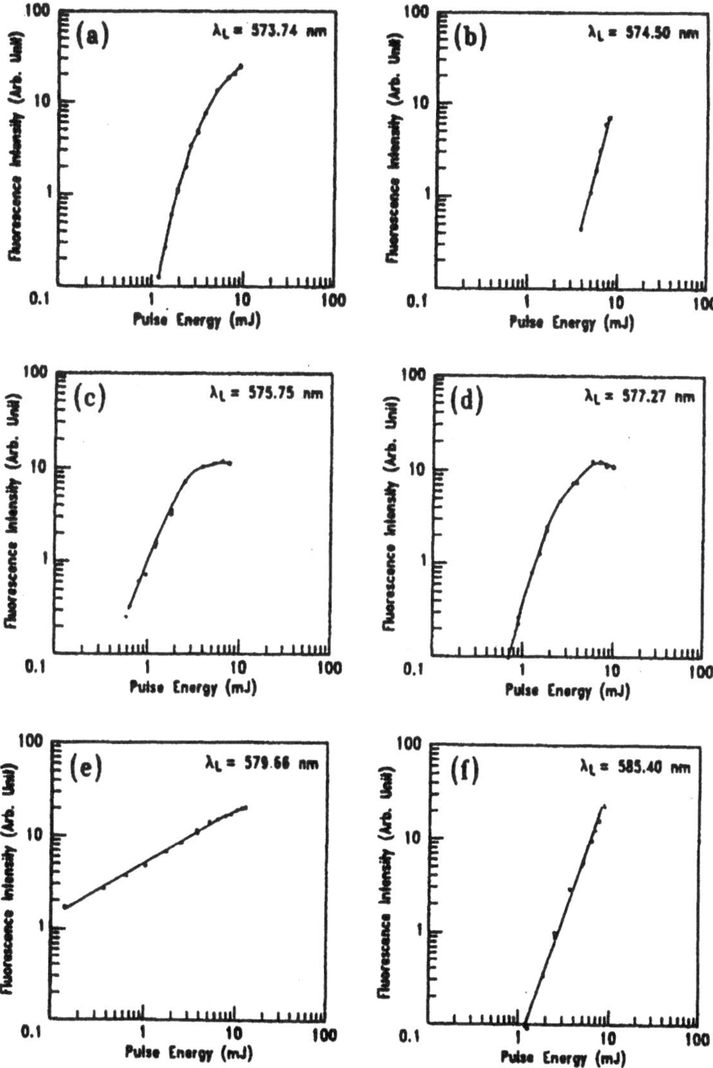

Figure 15. log-log plots of Ba D_1 line (6p, $^2P_{3/2} \rightarrow$ 6s, $^2S_{1/2}$) fluorescence intensity as a function of laser focal intensity at various laser excitation wavelengths: (a) 573.74 nm, four photon resonant ionization of Ba I; with slope, $K_{exp} = 3.53$; (b) 574.50 nm, four photon non-resonant ionization of I; $K_{exp} = 3.94$; (c) 575.75 nm, three photon resonant four photon resonance in Ba I; $K_{exp} = 2.30$; (d) 577.27 nm, three photon resonance in Ba II after photon non-resonant ionization of Ba I; $K_{exp} = 3.1$; (e) 579.66 nm, two photon four photon ionization of Ba I; $K_{exp} = 0.6$; (f) 585.40 nm, one photon resonance in Ba II after three photon non-resonant ionization of Ba I; $K_{exp} = 2.8$.

According to the perturbation theory of nonresonant multiphoton ionization discussed earlier in section 2, N-photon ionization rate W_N is given by $W_N = \sigma_N . I_L^N$, where σ_N is the generalized N–photon ionization cross section expressed in $cm^{2N} S^{N-1}$ units and I_L is the laser intensity which is directly proportional to laser pulse energy E_L. This consideration is valid when ionization is not saturated, i.e. when the number of ions produced are still small compared to the neutrals available. At saturation the ionization rate remains constant when the laser intensity is increased further. As anticipated the nonresonant four-photon ionization at $\lambda_L = 574.50$ nm (see Fig. 15a) shows $K_{exp} = 3.94$ confirming that the ionic fluorescence from 6P levels of Ba^+ is due to four-photon ionization of Ba atom. As discussed earlier the resonance at 585.40 nm is due to one-photon resonant process in Ba^+ ion (5d $^2D_{3/2} \rightarrow$ 6P $^2P_{3/2}$). The population of Ba^+ in 5d $^2D_{3/2}$ is due to three-photon nonresonant ionization. The observed experimental slope of 2.8 (see Fig. 15b) is in agreement with the theory of nonresonant ionization.

For R-photon resonant, N-photon ionization process, the rate of ionization is proportional to $(2R - N)^{th}$ power of the laser intensity if one stays on top of the resonance [23]. But this consideration is valid only under certain conditions; namely, (i) ionization is not saturated, (ii) the resonant state is more strongly coupled to the continuum than to the initial state and (iii) the transition from the resonant state to the continuum is saturated [24-26]. It is well established that in the presence of a strong laser field, the energy level shifts because of a.c. Stark effect. The shift is linear in intensity and is given by the relation $\delta = \alpha I$, where α is the a.c. Stark shift coefficient (see Sec. 2). The spatial inhomogeneities of the laser intensity in the ionization region results in spreading of the Stark shift thereby causing the broadening of the resonant level. In such a case the order of nonlinearity can take any value, varying from $2R - N$ to N [23, 25] including a fractional one, depending on the ratio between the a.c. Stark shift coefficient and the laser intensity broadening of the resonant state [24]. This seems to be the case for our observation in the two-photon resonant four-photon ionization $(2\omega + 2\omega)$ at the laser wavelength of 579.66 nm. Fig. 15c shows the slope of $K_{exp} = 0.6$ (whereas for this resonance $2R - N = 0$). Haugen and Othonos [27] have also reported the slope ~ 0.8 in the two-photon resonant four-photon ionization of Sr.

A minimum of three photons are required to each autoionizing state and four photons are required to populated the 6p states of Ba^+ in the laser wavelength range studied presently. As discussed earlier we did not observe the fluorescence from any ionic states higher than 6p. These facts suggest the autoionizing resonances we are observing at wavelengths 573.74 and 575.75 nm may be at the 3hv or 4hv position. The autoionizing state lying at higher photon absorption position would possibly have given the ionic fluorescence from higher ionic states (e.g. 7s state). The intensity dependence of I_F on E_L studied at the top of the resonances at 573.74 and 575.75 nm shows the slopes K_{exp} 3.53 and 2.30 respectively (see Fig. 16d and e). One observes for resonance at 575.75 nm (Fig. 15e) an early saturation at ~ 3 mJ/pulse compared to 6-7 mJ/pulse for the resonance at 573.74 nm (Fig. 15d). A comparison of the MPI spectrum obtained by collecting ions (see Fig. 12) with the spectrum obtained by monitoring the fluorescence intensity of D_1 and D_2 lines of Ba^+ (see Fig. 13) shows that the intensity of resonance at 573.74 nm in less than that of 575.75 nm in the spectrum obtained by collecting ions, whereas the intensity of 573.74 nm is more than that of 575.75 nm in the fluorescence spectrum. The autoionizing

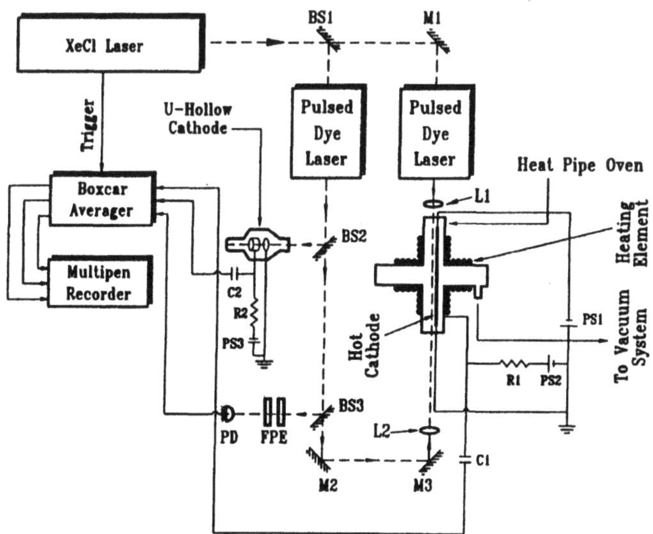

Figure 16. Schematic diagram of experimental set-up. BS1, BS2, BS3: Laser beam splitter; M1, M2, M3: Laser mirrors: L1, L2: Plano-convex lens (F = 30 cm); FPI: Fabry-Perot interferometer; PD: Photodiode; PSI: Regulated dc power supply for heating cathode; PS2: Regulated dc power supply for biasing voltage; PS3: Regulated dc high voltage power supply; R1, R2: Load resistor; C1, C2: Capacitor.

states lying in the 3hv position produce more ions than that of 4hv as a minimum number of three photons are required to ionize the Ba atom. On the other hand the same four-photon autoionizing resonance will give more intense D_1 and D_2 line fluorescence than that of three-photon autoionizing resonance as it has to absorb one more photon after resonance. These observations suggest that the resonances at 575.75 and 573.74 nm may be lying at 3hv and 4hv position respectively. Thus we can tentatively say that the position of these newly observed autoionizing resonances at 573.74 nm (17424.665 cm^{-1}) and 575.75 nm (17363.838 cm^{-1}) is at $4 \times 17424.665 = 69698.660$ cm^{-1} and $3 \times 17363.834 = 52091.502$ cm^{-1} respectively.

4.3   Summary

We have observed multiphoton ionization of barium atom at laser intensity of 10^{10} watts cm^{-2} in the range 570-608 nm. The technique presented to study multiphoton ionization process is rather simple and complementary to the ion and the electron spectroscopy studies of MPI. Atomic target vapor of moderate density allowed us to study the intrinsic strong-field process without significance interference from the collisional effects. The light-emission studies offer resolution of the J = 3/2 and 1/2 sublevels of the 6p state of Ba$^+$ which was not achieved in electron spectroscopic studies. We found that the J-sublevel population distribution is non-statistical and depends on the chosen path of excitation. The optical measurements are sensitive to rather weak channels, but no emission was observed from states higher than 6p ^{2}P$_{1/2, 3/2}$ of Ba$^+$ ion. This indicates that the multiphoton ionization occurs predominantly after three- or four-photon absorption. Without charge-selective ion detection we could tentatively identify the resonances in Ba$^+$ ion which may lead to the production of Ba^{2+}. Qualitative evidence is thus obtained for the two-step process in the double ionization of the Ba atom in the present intensity and laser wavelength range. We have also observed two new autoionizing levels at 69698.66 cm^{-1} and 52091.502 cm^{-1} [28].

5.   RESONANCE IONIZATION SPECTROSCOPY OF EUROPIUM ATOM

Resonance ionization spectroscopy (RIS) has proved to be a sensitive and most suitable experimental technique for studying high lying bound as well as Rydberg and autoionizing states of an atom. This technique, when using two-photon absorption, offers the advantage of studying the same parity states as that of ground state which are otherwise not accessible by conventional absorption spectroscopy technique. Moreover by using different stepwise excitation schemes, the analysis of the complex spectrum becomes conclusive. Information on some of the properties of the high lying atomic states, like the energy value, total angular momentum quantum number J, parity and excitation and ionization cross-sections are of interest not only from atomic structure theory but are also important practical for applications.

The neutral europium atom (Z = 63) has ground state electronic configuration $4f^7\,6s^2$, $^8S^0_{7/2}$. With half-filled f-shell the spectrum of neutral atom (Eu I) is relatively simpler compared to the complex spectra of other rare earths. But with the excitation of the f-electrons, the spectrum of Eu I becomes rather complicated. Information about high lying odd-parity bound states above 40000 cm^{-1} is very scanty and hardly 25 energy levels are known between 40000 and 45735 cm^{-1} (the ionization limit of Eu I) [29] although many levels are theoretically predicted in this region [30]. Very few laser spectroscopic studies have been undertaken for investigating the energy levels of Eu I. Laser resonance photoionization spectroscopy of excited and autoionizing states in the energy region 30600 - 36050 cm^{-1} and 45740 - 53270 cm^{-1} respectively has been reported by Zyuzikov et al. [31].

We have carried out investigations of high lying odd-parity bound energy levels of europium atom (Eu I) employing the technique of RIS using both single-color and stepwise two-color laser ionization schemes. New odd-parity energy levels of Eu I have been discovered in the region 40575 - 43410 cm^{-1}. The J values of these new odd levels have been uniquely assigned in most of the cases.

5.1   Experimental Technique

The experimental set-up is shown schematically in Fig. 16. The laser beam interacts with europium vapor in a heat-pipe oven and the photoions produced were detected by a thermionic diode. Various components of the experimental set up are described below in some detail.

We have used two-pulsed dye lasers (Lambda Physik, FL 3001 and FL 3002) pumped by an XeCl excimer laser (Lambda Physik, EMG 201 MSC) with the repetition rate around 10 Hz. The dye laser has a typical line-width of ~ 0.2 cm^{-1} and the pulse duration of 25 ns (FWHM). Coumarin-102 dye was used in the dye lasers to have the tunability in the wavelength range 459 to 522.5 nm. In two–color experiments the laser beams were counter propagating, parallel to the hot cathode of the thermionic diode, and were aligned to have maximum spatial overlap. A temporal delay of 10 ns was introduced between the two lasers.

The vapor of europium was generated in the crossed heat-pipe which was heated resistively by a regulated d.c. power supply. The temperature of the oven was monitored by a Chromel-Alumel thermocouple inserted inside the heat-pipe oven. The typical operational temperature of heat-pipe oven was around 640 °C and the

Eu vapor pressure is expected to be about 0.3 mtorr at this temperature. Argon at 5 torr pressure was used as a buffer gas.

The photoions produced by the laser atom interaction were detected by a thermionic diode having resistively heated hot cathode, design and construction of which has been already discussed in detail in section 3.1.2.

The calibration of the observed photoion peaks was carried out using optogalvanic (OG) spectra of uranium in a hollow cathode discharge (HCD) and also Fabry-Perot interferometer (FPI) fringes as frequency markers. A beam-splitter (BS2) was used to send a portion of the light from scanning laser into the open ended uranium/neon HCD lamp and another beam-splitter (BS3) to send the laser light through Fabry-Perot interferometer (see Fig. 16). Wavelength identification of OG spectrum was done using the atlas of U lines [32]. The free spectral range ($\Delta\sigma = 1/2t$) of the FPI was 1.366 cm^{-1} and the FPI fringes were recorded using a photodiode.

5.2   Results and Discussion

In the present investigation two sets of experiments have been carried out:
(a) Single-color resonance ionization spectroscopy in the dye tuning range $459 - 522.5$ nm corresponding to photon having energy in $19130 - 21780$ cm^{-1} range and (b) Stepwise two-color resonance ionization spectroscopy covering the energy region $40575-43410$ cm^{-1} (Fig. 17 b).

There are around 30 odd-parity energy levels of Eu I listed in the region $38260 - 43560$ cm^{-1}[29], so one expects as many number of two-photon resonances in the wavelength range $459 - 522.5$ nm ($19130 - 21780$ cm^{-1}) of the dye laser. However, we have observed many more resonances in Photoionization spectra in addition to the expected ones. In order to confirm the existence of the new odd-parity levels giving rise to so many additional resonances in single-color experiments and also to determine the J vales of these new levels, we have carried out two-color stepwise photoionization studies.

According to the selection rules of electric dipole transition, the probable J value of the energy levels accessible from the ground state ($J = 7/2$), by two-photon absorption via a virtual state, can be anything from $3/2$ to $11/2$. In the present experiment we have employed three excitation schemes which are given below. In all these schemes, the first step originates from the ground state of Eu I ($4f^7 6s^2$, $^8S^0_{7/2}$) and the laser excitation with three fixed frequencies populates the atoms in three levels (21761.26 cm^{-1}, $^8P_{9/2}$) and (21605.17 cm^{-1}, $^8P_{7/2}$) and (21444.58 cm^{-1}, $^8P_{5/2}$) of $4f^7$ 6s 6p configuration.

First step requires laser energy of $1 - 2$ mJ to saturate the transition. The second step is carried out by the tunable laser which covers the different energy regions of the odd-parity levels as shown on the extreme right of each of the schemes given below.

Scheme 1

$4f^7 6s^2\ {}^8S_{7/2} \xrightarrow{\ \lambda_1 = 459.4\ \text{nm}\ } 4f^7 6s6p\ {}^8P_{9/2} \xrightarrow{\ \lambda_2 = 459 - 522.5\ \text{nm}\ } E = 40890 - 43540\ \text{cm}^{-1}$

$J = 7/2,\ 9/2,\ 11/2$

Scheme 2

$4f^7 6s^2\ {}^8S_{7/2} \xrightarrow{\ \lambda_1 = 462.7\ \text{nm}\ } 4f^7 6s6p\ {}^8P_{7/2} \xrightarrow{\ \lambda_2 = 459 - 522.5\ \text{nm}\ } E = 40735 - 43385\ \text{cm}^{-1}$

$J = 5/2,\ 7/2,\ 9/2$

Scheme 3

$4f^7 6s^2\ {}^8S_{7/2} \xrightarrow{\ \lambda_1 = 466.2\ \text{nm}\ } 4f^7 6s6p\ {}^8P_{5/2} \xrightarrow{\ \lambda_2 = 459 - 522.5\ \text{nm}\ } E = 40575 - 43225\ \text{cm}^{-1}$

$J = 3/2,\ 5/2,\ 7/2$

With electric dipole selection rule of $\Delta J = 0,\ \pm 1$, the probable J values of the final odd levels are also indicated. These excitation schemes are also indicated. These excitation schemes are also depicted pictorially in Fig. 17B. A portion of the observed two-color photoionization spectrum obtained using excitation scheme 1 is shown in Fig. 18.

5.2.1 New Odd Parity Energy Levels of Eu I

The odd parity energy levels covered by the three excitation schemes shown above, lie in the region 40575 - 43410 cm^{-1}. A careful analysis was carried out of all the three sets of RIS spectra obtained by these three excitation schemes, and also the single-color photoionization spectrum. These analyses have resulted in the confirmation of the new energy levels of Eu I. With the application of three excitation schemes involving the intermediate levels with $J = 9/2$, 7/2 and 5/2 it was possible to assign unique J values for most of these energy levels. The highly excited odd-parity levels observed presently are given in Table 2. The energy value of the levels (in unit of cm^{-1}) are given in Column 2; the given value being the average of at least three measurements and with accuracy better than $\pm\ 0.4$ cm^{-1} in most of the cases. The next three columns give the positions of resonances observed in the respective stepwise excitation scheme leading to the identification of the particular energy levels. Our assignment of J is given in column 6. The earlier known energy levels in this region and their J values are given in column 7 and 8 respectively and are taken from [29].

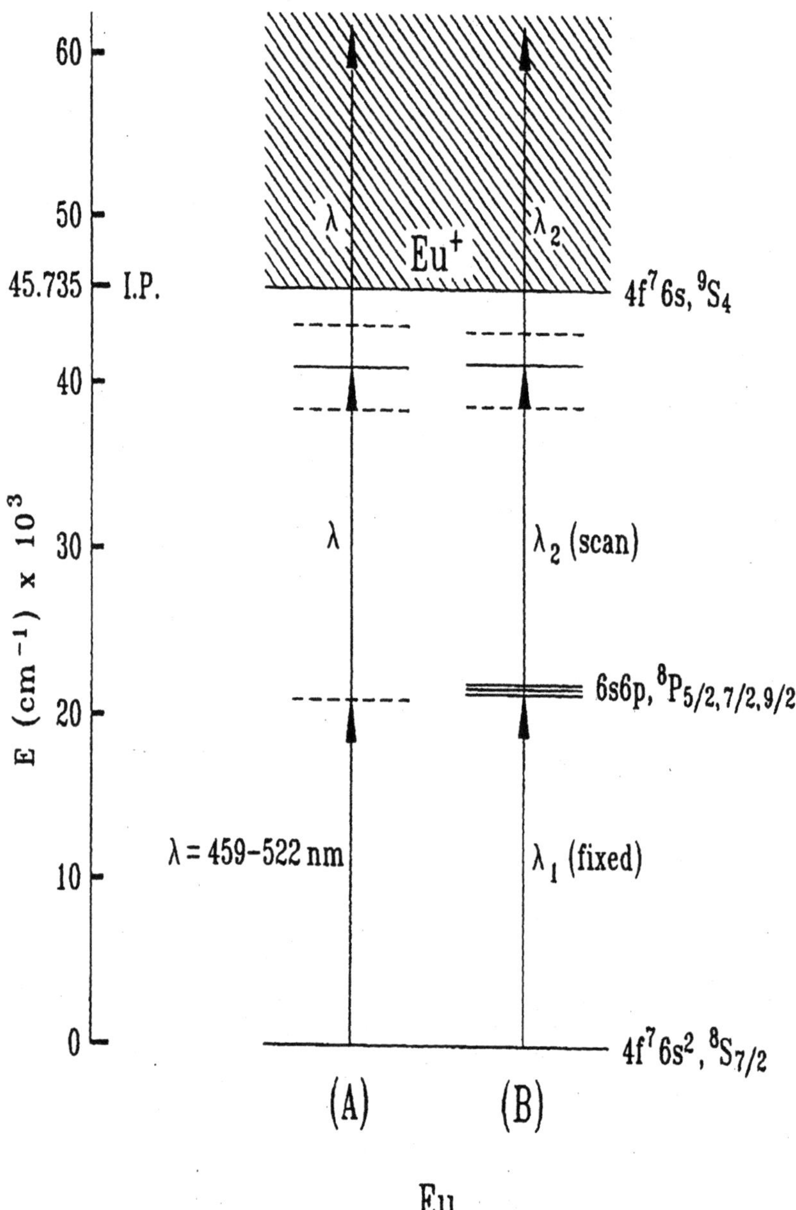

Figure 17. Energy diagram of Eu I indicating the excitation paths leading to odd-parity high-lying energy levels: (A) single-color excitation scheme, (B) two-color stepwise excitation scheme.

Table 2. Odd-parity energy levels of Eu I observed in the region 40575 – 43410 cm⁻¹. Under the excitation schemes (ES 1, 2, 3) we have indicated 'BR' and 'ND' 'BR' indicates that, expected resonances lying beyond the tuning range of the dye laser whereas, 'ND' indicates that, photoion peaks corresponding to this transition was not detected in this particular excitation scheme employed.

Present work					Earlier work[b]	
	Excitation schemes (ES)[a]					
E (cm⁻¹)	ES1 E1 = 21761.26 cm⁻¹ J = 9/2	ES2 E1 = 21605.17 cm⁻¹ J = 7/2	ES3 E1 = 21444.58 cm⁻¹ J = 5/2	J	E (cm⁻¹)	J
40576.2	BR	BR	19131.62	(7/2)	40576.00	7/2
40624.2	BR	BR	19179.62	5/2, 7/2	40624.10	5/2, 7/2
40650.6	BR	BR	ND	(see text)	40650.51	7/2
40759.6	BR	19154.43	19315.02	5/2, 7/2		
40764.2	BR	ND	ND	11/2	40764.08	9/2, 11/2
40768.3	BR	19163.13	ND	9/2	40768.31	9/2
40786.2	BR	19181.03	ND	9/2		
40814.8	BR	19209.63	19370.22	5/2, 7/2		
40854.1	19092.84	19248.93	ND	9/2	40854.11	9/2
40911.0	ND	19305.83	19466.42	5/2		
41174.8	19413.54	19569.63	ND	9/2	41174.74	9/2
41185.4	19424.14	19580.23	19740.82	7/2	41185.38	7/2
41201.1	ND	19595.93	19756.52	5/2	41201.21	5/2
41208.8	ND	ND	19764.22	3/2	41208.60	32
41272.0	19510.74	19666.83	19827.42	7/2		
41290.9	19529.64	19685.73	ND	9/2		
41332.1	ND	ND	19887.52	3/2		
41378.4	19617.14	ND	ND	11/2	41378.47	9/2, 11/2
41395.8	19634.54	ND	ND	11/2	41395.92	9/2, 11/2
41398.8	ND	ND	19954.22	3/2		
41443.9	19682.64	ND	ND	11/2	41443.0	11/2
41457.9	ND	19852.73	20013.32	5/2		
41516.2	19754.94	19911.03	20071.62	7/2		
41553.8	19792.54	19948.63	20109.22	7/2		
41673.0	ND	20067.83	20228.42	5/2		
41681.0	ND	ND	20236.42	3/2		
41685.0	19923.74	ND	ND	11/2		
41687.1	ND	20081.93	20242.52	5/2		
41722.1	ND	20116.93	20277.52	5/2		

	Present work				Earlier work[b]	
	Excitation schemes (ES)[a]					
E (cm⁻¹)	ES1 E1 = 21761.26 cm⁻¹ J = 9/2	ES2 E1 = 21605.17 cm⁻¹ J = 7/2	ES3 E1 = 21444.58 cm⁻¹ J = 5/2	J	E (cm⁻¹)	J
41728.2	19966.94	20123.03	20283.62	7/2		
41790.6	ND	ND	20346.02	3/2		
41845.2	20083.94	20240.03	ND	9/2		
42152.6	20391.34	ND	ND	11/2		
42244.2	20482.94	20639.03	20799.62	7/2		
42389.2	ND	20784.03	20944.62	5/2		
42401.6	ND	ND	20957.02	3/2		
42428.8	20667.54	20823.63	20984.22	7/2		
42453.1	ND	ND	21008.52	3/2		
42478.4	20717.14	20873.23	21033.82	7/2		
42525.1	ND	20919.93	21080.52	5/2		
42558.2	ND	20953.03	21113.62	5/2		
42575.3	20814.04	20970.13	21130.72	7/2		
42682.3	ND	21077.13	21237.72	5/2		
42696.7	ND	21091.53	21252.12	5/2		
42721.4	20960.14	21116.23	21276.82	7/2	42721.09[7]	9/2, 11/2
42743.3	20982.04	21138.13	21298.72	7/2		
42768.8	ND	ND	21324.22	3/2		
42790.6	ND	ND	21346.02	3/2		
42794.4	21033.14	21889.23	21349.82	7/2		
42848.3	ND	21243.13	21403.72	5/2		
42860.4	21099.14	21255.23	21415.82	7/2		
42881.8	21120.54	21276.63	21437.22	7/2		
42896.1	ND	ND	21415.82	3/2		
42937.0	ND	21331.83	21492.42	5/2		
42954.0	ND	ND	21509.42	3/2		
43029.9	21268.64	21424.73	21585.32	7/2		
43067.5	21306.24	21462.33	21622.92	7/2		
43100.6	21339.34	21495.43	ND	9/2	43100.66	7/2, 9/2
43165.3	ND	ND	21720.72	3/2		
43183.8	2122.54	21578.63	21739.22	7/2		
43233.3	21472.04	21628.13	BR	9/2, 7/2		
43238.2	21476.94	ND	BR	11/2		
43270.5	21509.24	ND	BR	11/2		
43281.5	21520.24	21676.33	BR	9/2, 7/2		

Present work				Earlier work[b]	
	Excitation schemes (ES)[a]				
E (cm^{-1})	ES1 E1 = 21761.26 cm^{-1} J = 9/2	ES2 E1 = 21605.17 cm^{-1} J = 7/2	ES3 E1 = 21444.58 cm^{-1} J = 5/2	J	E (cm^{-1})	J
43333.4	21572.14	21728.23	BR	9/2, 7/2		
43409.8	21648.54	BR	BR	11/2, 9/2, 7/2		

[a] The observed resonances, with E_1 as the first step, are given in column 3, 4 and 5 in units of cm^{-1}

[b] The values the energy level (E) and total angular momentum (J) are from [29]. The level marked with $^?$ is a doubtful one.

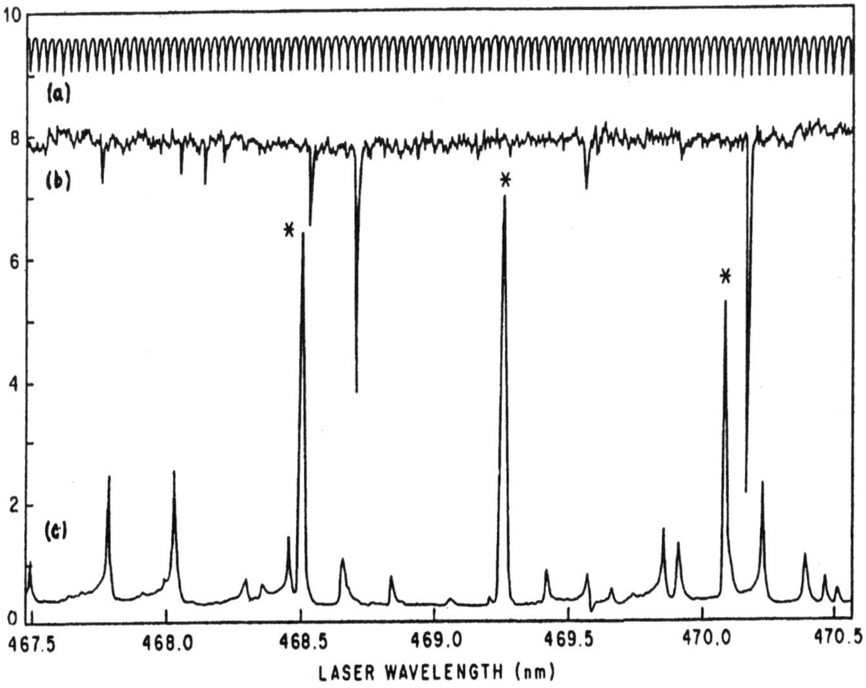

Figure 18. A portion of the observed photoionization spectrum (c) in two-color stepwise excitation of Europium atoms. For the first-step, the laser is tuned to the transition $4f^7\,6s^2,\ ^8S_{7/2} \xrightarrow{\ \lambda_1=459.4\,\text{nm}\ } 4f^7\,6s\,6p,\ ^8P_{9/2} \xrightarrow{\ \lambda_2=459-522\,\text{nm}\ } E = 40890 - 43540$ cm^{-1}; J = 7/2, 9/2, 11/2 (Scheme 1). Two-color photoionization peaks are marked with '*'. The optogalvanic lines of Uranium (b) and Fabry-Perot interferometer frequency markers 9a) are also shown.

The known odd configurations of the excited states of Eu I are $4f^7$ 6s nd (n = 5, 6, 7) $4f^7$ 6s ns (n = 7, 8), 4 f^7 $5d^2$ and $4f^7$ $6p^2$ [29]. For the second-step excitation from even parity levels (y $^8P_{5/2}$, y $^8P_{7/2}$ and y $^8P_{9/2}$) of $4f^7$ 6s 6p (see Fig. 18B), transitions to the levels of all these odd configuration are possible. Out of 19 known levels in the 40575 – 43410 cm^{-1} region listed by Martin et al. [29], there are only four levels at 41174.74, 41185.38, 41201.21 and 41208.60 cm^{-1} having respectively J values of 9/2, 7/2, 5/2 and 3/2 for which electronic configuration is known and they belong to f^6D^0 state of $4f^7(^8S^0)$ $6s(^7S^0)7d$ odd configuration. Wyart [33] has added the level at 41232.80 cm^{-1} with assigned value of J = 1/2. We have also observed these 4 levels with J = 9/2, 7/2, 5/2 and 3/2 (Table 2) and the observed photoionization resonances are quite strong. For rest of the levels, the electronic configuration is not known. One may get tempted to suggest probable electronic configurations of the observed levels on the basis of relative intensities of the photoionization resonances, but it is rather unsafe especially when it happens to be the complex energy levels of a rare-earth.

Wyart [30] has carried out calculation of $4f^7$ ($^8S^0$ $^6P^0$) 5d 6s and fitted the known levels, which belong nominally to the four terms a^{10} D^0, a^8 D^0, b^8 D^0 and a^6 D^0 of ($^8S^0$) 5d 6s (see also [33]). In his thesis [30] he has listed 12 calculated odd-parity energy levels of this configuration between 43000 and 44900 cm^{-1}. We could not find agreement with any of our values of energy levels except for 43333.4 cm^{-1} (J = 7/2, 9/2) which agrees well with calculated level E_{th}=43320 cm^{-1} (J = 7/2).

5.2.2 Comments on the Earlier Reported Energy Levels and J-value Assignment

The earlier studies ([29] and references there in) have provided 19 odd-parity energy levels, with only 9 levels having definite J assignments, in the region 40575 - 43410 cm^{-1} covered in our present investigations. The present two-step laser experiments have provided definite J values (Table 2) for four levels at 40764, 41378, 41395 and 43100 cm^{-1} which were earlier assigned two possible J values [29]. There are three doubtful levels at energies 41037.58, 42721.09 and 43212.06 cm^{-1}listed in [29] and out of these three levels we have observed only one level at 42721.4 $cm^{-1}$and the other two levels could not be observed. However, the J value for the level at 42721.4 cm^{-1} is found to be 7/2 instead of 9/2, 11/2 as listed by Martin et al. [29].

Corresponding to the excitation to the level at 41631.37 cm^{-1}, a weak photoionization resonance was observed in single-color experiment at ν = 20815.7 cm^{-1} (2ν = 41631.4 cm^{-1}); we also observed a strong resonance in two-color experiment at 20026.0 cm^{-1} from a level at 21605.17 cm^{-1}

(J = 7/2) (scheme 2, see section 5.2). But somehow we did not observe any corresponding resonance when the excitation was made from 21761.26 cm^{-1} (J = 9/2) and 21444.58 cm^{-1} (J = 5/2) level; so the level at 41631.4 cm^{-1} is not included in Table 2.

The known level at 40650.51 cm^{-1} has J = 7/2 and in our single-color experiment we observed a resonance at ν = 20325.3 cm^{-1} (2ν = 40650.6 cm^{-1}) which could be a 2-photon excitation from ground state to this odd level; the observed intensity also supporting the 2-photon path. But we did not observe any photoions peak when excitation was made using the level at 21444.58 cm^{-1}, J = 5/2 (scheme 3, see section 5.2). The absence of the photoion peak does not confirm the earlier assignment of J = 7/2 for the level at 40650.51 cm^{-1}. Scheme 1 and 2, which would confirm J assignment, could not be used for reaching this level as the laser wavelength required was out of the tuning range of the dye used presently. In the view of this, for the time being we have not made any assignment of J value for this level (Table 2).

5.2.3 Rydberg Series and Autoionizing states of Europium Atom

Recently we have undertaken studies to explore the high-lying bound states, Rydberg states and the autoionizing states of Eu I in the region 43510 - 47270 cm^{-1}; the first ionization potential (IP) of Eu I being at 45735 cm^{-1}. The experimental set up is similar to the one described in detail elsewhere [34]. Coumarin-120 and Slilbene-3 dyes were used to get tunability in the region 408 - 461 nm (corresponding to the photon energy in 21705 - 23635 cm^{-1} region). We have observed more than 500 resonances in the photoionization spectrum. Most of these are expected to be two–photon resonance to odd parity high-lying bound, Rydberg and autoionizing states of Eu I in the region 43410 – 47270 cm^{-1}. The observed Rydberg series below the first ionization potential is shown in Fig. 19. The photoionization spectrum beyond 45735 cm^{-1} (the first IP of Eu I) shows broad as well as sharp autoionizing resonances. The autoionizing resonances can be identified with their asymmetric Fano profile. Rydberg series converging to the ground state 4f^7 (^{8}S^0) 6s, ^{7}S$_3$ (at 1660.21 cm^{-1} above first IP of Eu I) of Eu$^+$ have been also observed.

In order to confirm the existence of many new odd parity states of Eu I and also uniquely assign total momentum J to these new energy levels we have carried out stepwise two-color experiments using all the three excitation schemes mentioned in the Sec. 5.2. We have also identified a number of Rydberg series below the first ionization limit. The important results obtained from the analysis of our exhaustive data will be presented in detail somewhere else.

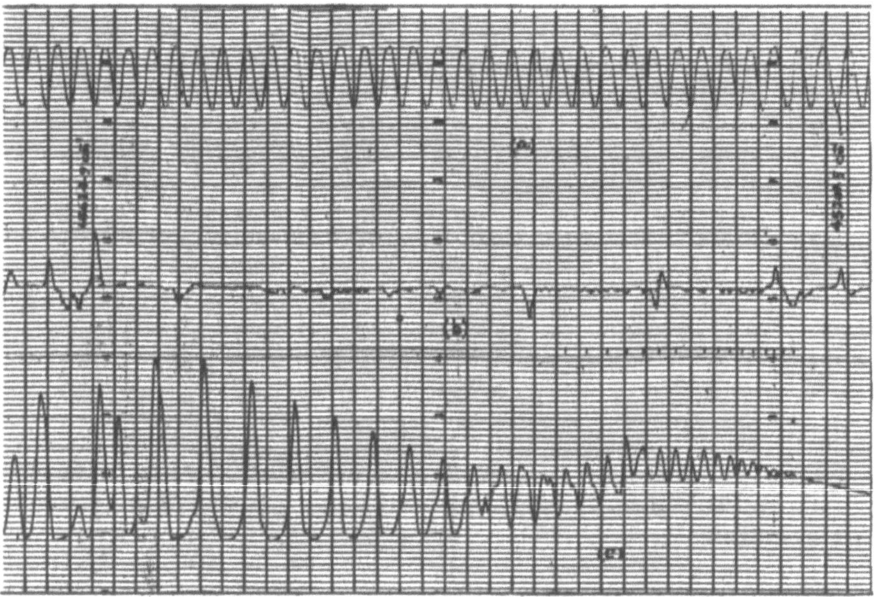

Figure 19. Rydberg series observed in the single-color two photon ionization spectrum of Eu I in the region 45625-45710 cm^{-1}. The first ionization limit of Eu I is at 45735 cm^{-1}. The Fabry-Perot interferometer frequency markers (a) and optogalvanic lines of Uranium (b) are also shown.

5.3   Summary

The present investigations were carried out on europium atoms in a heat-pipe oven thermionic diode system using the technique of resonance ionization spectroscopy with pulsed tunable dye lasers. Single-color and stepwise two-color experiments with three different excitation schemes provided a rich data. Careful analyses of the photoionization spectra has resulted in identification of 52 new odd-parity energy levels of Eu I in the region 40575 - 43410 cm^{-1} with definite J value assignment to 46 of these levels [34]. In addition, out of 19 levels reported by earlier workers in the region of our investigation, 14 levels have been confirmed and also unique J values have been assigned to 4 levels which were having two possible assigned values of J.

The additional data provided on the energy levels of Eu I in the region 40575 - 43410 cm^{-1} may be helpful in improving the accuracy of parametric calculations of the high lying odd-parity energy levels of Eu I as most of the known levels in this region are presently without any configuration assignments.

The recent RIS studies on Eu atom carried out by us to explore the energy levels of Eu atom (Eu I) in the region 43410 - 45735 cm^{-1} have resulted in identification of more than 80 high-lying odd parity energy levels of Eu I with unique assignment of J quantum number for most of these levels. Rydberg series below the first ionization potential of Eu I (45735 cm^{-1}) have been observed and analyzed. Autoionizing states and Rydberg series have been also observed above 45735 cm^{-1} providing new information in the energy levels of Eu I in 45735 - 47270 cm^{-1} region.

ACKNOWLEDGMENTS

The authors thank Dr. A. P. Roy, Head Spectroscopy Division for his interest in this work. Thanks are also due to Dr. M. A. N. Rizvi for his generous help in some of the experimental studies reported here.

REFERENCES

1.　(a) J. H. Hall, E. J. Robinson and L. M. Branscomb, Phys. Rev. Lett. **14**, 1013 (1965).

　　(b) S. L. Chin and P. Lambopoulos (eds.), *Multiphoton Ionization of Atoms* (New York: Academic) (1984).

　　(c) G. Mainfray and G. Manus, Rev. Prog. Phys. **54**, 1333 (1991).

　　(d) M. Gavrila (ed.), *Atoms in Intense Radiation Fields, Adv. At. Mol. Opt. Phys.* (1991).

　　(e) F. H. M. Faisal, *Theory of Multiphoton Processes,* Plenum (1987).

　　(f) K. Brunett, V. C. Reed and P. L. Knight, J. Phys. B: At. Mol. Opt. Phys. **26**, 561 (1993).

2.　P. Agostini, F. Fabre, G. Mainfray, G. Petite and N. Rahman, Phys. Rev. Lett. **42**, 1127 (1979).

3.　A. L'Huillier, L. A. Lompre', G. Mainfray and C. Manus, Phys. Rev. Lett. **48**, 1814 (1982).

4.　S. Suckewere and C. H. Skinner, Comments At. Mol. Phys. **30**, 331 (1995).

5.　(a) S. G. Nakhate, Ph.D. Thesis, University of Bombay (1998).

　　(b) S. G. Nakhate, S. A. Ahmad, A. Venugopalan, P. M. Rao and G. D. Saksena, Proc. *National Symposium on At. Mol. Phys., Bombay, India,* p.166 (1992).

6.　G. M. Grover, T.P. Cotter and G. F. Erickson, J. Appl. Phys. **35**, 1990 (1964).

7.　C. R. Vidal and J. Cooper, J. Appl. Phys. **40**, 3370 (1969).

8. K. H. Kingdon, Phys. Rev. **21**, 408 (1923).

9. G. Hertz, Z. Phys. **18**, 307 (1923).

10. P. D. Foote and F. L. Mohler, Phys. Rev. **26**, 195 (1925).

11. K. Niemax, Appl. Phys. **B38**, 147 (1985).

12. R. Beigang, W. Makat and A. Timmermann, Optics Commn. **49**, 253 (1984).

13. I. S. Aleksakhin, I. P. Zapesochny, JETP Lett. **26**, 11 (1977).

14. I. S. Aleksakhin, N. B. Delone, I. P. Zapesochny and V. V. Suran, Sov. Phys. JETP **49**, 447 (1979).

15. N. B. Delone V. V. Suran and B. A. Zon, *Multiphoton Ionization of Atoms*, ed S. L. Chin and P. Lamropoulos (Orlando, FL: **Academic**), p235 (1984).

16. P. Agostini and G. Petite, J. Phys. B: At. Mol. Phys. **17**, L811 (1984).

17. J. L. Dexter, S. M. Jaffe and T. F. Gallagher, J. Phys. B: At. Mol. Phys. **18**, L735 (1985).

18. I. I. Bondar, M. I. Dudich and V. V. Suran, Sov. Phys.-JETP **63**, 1142 (1986).

19. I. I. Bondar, N. B. Delone, M. I. Dudich and V. V. Suran, J. Phys. B: At. Mol. Opt. Phys. **21**, 2763 (1988).

20. Y. Zhu, R. R. Jones, W. Sandner, T. F. Gallagher, P. Camus, P. Pillet and J. Boulmer, J. Phys. B: At. Mol. Opt. Phys. **22**, 585 (1989).

21. S. G. Nakhate, M. A. N. Razvi and S. A. Ahmad, Ind. J. Phys. **66B**, 437 (1992).

22. C. E. Moore, *Atomic Energy Levels* **vol 2** NBS Circular No. 467 (Washington DC: US Govt. Printing Office), p132 (1958).

23. P. Camus, M. Kompitsas, S. Cohen, C. Nicolaides, M. Aymar, M. Crance and P. Pillet, J. Phys. B: At. Mol. Opt. Phys. **22**, 445 (1989).

24. G. Petite, J. Morellec and D. Normand, J. Physique **40**, 115 (1979).

25. M. Crance, *Multiphoton Ionization of Atoms* ed S. L. Chin and P. Lamropoulos (Orlando, FL: Academic), p65 (1984).

26. Y. Gontier and M. Trahin, *Multiphoton Ionization of Atoms* ed S. L. Chin and P. Lamropoulos (Orlando, FL: Academic), p49 (1984).

27. H. K. Haughen and A. S. Othonos, Phys. Rev. **A39**, 3392 (1989).

28. S. G. Nakhate, S. A. Ahmad, M. A. N. Razvi and G. D. Saksena, J. Phys. B: At. Mol. Opt. Phys. **24**, 4973 (1991).

29. W. C. Martin, R. Zalubas and L. Hagan, *Atomic Energy Levels of the Rare Earth elements* NSRDS-NBS 60 (Washington DC, US Government Printing Office) p 185 (1978).

30. J. F. Wyart, Ph.D. Thesis, Univ. Paris-Sud, Orsay, 194 pp. (1973).

31. A. D. Zyuzikov, V. S. Letokhov, V. I. Mishin and V. N. Fedoseev, Opt Spectrosc. (USSR) **63**, 572 (1987).

32. B. A. Palmer, R. A. Keller and R. Engleman Jr., *Los Alamos Rep. No. LA-8251-MS* (1980).
33. J. F. Wyart, Phys. Scr. **32**, 58 (1985).
34. S. G. Nakhate, M. A. N. Razvi, G. L. Bhale and S. A. Ahmad, J. Phys. B: At. Mol. Opt. Phys. **29**, 1439 (1996).

Radiative Lifetime Measurements and Study of Perturbed Electronic States of NO_2

K. P. Subramanian, V. Sivakumaran and Vijay Kumar

Physical Research Laboratory
Navrangpura, Ahmedabad - 380009, India

Mixing of ro-vibrational levels in the excited states of certain molecules leads to elongation of average lifetime of excited states. The relaxation of pure adiabaticity condition of Born-Oppenheimer states allows mixing of levels in the excited state. The extent of perturbation on the emitting level by the nearby lying electronic states is ro-vibrational and is quantum number specific. This introduces variability in the anomalous lifetime as a function of the excitation wavelength. For certain wavelengths, emissions from pure unperturbed states also are observed. The radiative lifetime measurement of NO_2 in the excitation region 465-490 nm is presented in this paper.

1.   INTRODUCTION

In molecules like SO_2 and NO_2, it has been found that the observed fluorescence lifetimes are much longer than those expected from the integrated absorption coefficient [1, 2]. In these molecules the rotational and vibration levels are generally perturbed exhibiting a complex absorption and fluorescence spectra in the uv and visible region. This discrepancy in lifetime of excited state, known as lifetime anomaly, was first explained by Douglas [3] on the basis of interelectronic level mixing. The electronic-vibrational (E-V) or electronic-vibrational-rotational (E-V-R) interaction between two levels allows a particular level in the excited level to interact with many levels of the highly lying ro-vibrational levels of the ground state. If the perturbing state has a high density of suitable vibrational levels and the excitation energy is large compared to the energy spacing between these interacting levels, then the lifetime of the perturbed level will be considerably influenced. The strength of interaction varies with the rotational and vibrational quantum numbers of the excited state, and therefore, the lifetimes of even closely spaced levels may differ. Depending up on the strength of interaction and level mixing, the total fluorescence may contain emissions from perturbed and unperturbed levels whose lifetimes are vastly different. This would make the decay curve of fluorescence a multi-exponential curve.

The perturbation of the electronic levels and the lifetime anomaly in NO_2 is being discussed in this paper. The ground state of NO_2 is $\tilde{X}\,^2A_1$, with bond angle 134.1°. The absorption spectrum of NO_2 in the near ultraviolet and visible is extremely complex and most of the regions has no apparent regularity in rotational and vibrational structure. NO_2 fluoresces when irradiated with light of wavelength above 397.9 nm (dissociation limit). The fluorescence spectrum lies in the region from the exciting wavelength to above 800 nm.

Atherton et al [4], based on polarization studies of NO_2 in the region 410-420 nm, reported $^2B_2 \leftarrow {}^2A_1$ transition and suggested the possibility of a lower energy $^2B_1 \leftarrow {}^2A_1$ transition to overlap with $^2B_2$ in this region. Steven et al [5], from the rotational analysis of the fluorescence spectra of NO_2 had found the evidence of two excited states $^2B_2$, $^2B_1$ in the region 593.4 to 594 nm. They concluded that the $^2B_2$ has lifetime of 30 ± 5 μ sec and the $^2B_1$ has a lifetime of $115 \pm 10\,\mu$ sec.

Sackett and Yardley [6] observed several unperturbed levels with short lifetime from $^2B_1$ state in the region 454.4-455 nm. Similar observation was done by Solarz and Levy [7] at 488 nm. Paech et al [8] measured collision free lifetime of NO_2 excited by a tunable laser near 488 and 514.5 nm. They observed three different lifetimes of fluorescence (3, 28 and 75 μ sec) and inferred that the initially found $^2B_1$ state crosses rapidly to another state $^2B_2$, with high level density. The $^2B_2$ state can have two different lifetimes (28 and 75 μ sec) depending on the extent of interaction with the ground state (vibronic coupling). The short component from $^2B_1$ was primarily determined by the rate of internal conversion from $^2B_1$ to $^2B_2$.

Mulliken [9], Pearse and Gaydon [10] and Walsh [11] carried out calculation on the energy levels of NO_2 and concluded that $^2B_1$ and $^2B_2$ are the excited states above the ground state $^2A_1$ and were responsible for the visible spectrum. They suggested that $^2B_2$ is the state primarily responsible for fluorescence in visible spectrum which is stronger than $^2B_1 \leftarrow {}^2A_1$ transitions. Gangi and Burnelle [12] did extensive configuration interaction calculation to compute the potential energy surface of NO_2 and placed $^2B_1 \leftarrow {}^2A_1$ and $^2B_2 \leftarrow {}^2A_1$ in the visible region and the forbidden $^2A_2$, $^4B_2$ and $^4A_2$ states somewhat higher in energy. The computed potential energy curves for different bond angles are shown in Figure 1.

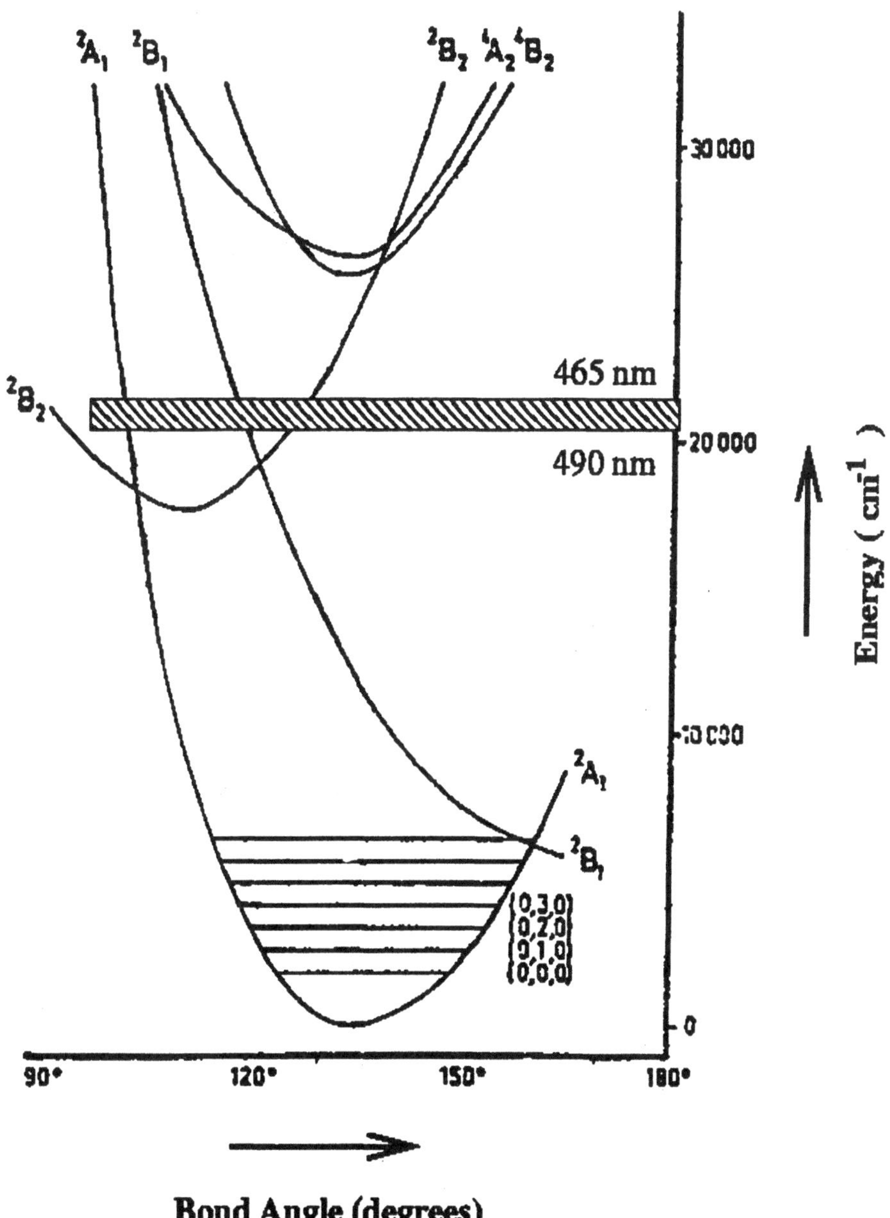

Bond Angle (degrees)

Figure 1. Potential energy diagram (bond angle Vs. energy) of NO_2 due to theoretical calculations by Gangi and Burnelle [12]. The shaded region on the energy axis shows the region of excitation studied in the present work.

Herzberg [13] pointed out that the vibronic levels of the $^2\Pi_4$ state of the linear NO_2 molecule undergoes a Renner-Teller splitting to form the $^2A_1$ ground state and $^2B_1$ excited state upon bending. All the wave functions contain contribution from both states and may be assumed to contain only small contribution from the $^2B_1$ electronic state. While an integration of the absorption intensity should be capable of measuring essentially the total oscillator strength of the $^2B_1 \leftarrow {}^2A_1$ transition, it will be distributed among the transitions between various vibronic level more evenly and will be smaller than expected in Born-Oppenheimer approximation leading to a low estimated lifetime than those observed experimentally.

We report here systematic measurement of fluorescent lifetime in NO_2 molecule at room temperature. The measurements are made in the excitation wavelength region 465-490 nm at an interval of 0.2 nm. The measurements are done at cell gas pressures 1, 3, 5, 15, 51, 81 and 110 mTorr. In all previous studies, such measurements have been carried out at a few selected excitation wavelengths. The present work would be particularly important to understand the dynamical nature of the emission bands with regard to the variation in the intrinsic component of lifetime.

2.   EXPERIMENTAL SETUP

The method of pulsed excitation was used to study the fluorescence lifetime in NO_2. The excitation of the NO_2 gas inside a stainless steel fluorescence chamber was achieved by an excimer laser pumped tunable dye laser (Lambda Physics LPX 110 and LPD 3000). Using appropriate collection optics, the fluorescence signal was fed to a fast photomultiplier and a digital storage oscilloscope (Tektronix TDS 540) having large sampling rate. The pulse width of the excitation source was about 8 nsec, which was considerably smaller than the lifetimes analyzed (typically, a few μ sec) and thus the effect of convolution of the source function on the decay curve was safely ignored. The excitation source was covering the spectral range from 465 to 490 nm with an instrumental resolution of 0.01 nm. Lower energies of the laser were used in the present experiment ranging from a few μ J to 10 mJ. The laser beam was 2 mm in diameter and the intensity distribution within the beam was nearly Gaussian. The laser beam was expanded using a 4-X beam expander placed before a 5% beam splitter. A photodiode monitored the laser light reflected at the beam splitter and its signal was used as a time-marker in time-resolved studies. The laser beam transmitted by the beam splitter was then allowed to interact with the target molecule in the fluorescence chamber. The digital fluorescence decay traces obtained during the experiment were transferred to PC and then to IBM RS6000 for off-line analysis.

It is known that NO_2 forms its dimer even at room temperature and it is found that the dimer concentration increases as the temperature is lowered.

While doing the experiment, it was made sure that the dimer concentration at room temperature did not affect the lifetime measurements. Amoruso et al [14] has given a method to estimate the partial pressure of N_2O_4 in NO_2 at different temperatures and pressures and this method was adopted to estimate the dimer concentration in the present experiment. It has been estimated that the maximum error introduced in the pressure measurement due to this source is about 0.3%.

3.   METHOD

Figure 2(a) shows a typical NO_2 fluorescence decay curve at an excitation photon wavelength of 474 nm obtained at a gas pressure of 5 mTorr. The decay curve has been obtained after averaging over a large number of laser shots. The decay curves are subjected to multi-exponential curve fitting procedure to obtain single or multiple lifetimes of the excited state.

For the decay of total fluorescence of a multi-level system, ignoring the convolution effects of the excitation function, the decay could be expressed as

$$I(t) = \sum_{i=1}^{N} a_i \exp(-t/\tau_i) \qquad (1)$$

where a_i and τ_i are the amplitudes and average fluorescence lifetimes of the N decaying levels. An exponential sum-fitting programme given by Wiscombe and Evans [15] using the least square technique was used to fit the experimental points on the fluorescence decay curve. No trial values were required for the sum-fitting programme. The fit based on the above programme is shown in Fig. 2(a) by a solid line. The residue of the fit is plotted in Fig. 2(b). The autocorrelation of the residue [16] after the fit was used to validate the results obtained from the numerical fitting. The residual autocorrelation curve in the time domain is shown in Fig. 2(c). Pure white noise of the autocorrelation of the residue as shown in Fig. 2(c) indicates that the assumed fit parameters were acceptable. The lifetimes estimated by this method for the typical decay curve shown in Fig. 2(a) are 27.4 and 70.3 μ sec.

Proper tests were performed for convergence of lifetime values with varying record length chosen for omission due to blending of scattered light. Further, for the exact determination of proper baseline, a 25 μ sec long data was stored before the start of the excitation pulse. The contribution for this

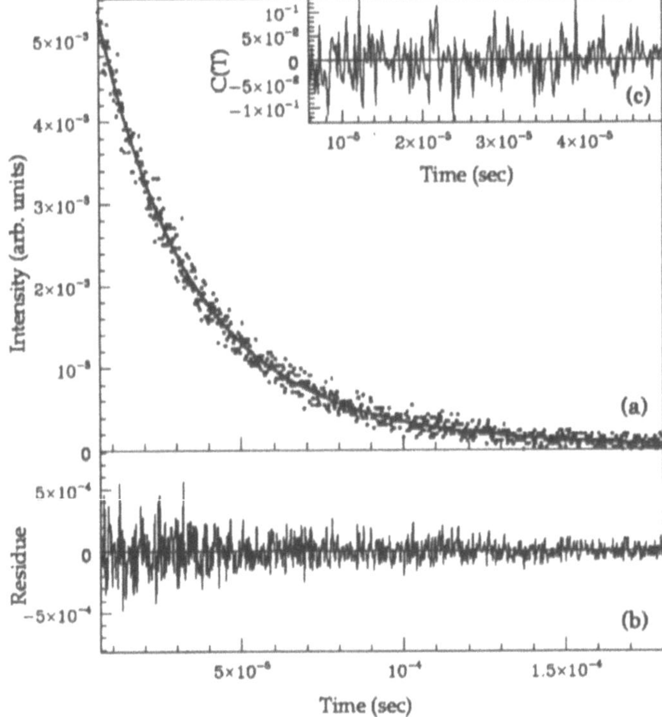

Figure 2. Fluorescence decay curve showing details of exponential fitting. (a) The observed and the bi-exponential fitting with $\tau_1 = 70.3\ \mu$ sec, $\tau_2 = 27.4\ \mu$ sec at excitation wavelength 474.0 nm. (b) The residual plot of the fitted data. (c) The autocorrelation plot of the residues.

baseline was from the dark current of the photomultiplier which was found to be constant in all runs.

The most probable error in the measurement of radiative lifetimes in the present experiment can be calculated from the two main contributing errors, i.e. the statistical error and the error in curve fitting. Since in the present work, the waveform of the fluorescence decay curve is sampled using a digital sampling oscilloscope, the statistical error would follow a Gaussian distribution. This error has been estimated to vary from ± 1.5 to ± 3% for different pressure regions, but the maximum error of ± 3% has been used to calculate the most probable error. The error in curve fitting has been found to be different for different lifetimes. For lifetimes varying from 1 to $3\ \mu$ sec, this error has been found to vary from ± 4 to ± 20% where as the error varies from ± 0.1 to ± 2.6% and ± 0.001 to ± 2% for lifetimes varying from 20 to 40 μ sec and 70 to 120 μ sec respectively. The maximum error of ± 20%, ± 2.6% and ± 2% has been taken for three lifetime groups to calculate the most probable error.

4.   RESULTS AND DISCUSSION

Excitation spectra were indirectly constructed by measuring the area under decay curves at different wavelengths. Such spectra were constructed for measurements at all cell gas pressures. These spectra were corrected for shot-to-shot intensity variations of the laser light and the quantum efficiency of the dye in use at different photon wavelengths. The excitation spectra recorded at pressures 1, 3, 5, 15, 51 and 81 mTorr are shown in Fig. 3. All these measurements have been carried out at room temperature and at large excitation wavelength intervals of 0.2 nm. Because of this, only low resolution excitation spectra with broad feature could be observed.

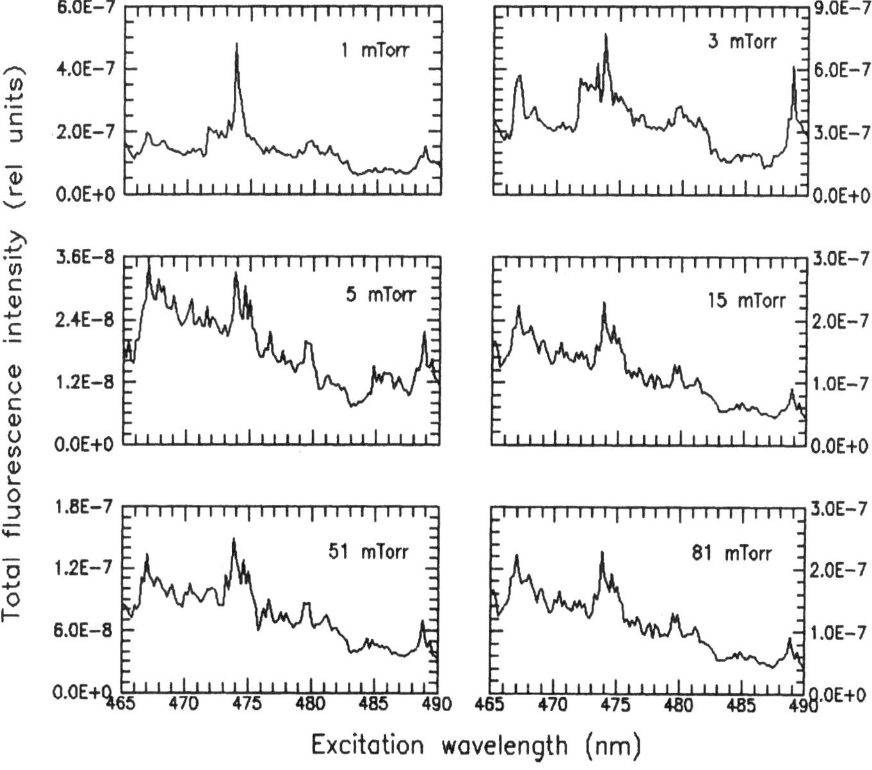

Figure 3. Fluorescence excitation spectra of NO_2 at different gas pressures.

The lifetime spectra of the static component at 1, 3, 5 and 15 mTorr pressures are shown in Fig. 4(a) where as the spectra at 51, 81, 110 mTorr pressures are shown in Fig. 4(b) as a function of excitation photon wavelengths ranging from 465 to 490 nm. The spectra constructed from the observed fluorescence decay curves at these pressures could be fitted only with a single exponential. The lifetime was found to vary between 26-40

μ sec in this spectral region expect at a few excitation wavelengths mentioned below. This component has been found to be present at all pressures; that is why it has been named as the static component. At excitation wavelengths between 486 and 488 nm and also in other small wavelength regions, the values of lifetime have been found to be larger than the highest value 40 μ sec of the static component obtained at other excitation wavelengths. The values of the static component obtained in the present work are more or less in agreement with those reported by Paech et al [8] at 488 and 514.5nm. Their high resolution measurements gave the values varying from 26-30 μ sec. When excited with a 0.1 μ sec laser beam at 488 nm, Paech et al [8] reported a value of 28 μ sec for the radiative lifetime. Varying the pulse width, they could obtain the other components of lifetime also even at very low pressures. However, this could not be verified in our experiment as it was not possible to vary the pulse width of excitation. The long lifetime components could, therefore, not be observed at 1 and 3 mTorr pressures.

The average time between collision decreases with increase in pressure and therefore, it will have an effect on the observed lifetime due to collisional quenching. The lifetime spectra (Fig. 4(a) and 4(b)) at different pressures give an indication that the quenching of the emitting states of NO_2 due to collisional effects are less important below 5 mTorr. This can be ascertained from the excitation spectrum at 5 mTorr pressure shown in Fig. 3. The excitation spectrum shows rapid changes in fluorescence intensity with excitation wavelength. This could be explained by the fact that lifetime components other than the static component are also present at 5 mTorr pressure. Again, the general shape of the excitation spectrum at 5 mTorr is more or less similar to those shown at higher pressures, viz. 15, 51, 81 and 110 mTorr. Thus, it could be safely concluded that even though the electronic state rearrangement are initiated through collisions resulting in redistribution of energy of the excited state, its effects on quenching of the states are pronounced at pressures larger than 5 mTorr. They are weak at pressures smaller than or equal to 5 mTorr. This is in agreement with the observations of Donnelly and Kaufman [17].

A larger component of lifetime in the range of 70-120 μ sec was obtained at three excitation wavelengths (466.8, 473.8 and 474 nm) at a pressure of 5 mTorr only. The strength of these components varied from 3.7 to 12.4% of the total in terms of amplitudes, a_i, defined in equation (1). A similar component of 80 μ sec was also found in the present experiment at 482.4 nm at 51 mTorr pressure. But this may not be reliable because the

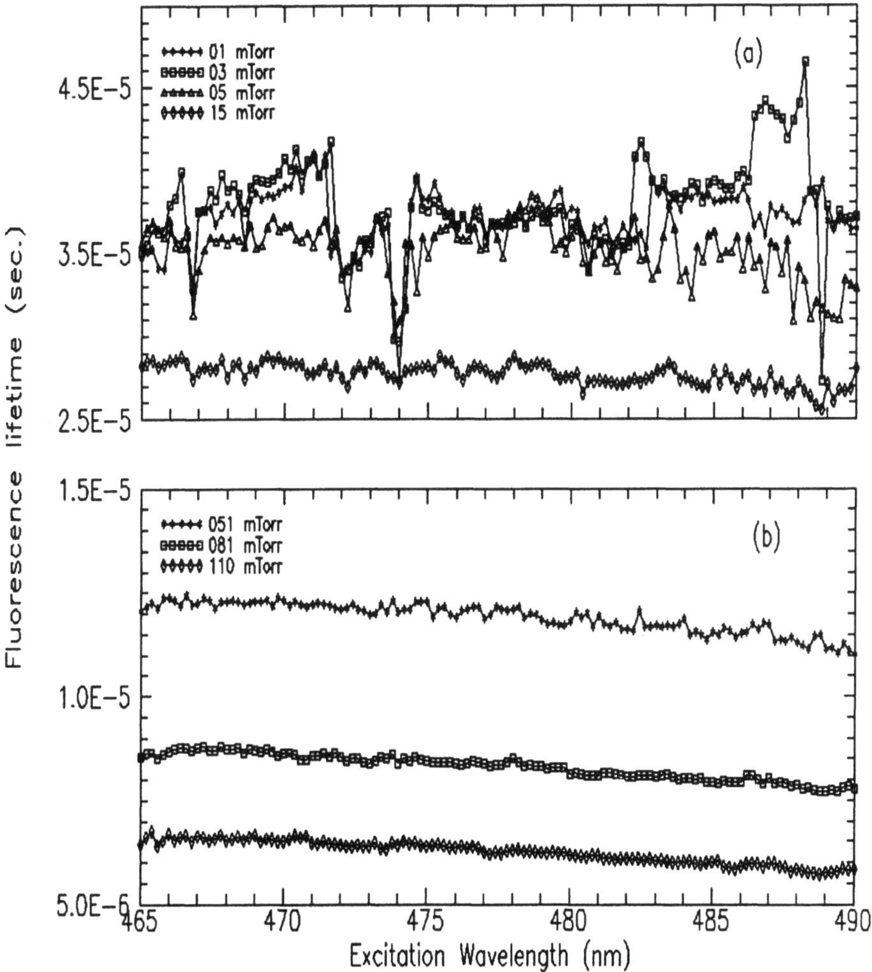

Figure 4. Fluorescence lifetime spectra of the static component in NO_2 at different gas pressures.

amplitude was found to be less than 0.1% of the total signal. Paech et al [8] also obtained a longer lifetime component varying from 73-105 μ sec at two excitation wavelengths, 488 and 514.5 nm under collision-free conditions. The larger component was also reported by Donnelly and Kaufman [17] at low pressures ($\leq$ 0.1 mTorr) which varied from 88 to 120 μ sec as the excitation wavelength was increased from 473 to 562 mm.

The shorter component of lifetime in the range of 1-3 μ sec appears more frequently than the longer component in the present experiment. The lifetimes of these components derived from the fluorescence decay curves of NO_2 at different excitation wavelengths and at different pressures are given in Table 1. At higher pressures, the measured lifetime was found to decrease due to collisional quenching. At pressures 81 and 110 mTorr, the lifetime decreases to about 0.1μ sec; however, their frequency of appearance at different excitation wavelengths up to 81 mTorr pressure was found to increase as the pressure is increased. At 110 mTorr, collision frequency becomes so large that the collision time is of the same order as the fluorescence lifetime; therefore, the frequency of appearance decreases at this pressure. This shorter component of lifetime was also studied by Sackett and Yardley [6], Solarz and Levy [7] and Paech et al [8] only at a few limited excited wavelengths. The range of lifetimes obtained by these researchers was from 1-3 μ sec and no pressure dependence of this component was studied by either of these three groups in contrast to the work carried out in the present experiment.

The mechanism suggested by Paech et al [8] to explain the three widely different lifetimes obtained in different experiments discussed above is consistent with the results obtained in the present experiment. The mechanism implies that the upper excited states are responsible for the absorption of laser photons and the subsequent fluorescence emissions have to be different states which could be connected by internal conversion processes. It was reported that the primary excited state in the $^2B_1$ electronic configuration decays mainly by radiationless transitions into a manifold of densely lying levels which can be described by linear combinations of $^2B_2$ and $^2A_1$ vibronic states. The primary transition which results in fluorescence from $^2B_1$ state to the ground state of NO_2 is supposed to belong to the group having lifetimes varying from 1-3 μ sec where as the other two transitions having lifetimes 20-40 and 70-120 μ sec occur from $^2B_2$ and $^2B_1$ vibronic states.

From the above studies, it is evident that the fluorescence lifetime in NO_2 exhibit considerable variation when excited with narrow band wavelengths. The anomaly in lifetime and its variability is attributed to the perturbation of the emitting levels by adjoining excited states. It may be stated here that one would learn more about the mechanisms for populating the emitting states if the lifetime measurements are carried out at the single fluorescing bands produced in the dispersed fluorescence spectra. Such a study would help to identify unambiguously the upper emitting levels of NO_2.

Table 1. Lifetimes 1-3 μ sec present in the fluorescence decay of NO_2 at different excitation wavelengths and at different pressures.

λ_{ex}(nm)	5 mT	15 mT	51 mT	81 mT	110 mT
466.8	-----	-----	-----	0.24	-----
467.0	-----	1.4	-----	-----	-----
472.0	-----	-----	-----	0.25	-----
472.2	1.1	-----	-----	0.31	-----

λ_{ex}(nm)	5 mT	15 mT	51 mT	81 mT	110 mT
472.4	-----	-----	-----	0.21	-----
473.8	2.6	-----	-----	0.20	-----
474.0	-----	1.8	-----	0.36	0.18
474.6	-----	-----	-----	0.18	-----
475.2	-----	-----	-----	0.16	-----
476.8	-----	-----	1.0	0.17	-----
477.6	-----	-----	-----	0.12	-----
479.6	-----	-----	-----	0.24	-----
479.8	-----	-----	-----	0.42	-----
480.0	-----	-----	-----	0.29	-----
480.4	-----	-----	-----	0.38	-----
480.6	-----	1.1	-----	0.21	-----
480.8	-----	-----	-----	0.29	0.12
481.0	-----	-----	1.5	0.15	0.12
481.2	-----	-----	-----	0.14	0.13
481.4	-----	-----	-----	0.16	-----
481.6	-----	-----	-----	0.29	-----
481.8	-----	-----	-----	0.23	-----
482.0	-----	-----	-----	0.25	-----
482.4	-----	-----	-----	0.28	-----
482.6	-----	-----	-----	0.22	-----
482.8	-----	-----	-----	0.16	-----
484.2	-----	-----	-----	0.13	-----
484.4	-----	-----	-----	0.35	-----
484.6	-----	-----	-----	0.26	-----
484.8	-----	-----	-----	0.14	-----
485.2	-----	-----	-----	0.11	0.16
485.6	-----	1.0	-----	-----	0.11
485.8	-----	-----	-----	0.72	-----
486.0	-----	-----	-----	0.50	-----
486.6	-----	-----	1.7	-----	0.20
486.8	-----	-----	-----	0.58	0.13
487.2	-----	-----	-----	0.48	-----
487.6	-----	-----	-----	0.33	-----
487.8	-----	-----	-----	0.40	-----
488.0	-----	-----	-----	0.37	-----
488.2	1.1	-----	-----	0.43	0.20
488.4	-----	1.5	-----	0.56	0.19
488.6	-----	2.6	1.7	0.66	0.32
488.8	-----	2.6	-----	0.58	0.41
489.0	-----	1.9	1.2	0.50	0.53
489.2	-----	-----	-----	0.55	-----
489.4	-----	-----	-----	0.64	0.34
489.6	-----	-----	-----	0.29	-----
489.8	1.2	1.7	1.2	0.44	0.29
490.0	-----	-----	1.2	0.43	0.20

REFERENCES

1. R. Kullmer and W. Demtroeder, Chem. Phys., **92**, 423 (1985).
2. L. E. Brus and J. R. McDonald, Chem. Phys. Lett., **21**, 283 (1973).
3. A. E. Douglas, J. Chem. Phys., **45**, 1007 (1966).
4. N. M. Atherton, R. N. Dixon and G. H. Kirby, Trans. Faraday Soc., **60**, 1688 (1964).
5. C. G. Stevens, M. W. Swagel, R. Wallace and R. N. Zare, Chem. Phys. Lett., **18**, 465 (1973).
6. P. B. Sackett and J. T. Yardley, Chem. Phys. Lett., **9**, 612 (1971).
7. R. Solarz and D. Levy, J. Chem. Phys., **60**, 842 (1974).
8. F. Paech, R. Schmiedl and W. Demtroder, J. Chem. Phys., **63**, 4369 (1975).
9. R. S. Mulliken, Rev. Mod. Phys., **14**, 204 (1942).
10. R. W. B. Pearse and A. G. Gaydon, *The identification of molecular spectra*, Chapman and Hall, London (1950).
11. A. D. Walsh, J. Chem. Soc., 2266 (1953).
12. R. A. Gangi and L. Burnelle, J. Chem. Phys., 55, 843 (1971).
13. G. Herzberg, *Molecular spectra and molecular structure III*, Van Nostrand, Princeton, N.J, 1966.
14. A. Amoruso, L. Crescentini, G. Fiocco and M. Volpe, J. Geophys. Res., **98**, 16857 (1993).
15. W. J. Wiscombe and J. W. Evans, J. Comp. Phys., **24**, 416 (1977).
16. A. Grinvald and I. Z. Steinberg, *Analytical Biochemistry*, **59**, 583 (1974).
17. V. M. Donnelly and F. Kaufman, J. Chem. Phys., **69**, 1456 (1978).

Electron Exchange Model Potential: Application to Positronium Scattering from Atoms and Molecules

P. K. Biswas

Instituto de Fisica Teorica,
Universidade Estadual Paulista, 01405-900 Sao Paulo, Brazil.

We present a one-parameter nonlocal model potential for electron exchange in complex systems described by Hatree Fock functions, and use it in the ab initio framework of Close Coupling Approximation (CCA). We check our scheme in electron-hydrogen and electron-helium systems, and perform coupled-channel calculations for elastic and inelastic scattering of positronium (Ps) by H, He, and H_2. Results for binding and resonances in Ps-H are compared with variational predictions and cross sections for Ps-He, and Ps-H_2 are compared with experiments.

1.　　INTRODUCTION

Positronium scattering studies now play an important role in the domain of both Physics and Chemistry, specially in the study of condensed matter physics and surface physics [1, 2, 3]. The neutral character of Ps, vis-a-vis the intimate interaction affinity of its positron with matter, makes it attractive as a probe. Ps scattering studies provides a great deal of interest to the theoreticians also owing to the underlying charge and mass symmetry of a Ps atom. Due to this symmetry in Ps, all its elastic and even-parity transition direct potentials with any target are exactly zero and the short-ranged exchange correlation dominates its low and medium energy scattering processes [4, 5, 6, 7]. The dominance of this short range interaction causes serious trouble towards the convergence of any coupled-channel fromalism with a truncated basis. Recently, pseudostate technique has been developed by the Belfast group to cope with this kind of problem.

Trends in Atomic and Molecular Physics,
Edited by Sud and Upadhyaya. Kluwer Academic/Plenum Publishers, New York, 2000.

They have got very successful results for e^+-H and e^+-He systems [8]. However, for Ps scattering, the R-matrix calculational scheme using as much as 22 coupled states in the simplest Ps-H system has exhibited the convergence difficulty while reporting the resonance and binding energies [9]. In ortho-Ps-He scattering, the galloping difference between the measured and theoretical values for low energy elastic cross sections and Zeff [10, 11] also contributes to this view of convergence difficulty for the exchange dominated Ps scattering.

To find an alternate solution to the nonconvergence problem and also to keep the calculational scheme tractable, we have suggested the use of a nonlocal electron-exchange model potential [5, 6, 7] in the ab initio framework of CCA. Two versions of this potential were suggested: one is time-reversal-symmetric and the other nonsymmetric. The symmetric potential developed subsquently has been found to be very reliable for all kind of investigations. Here we employ this potential in the exchange kernel of the coupled Lippmann-Schwinger equation and perform quantum coupled coupled-channel calculations for the description of Ps scattering from H, He, and H_2 targets.

2. THEORY

We consider the scattering of the long-lived ortho-Ps(1s) atom. The total wave function of the Ps-target system may be expanded in terms of Ps and target eigenstates as [9]

$$\Psi^{\pm}(\mathbf{r}_1, ..., \mathbf{r}_N; \mathbf{r}_{N+1}, \mathbf{x}) = \sum_{\nu}\sum_{\mu} \left[F_{\nu\mu}(\rho_{N+1})\, \chi_{\nu}(t_{N+1})\, \phi_{\mu}(\mathbf{r}_1, ..., \mathbf{r}_j, ...\mathbf{r}_N) \right.$$

$$\left. + \sum_{j=1}^{N} (-1)^{S_{N+1,j}}\, F_{\nu\mu}(\rho_j)\, \chi_{\nu}(t_j)\, \phi_{\mu}(\mathbf{r}_1, ..., \mathbf{r}_{N+1}, ...\mathbf{r}_N) \right],$$

$$(1)$$

where antisymmetrization with respect to Ps- and target-electron coordinates has been made and $S_{N+1,j}$ is the total spin of the Ps electron $(N+1)$ and target electron (j). Here $\rho_i = (\mathbf{x} + \mathbf{r}_i)/2$, $t_i = (\mathbf{x} - \mathbf{r}_i)$, where $\mathbf{r}_i$, $i = 1, ..., N$ denote the target electron coordinates, and $\mathbf{r}_{N+1}$ and $\mathbf{x}$ are the electron and positron coordinates of Ps; ϕ_{μ} and χ_{ν} denote the target and Ps wave functions, $F_{\nu\mu}$ is the continuum orbital of Ps with respect to the target. The parameters ϵ'', and Υ'' are the energies of the intermediate Ps and target states, and $\mathcal{E}$ is the total parametric energy of the Ps-target system.

Due to the exchange of the Ps electron with a target electron, the ortho-Ps atom may change to a para-Ps atom, if the exchange occurs between unlike spins. However, without a target excitation this possibility arises only in the case of a open-shell atom (here H). So for Ps-H, both the electronic singlet and triplet amplitudes govern the scattering. For Ps scattering by closed-shell atoms, the ortho to para conversion of Ps is not possible without a target excitation. Therefore, for target-elastic Ps scattering by He, and H_2 only the exchange amplitudes consisting of the two-electron triplet channels will contribute to the scattering. Projecting for a final state and spin averaging, the resulting momentum-space Lippmann-Schwinger scattering equation can, in general, be written as [4, 5, 12, 13]

$$f^{\pm}_{\nu'\mu',\nu\mu}(\mathbf{k}_f, \mathbf{k}_i) = B^{\pm}_{\nu'\mu',\nu\mu}(\mathbf{k}_f, \mathbf{k}_i)$$

$$- \frac{1}{2\pi^2} \sum_{\nu'} \sum_{\mu'} \int d\mathbf{k}'' \frac{B^{\pm}_{\nu'\mu',\nu''\mu''}(\mathbf{k}_f, \mathbf{k}'') f^{\pm}_{\nu''\mu'',\nu\mu}(\mathbf{k}'', \mathbf{k}_i)}{\varepsilon - \epsilon''_{\nu} - \Upsilon''_{\mu} - k''^2/4 + i0} \quad (2)$$

where, $f^{\pm}$ are the symmetrized final amplitudes. Due to spin averaging, both these amplitudes will contribute to the scattering process for Ps-H [4, 12], while for Ps-He [5, 13] and Ps-H2 [6] only the f^- channels will contribute for target elastic processes. In the present coupled channel scheme we shall describe only the target elastic processes so that the summation over μ'' is dropped and both of μ and μ' will represent the ground state of the target. For Ps-H, the input potentials to the above equation reduces to

$$B^{\pm}_{\nu'\nu}(\mathbf{k}', \mathbf{k}) = B^{D}_{\nu'\nu}(\mathbf{k}', \mathbf{k}) \pm B^{E}_{\nu'\nu}(\mathbf{k}', \mathbf{k});$$

whereas for Ps-He and Ps-H_2, $B^{+}_{\nu'\nu}(\mathbf{k}', \mathbf{k}) = 0$ and

$$B^{-}_{\nu'\nu}(\mathbf{k}', \mathbf{k}) = B^{D}_{\nu'\nu}(\mathbf{k}', \mathbf{k}) - B^{E_j}_{\nu'\nu}(\mathbf{k}', \mathbf{k});$$

where, B^D is the direct Born amplitude and B^E is the exchange amplitude [12] and B^{E_j} is the same [13] for the exchange of the Ps-electron with the j-th target electron (j = 1, 2).

The most crucial aspect of Ps scattering is that B^D, becomes exactly zero when $\nu \to \nu'$ represents an elastic or even parity transition of Ps; irrespective of any transition of target-states ($\mu \to \mu'$). So, upto lowest order the solution of the coupled LS-equation is governed solely by B^E (or B^{E_j}), the nonorthogonal exchange potential. The dominance of this short range exchange force hinders the convergence of any calculational scheme. To remedy this, we replace this exchange kernel by suitable model exchange potential generated from the $1/r_{12}$ term of the antisymmetrized part of the amplitude. For Ps-H scattering, in general B^E is derived from:

$$B^E_{\nu'\mu'\nu\mu}(\mathbf{k_f}, \mathbf{k_i}) = -\frac{1}{\pi}\int d\mathbf{r_1}\, d\mathbf{r_2}\, d\mathbf{x}\, e^{-i\mathbf{k_f}\cdot(\mathbf{x}+\mathbf{r_1})/2}\chi^*_{\nu'}(\mathbf{x}-\mathbf{r_1})\,\phi^*_{\mu'}(\mathbf{r_2})\cdot\frac{1}{r_{12}}$$

$$\times\,\chi_\nu(\mathbf{x}-\mathbf{r_2})\,\phi_\mu(\mathbf{r_1})e^{i\mathbf{k_i}\cdot(\mathbf{x}+\mathbf{r_2})/2}.$$

where we have shown the target lebels (μ, μ'). In the present investigation, both of them will be considered at their ground state and these indices will be dropped finally. To obtain the model potential, we take momemtum space expansion of the above amplitude and assign quantum average values of energy (multiplied by twice the reduced mass) for the (momentum)^2 terms for the Ps and H electrons. Detailed mathematical scheme is worked out in reference [5] and here we quote only the final results:

$$B^E_{\nu'\mu', \mu\nu}(\mathbf{k_f}, \mathbf{k_i}) = \frac{4(-1)^{l'+l}}{\langle D\rangle}\int\phi^*_{\mu'}(\mathbf{r_2})\exp(i\mathbf{q}\cdot\mathbf{r_2})\phi_\mu(\mathbf{r_2})\,d\mathbf{r_2}$$

$$\times\int\chi^*_{\nu'}(\mathbf{t_2})\exp(i\mathbf{q}\cdot\mathbf{t_2}/2)\chi_\nu(\mathbf{t_2})\,d\mathbf{t_2}, \tag{4}$$

where $\mathbf{q} = \mathbf{k_i} - \mathbf{k_f}$; $\mathbf{k_i}$ and $\mathbf{k_f}$ are the initial and final momenta of Ps with respect to the H-atom. The indices l and l' are the orbital angular momenta of the initial and final Ps-states. The function $\langle D\rangle$, for symmetric form of the potential, is given by

$$\langle D\rangle = (k_i^2 + k_f^2)/8 + C^2\left[(\alpha_\mu^2 + \alpha_{\mu'}^2)/2 + (\beta_\nu^2 + \beta_{\nu'}^2)/2\right] \tag{5}$$

where α_μ^2, $\alpha_{\mu'}^2$ correspond to the initial- and final-state ionization energies of the H atom in (here both are in ground state) and β_ν^2, $\beta_{\nu'}^2$ correspond to those of the Ps-atom. For, non-symmetric form of the potential, $\langle D\rangle = k_f^2/4 + \alpha_\mu^2 + \beta_\nu^2$ or $\langle D\rangle = k_i^2/4 + \alpha_{\mu'}^2 + \beta_\nu^2$ and the factor $(-1)^{l+l'}$ in the expression for B^E is replaced by $(-1)^{l'+1}$ or $(-1)^{l=1}$, respectively [5]. The parameter $C = 1$varies from unity to facilitate tuning of the assign average values of the (momentum)^2.

For He, the space part of the HF wave function [14] is given by $\Psi(\mathbf{r_1}, \mathbf{r_2}) = A[\phi_1(\mathbf{r_1})\phi_2(\mathbf{r_2})]$, where A is the antisymmetrization operator. The position vectors of the electrons are $\mathbf{r_j}, j = 1, 2$ and ϕ's have the form: $\phi_j(\mathbf{r_j}) = \sum_{k}a_{kj}\phi_{kj}(\mathbf{r_j})$. For target elastic channels, the spin-averaged exchange amplitude entering the coupled LS-equations are equivalent to considering exchange from either of the two 1s-electrons (r_j, $j = 1, 2$) (from the space part of the wave function) and is thus derived from

$$B_{v'v}^{E_j}(\mathbf{k}_f, \mathbf{k}_i) = -\frac{1}{\pi} \int d\mathbf{r}_1\, d\mathbf{r}_2\, d\mathbf{r}_3\, d\mathbf{x}\, e^{-i\mathbf{k}_f\cdot(\mathbf{x}+\mathbf{r}_2)/2}\, \chi_{v'}^*(\mathbf{x}-\mathbf{r}_2)\phi_0^*(\mathbf{r}_1, \mathbf{r}_3)\frac{1}{r_{23}}$$

$$\times\; \chi_v(\mathbf{x}-\mathbf{r}_3)\phi_0(\mathbf{r}_1, \mathbf{r}_2)e^{i\mathbf{k}_i\cdot(\mathbf{x}+\mathbf{r}_3)/2}.$$

$$(6)$$

and is reduced to [5]

$$B_{v'v}^{E_j}(\mathbf{k}_f, \mathbf{k}_i) = \sum_{\kappa\kappa'} \frac{4a_{\kappa j}a_{\kappa' j}(-1)^{l+l'}}{\langle D_{\kappa\kappa'}\rangle} \times \int \phi_{\kappa' j}^*(\mathbf{r}_j)\exp(i\mathbf{Q}\cdot\mathbf{r}_j)\phi_{\kappa j}(\mathbf{r}_j)\, d\mathbf{r}_j$$

$$\times\; \int \chi_{v'}^*(\mathbf{t}_j)\exp(i\mathbf{Q}\cdot\mathbf{t}_j/2)\,\chi_v(\mathbf{t}_j)\, d\mathbf{t}_j.$$

$$(7)$$

with

$$\langle D_{\kappa\kappa'}\rangle = (k_i^2 + k_f^2)/8 + C^2\left[(\alpha_{\kappa j}^2 + \alpha_{\kappa' j}^2)/2 + (\beta_v^2 + \beta_{v'}^2)/2\right],$$

for the symmetric form of the potential. The non-symmetric form of $\langle D_{\kappa\kappa'}\rangle$ is similar, as described in the case of Ps-H earlier. Here $\alpha_{\kappa j}^2$ corresponds to the binding energy of the electron represented by the slater orbital $\phi_{\kappa j}(\mathbf{r}_j)$ [14].

For molecular target, the Slater-type orbitals will have additional $\vec{R}$ dependence but the modeling scheme is exactly the same [6]. The present scheme can also be extended to scattering by other composite projectiles and to electron-impact cases [5]. In a previous work [5], the model exchange potential for electron impact cases have been published and the scheme is verified for e-H and e-He cases and we do not repeat them here.

3.   NUMERICAL PROCEDURES

In the present investigation, we consider Ps scattering from atoms and molecules. We use exact wave function for the Ps and H atom and for He we use the Hatree-Fock wave functions of Clemmenti and Roetti [14]. For H_2, we use the self-consistent field molecular orbitals of ref [15] and employ a fixed nuclei approximation. After a partial-wave decomposition of the three-dimensional LS equation, the coupled equations for the two different

414

channels are solved by the method of matrix inversion [12, 13]. For the
molecular target, the angular dependence of the nuclear axis has been treated
as parameters (θ_R, ϕ_R) and we solve the corresponding partial wave coupled-
equations for arbitrary orientations of the molecule. Scattering parameters
are finally, numerically averaged over molecular orientations. Maximum
number of partial waves included in the calculation is fifteen (L = 0 - 14).
Contribution of higher partial waves to the cross sections was included by
corresponding partial wave Born terms.

4.   RESULTS AND DISCUSSIONS

We start our discussion with the simplest Ps-H system. Here, we shall
use the symmetric form of the potential. We first present a study of the
effect of the variation of the parameter C of the exchange potential using a
three-Ps-state model with Ps(1s,2s,2p) states. We start our discussion with
the singlet S-wave resonance.

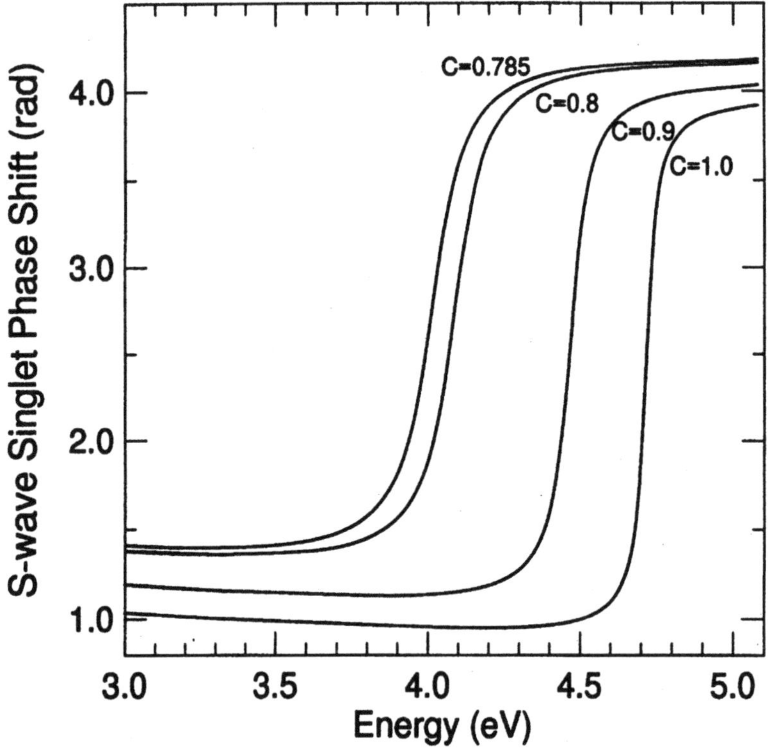

Figure 1. Variation of S-wave singlet phase shifts (in radian) for Ps-H scattering using 3-Ps-
state model for different values of the parameter C of the model potential

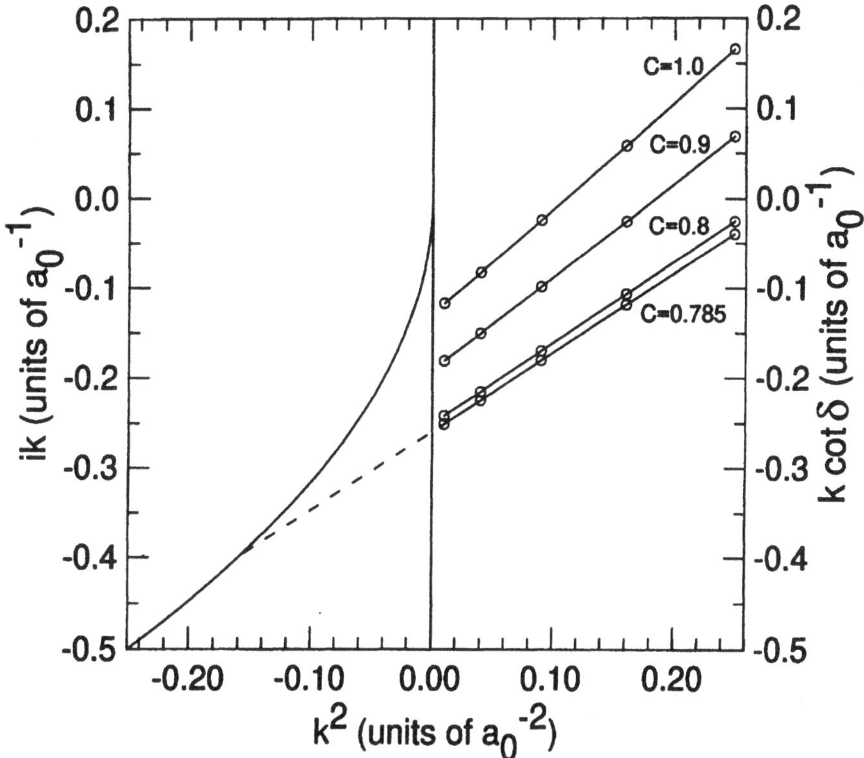

Figure 2. $k \cot \delta$ and ik (in a_0^{-1}) versus k^2 (in a_0^{-2}). The corresponding abscissa of the meeting point of negative energy extrapolation of $k \cot \delta$ curve with ik curve represents the binding energy of the PsH system for breaking into Ps and H in a.u. (Energy = 6.8 k^2 eV).

In Fig. 1 we plot the S-wave phase shift at different energies which illustrate the resonance pattern obtained with different values of C. The resonance position shifts monotonically towards lower energies with decreasing of C from unity. We have shown it in steps where the value of C is varied from unity to 0.785. The resonance position matches with the accurate prediction of 4.01 eV [16] for $C = 0.785$. In Fig. 2, we plot $k \cot \delta$ versus k^2 for the corresponding low-energy S-wave phase shifts δ. Figure 2 demonstrates how the improvement in the resonance position simultaneously improves the Ps-H binding energy. For $C = 1.0$ the resonance and binding energies are 4.715 eV and 0.165 eV, respectively; for $C = 0.9$ the corresponding energies are 4.470 eV and 0.445 eV, respectively. At $C = 0.785$, while the resonance position is correctly fitted to 4.01 eV (Fig. 1), we obtain an approximate binding energy of 1.02 eV from a linear extrapolation as in Fig. 2, and 0.99 eV with more precise fitting considering

416

second order corrections, compared to the accurate prediction of
1.0598 - 1.067 eV [17]. This behavior of the low-energy phase shifts, which
yields simultaneously the Ps-H resonance and binding energies, exhibits the
reliability of the present scheme. Previous best calculations using
stabilization method predicts 4.455 eV and 0.672 eV for the resonance and
binding energies, while that using a 22-coupled-pseudostate R-matrix
approach predicts those as 4.55 eV and 0.634 eV, respectively.

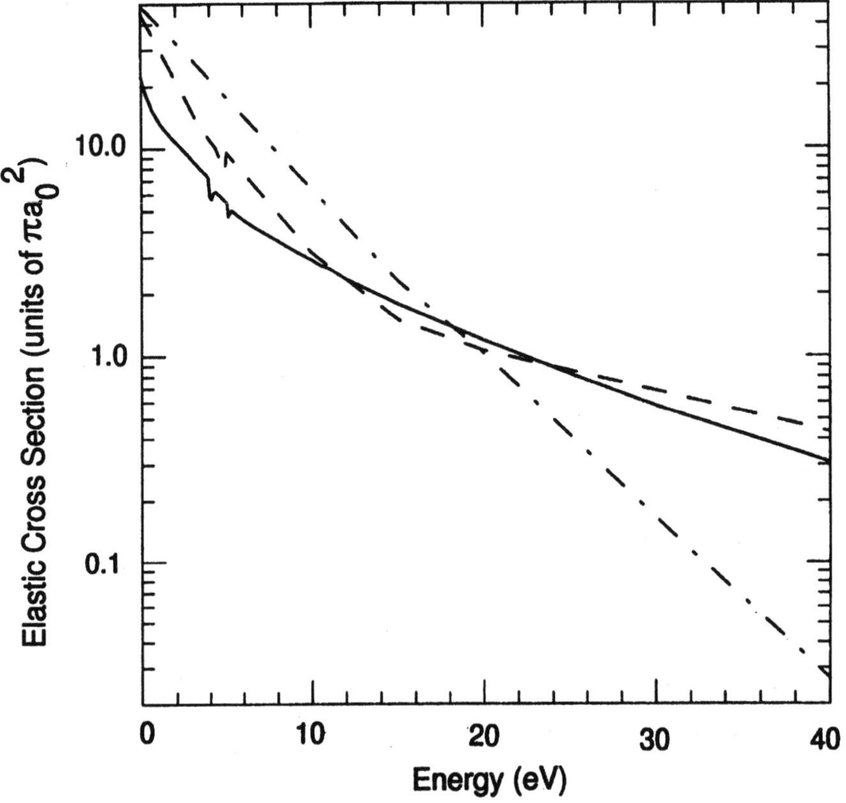

Figure 3. Elastic cross sections (in πa_0^2) against incident Ps energies (in eV). Solid line:
present result; Dashed line: Campbell et al [9]; Dashed-dotted line: Sinha et al
[12].

In Figs. 3 and 4 we demonstrate the elastic and target-elastic total cross
sections for Ps-H, respectively, along with other theoretical results. Elastic
and Ps(2s) and Ps(2p) excitation cross sections are obtained from the 3-Ps-
state CCA model while higher excitations and ionizations of Ps are
evaluated using a first Born approximation with present exchange. Present
elastic cross sections grossly disagree with other two calculations at low

energies. This was quite expected from the converged behaviour of the present low energy phase shifts as indicated by the resonance and binding energies. At medium energies, present results grossly deviate from the 3-state calculation of Sinha et al while it remains very close to the 22-state predictions which is obviously better converged than the results of Sinha et al.

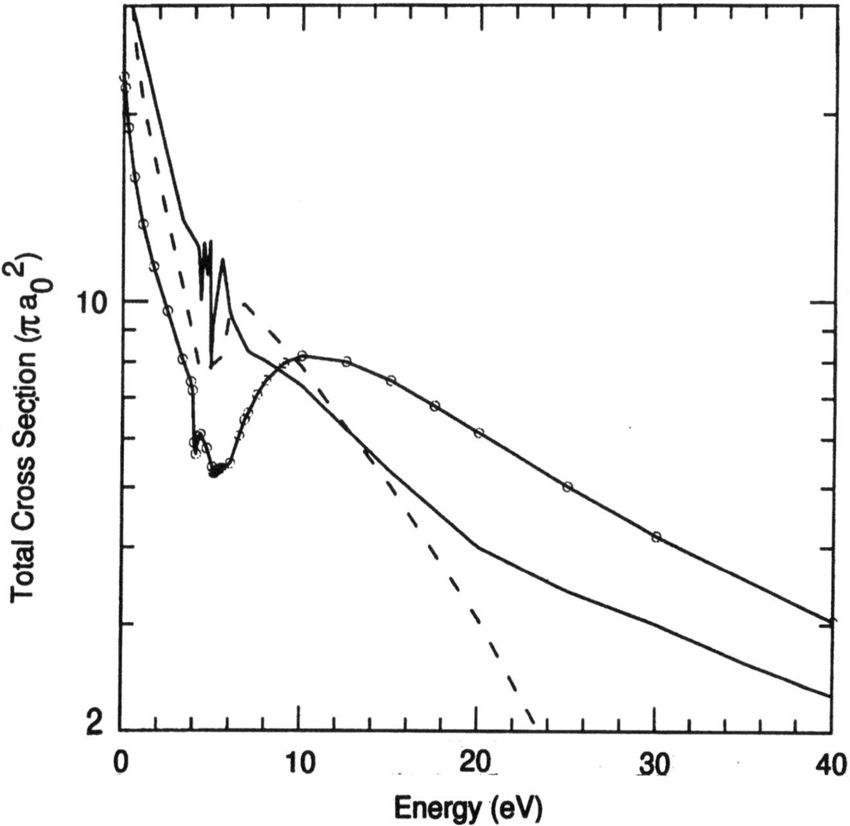

Figure 4. Target-elastic total cross sections (in πa_0^2) against incident Ps energies (in eV). Solid line with open circles: present result; Solid line: Campbell et al [9]; Dashed line: Sinha et al [12]

In Fig. 4 we plot the target-elastic total cross sections for the system obtained from the 3-Ps-state calculation plus Born approximation applied for the $n \geq 3$ higher excitations and ionization of Ps. Present results demonstrate the feature of a minimum near the first excitation threshold of Ps and a maximum thereafter. This feature is well established in experimental Ps scattering studies from He and H_2 [18].

418

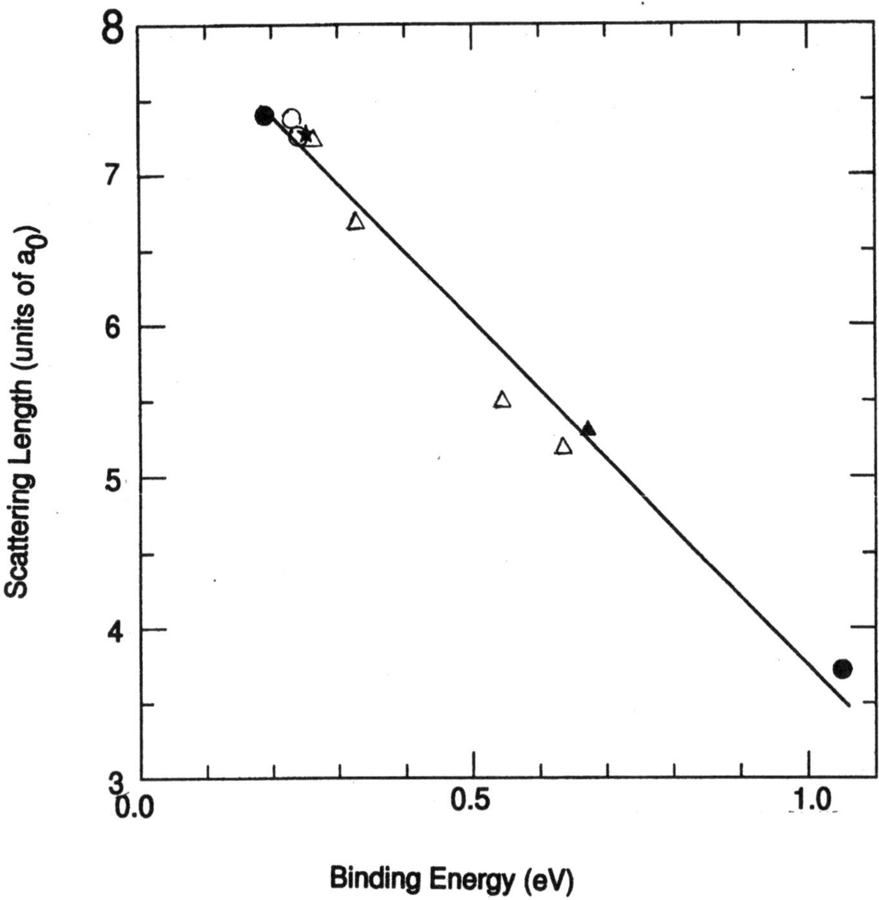

Figure 5. S-wave singlet scattering length versus PsH binding energy of different calculations. Open triangels: Ref.[9]; Solid triangle: Ref.[20]; Open circles: Ref.[12]; Solid circle: Present results Solid star: Ref.[19].

In Fig. 5, we plot an interesting feature of the low energy Ps-H system. We plot the singlet scattering length versus the binding energy obtained in different calculations. Due to the presence of an effective range expansion, these two were expected to be correlated and the figure clearly demonstrates this behaviour. The convergence of binding energy and scattering length predictions from static exchange, 3-Ps-state, 9-state, and 22-state follows the straight line correlation curve obtained with different results. The predictions by Drachman [20] using stabilization method also alingned in this correlation and appeared to be slightly better converged than the 22-state prediction of Campbell et al [9] as also apprehended by them. Present result, which claim to be better converged than all previous predictions also lie along this correlation curve.

Next, we describe our results for Ps-He and Ps-H$_2$ systems. In these cases, we shall describe our results with the previously proposed non-symmetric form of the potential. Results with symmetric forms are under preparation and will be published elsewhere, shortly. We exhibit the results for Ps-He system obtained using a 3-Ps-state coupling for elastic and Ps(2s), Ps(2p) excitations and using Born for n $\geq$ 3 excitations and ionization of Ps.

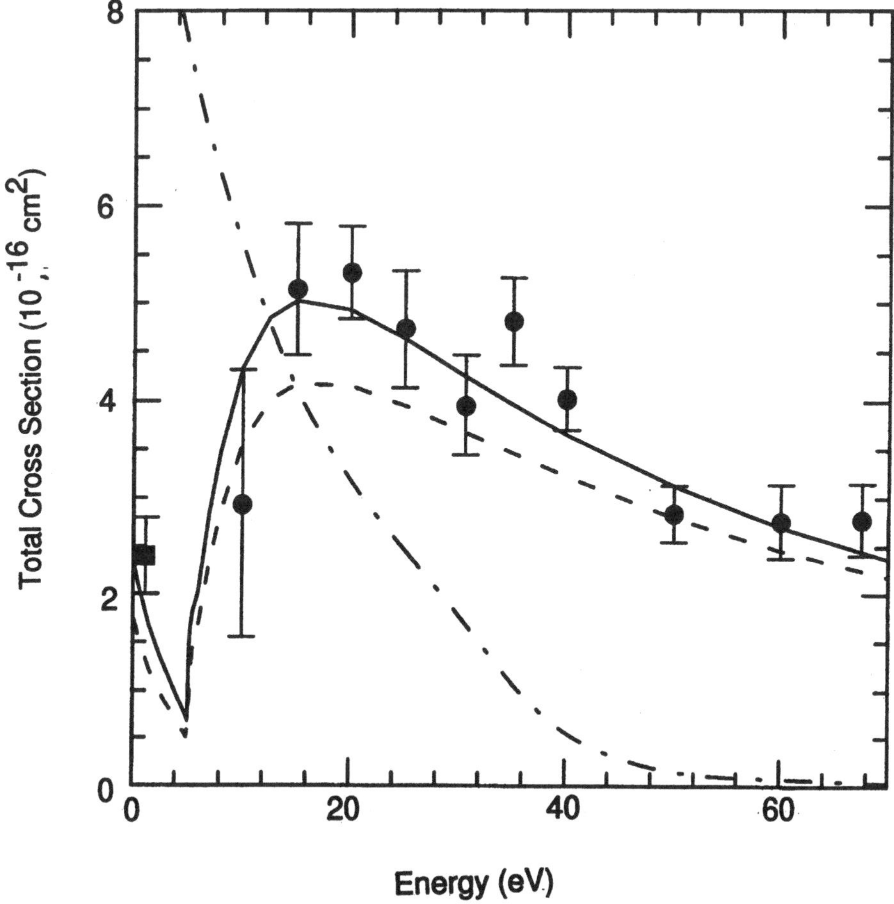

Figure 6. Total cross sections for Ps-He. Solid line: present results with C = 0.88; Dashed line: results with $C = 1.0$ (unchanged); Solid circle with error bars: Measured data of Ref. [18]; Solid square with error bar: Measured data of Ref. [21]; Dashed-dotted line: calculation of Ref. [13].

In Fig. 6 we plot the target-elastic total cross section so obtained along with the experimental data. The encouraging agreement of the cross sections from low to medium energies are obtained with a value of the

420

parameter $C = 0.88$. The salient feature of this calculation is that it bridges the two sets of experiments, performed at low and medium energies, with the prediction of a minimum near the Ps-excitation thresholds (5.1-6.8 eV). This feature arising from the opening up of an infinite number of excitation and ionization channels of Ps in between 5.1-6.8 eV, is although obtained for the first time [22] in Ps scattering, is well known in positron-atom scattering [8] arising due to opening up of Ps-formation channels. All previous calculations, except the unpublished work of Peach [22], do not provide such a feature and also do not satisfy the experimental data.

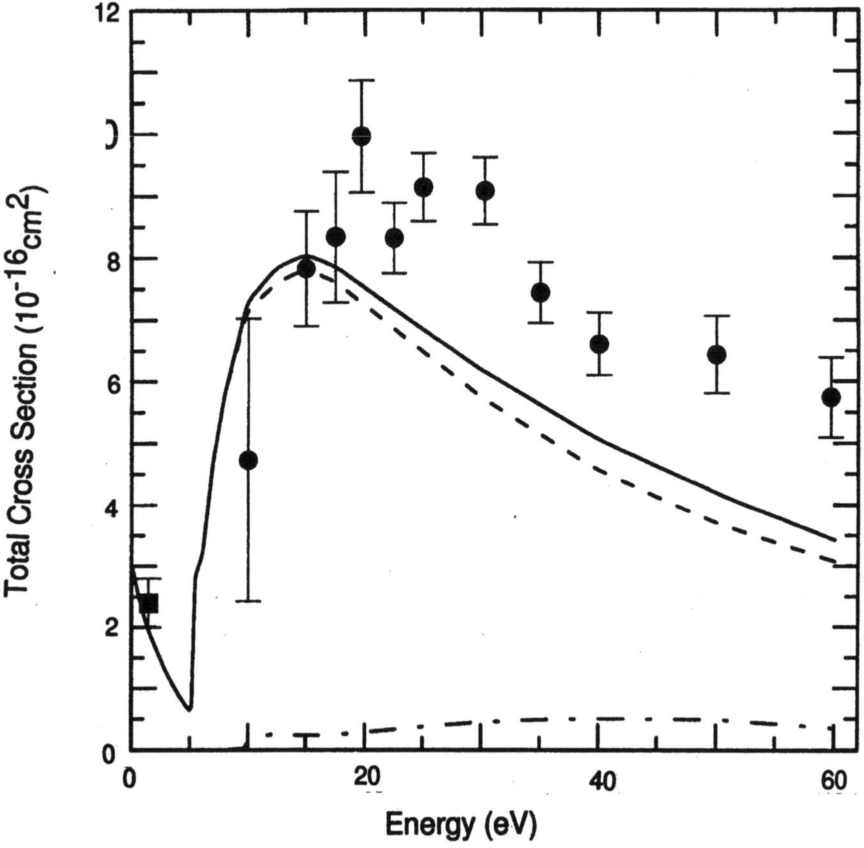

Figure 7. Total cross sections for Ps-H$_2$. Solid line: Total cross section; Dashed line: Target-elastic total cross section; Dashed-dotted line: Target inelastic cross sections for $X^1\Sigma_g^+ \to b^3\Sigma_u^+$ and $X^1\Sigma_g^+ \to B^1\Sigma_u^+$ excitations [6] Solid circle with error bars: Measured data of Ref. [18]; Solid square with error bar: Measured data of Ref. [21].

Finally, in Fig. 7, we plot our results for the Ps-H$_2$ system. Here, apart from presenting the target-elastic cross section evaluated as in Ps-He case, we also add with it the cross sections for the $X^1\Sigma_g^+ \to b^3\Sigma_u^+$ and

$X^1\Sigma_g^+ \to B^1\Sigma_u^+$ excitations of the hydrogen molecule calculated using first Born with present exchange [6]. The agreement of the theoretical total cross section with the measurement is very encouraging from the fact that only two low lying target inelastic processes have been added to the scheme. The total contributions from these two target inelastic channels is indicative that the remaining significant excitation channels of H$_2$ might improve the total cross section to closely agree with the measured data. This total cross section is, however, obtained using our previously suggested non-time-reversal symmetric form of the exchange potential. We are in the process of reinvestigating this problem with the time reversal symmetric form of the model exchange potential.

5.   CONCLUSION

In conclusion, we present a nonlocal model potential for electron exchange in complex systems and perform coupled-channel calculations for elastic and inelastic scattering of positronium (Ps) by H, He and H$_2$. Present calculational reproduces the variational findings for the resonance and binding in Ps-H. Elastic and total cross section for Ps-He and Ps-H$_2$ are also in good agreement with experiment. The tractibility and reliability of the hybrid scheme in the ab initio framework opens the possibility of having realistic predictions on complex systems.

The work is supported by the Fundação de Amparo á Pesquisa do Estado de São Paulo of Brazil.

REFERENCES

1.   D. M. Schrader and Y. C. Jean, Positron and Positronium Chemistry (Amsterdam:Elsevier), 1988; D. M. Schrader, Nucl. Inst. Methods, B 143, 112 (1998).

2.   H. Weber et al, Phys. Rev. Lett., 61, 2542 (1988); S. Tang and C. M. Surko, Phys.Rev. A 47, R743 (1993).

3.   M. Charlton and G. Laricchia, Comm. At. Mol. Phys. 26, 253 (1991).

4.   P. K. Biswas and S. K. Adhikari, J. Phys. B 31, 3147 (1998).

5. P. K. Biswas and S. K. Adhikari, Phys. Rev. A 59, 363 (1999).

6. P. K. Biswas and S. K. Adhikari, J. Phys. B 31, L315 (1998); ibid, L737 (1998).

7. S. K. Adhikari and P. K. Biswas, Phys. Rev. A 59, 2058 (1999).

8. A. A. Kernoghan, M. T. McAlinden, and H. R. J. Walters, J. Phys. B. 28, 1079 (1995).

9. C. P. Campbell, M. T. McAlinden, F. G. R. S. MacDonald, and H. R. J. Walters, Phys. Rev. Lett. 80, 5097 (1998).

10. S. K. Adhikari, P. K. Biswas, and R. A. Sultanov, Phys. Rev. A, to appear as brief report, vol. 59 (1999).

11. M. I. Barker and B. H. Bransden, J. Phys. B 1, 1109 (1968); ibid 2, 730 (1968).

12. P. K. Sinha, P. Chowdhuri, and A. S. Ghosh, J. Phys. B 30, 4643 (1997).

13. N. K. Sarkar. and A. S. Ghosh, J. Phys. B 30, 4591 (1997).

14. E. Clemmenti, and C. Roetti, At. data Nucl. data Tables 14, 177 (1974).

15. R. S. Mulliken and W. C. Ermler, 'Diatomic molecules: Results of Abinitio Calculations' (New York: Academic), pp.47 (1977).

16. Y. K. Ho, Phys. Rev. A 17, 1675 (1978).

17. A. M. Frolov, and V. H. Smith, Phys. Rev. A 55, 2662 (1997) and references therein.

18. A. J. Garner, G. Laricchia, A. Ozen, J. Phys. B 29, 5961 (1996); N. Zafar, G. Laricchia, M. Charlton, and A. Garner, Phys. Rev. Lett., 76, 1595 (1996).

19. S. Hara and P. A. Fraser, J. Phys. B 8 L472 (1975).

20. R. J. Drachman and S. K. Houston, Phys. Rev. A 12, 885 (1975); ibid, bf 14, 894 (1976).

21. M. Skalsey, J. J. Engbrecht, R. K. Bithell, R. S. Vallery, D. W. Gidley, Phys. Rev. Lett., 80, 3727 (1998).

22. G. Peach, (unpublished), as quoted in [18]

INDIAN SOCIETY FOR ATOMIC AND MOLECULAR PHYSICS (ISAMP)

President	*Dr. S. A. Ahmad*
	Spectroscopy Division
	Bhabha Atomic Research Centre
	Trombay, Mumbai - 400 085.
Vice President	*Prof. Krishan K. Sud*
	Department of Physics
	College of Science Campus
	M. L. Sukhadia University,
	Udaipur - 313 001.
Secretary	*Prof. E. Krisnakumar*
	Tata Institute of Fundamental Research
	Mumbai - 400 005.
Treasurer	*Mr. I. A. Prajapati*
	Physical Research Laboratory
	Navrangpura, Ahmedabad - 380 009.
Members	*Prof. A. N. Tripathi*
	Department of Physics
	University of Roorkee
	Roorkee - 247 667.
	Prof. K. L. Baluja
	Department of Physics and Astrophysics
	University of Delhi
	Delhi - 100 007.
	Prof. K. N. Joshipura
	Department of Physics
	Sardar Patel University
	Vallabh Vidyanagar - 388 120.

Prof. Kanika Roy

Department of Theoretical Physics
Indian Association for the Cultivation of Science
Jadavpur, Calcutta - 700 032

Prof. Naresh Chandra

Department of Physics and Meteorology
Indian Institute of Technology
Kharagpur - 721 302

Dr. Y. N. Tiwari

Department of Physics
Pachhunga University College
North-Eastern Hill University
Aizawal - 796 001

Ultra short pulse excited plasmas, 3